IUTAM Symposium on Multi-Functional Material Structures and Systems

IUTAM BOOKSERIES

Volume 19

Aims and Scope of the Series

The IUTAM Bookseries publishes the proceedings of IUTAM symposia under the auspices of the IUTAM Board.

For other titles published in this series, go to
www.springer.com/series/7695

B. Dattaguru • S. Gopalakrishnan
V.K. Aatre
Editors

IUTAM Symposium on Multi-Functional Material Structures and Systems

Proceedings of the IUTAM Symposium on Multi-Functional Material Structures and Systems, Benguluru, India
December 10–12, 2008

Springer

Editors
Prof. B. Dattaguru
Dept. Aerospace Engineering
Indian Institute of Science
Benguluru-560012
India
datgur@aero.iisc.ernet.in

Dr. V.K. Aatre
Dept. of Electrical Communication
Engineering
Indian Institute of Science
Benguluru-560012
India
vkaatre@yahoo.co.in

Prof. S. Gopalakrishnan
Dept. of Aerospace Engineering
Indian Institute of Science
Benguluru-560012
India
krishnan@aero.iisc.ernet.in

ISSN 1875-3507 e-ISSN 1875-3493
ISBN 978-90-481-3770-1 e-ISBN 978-90-481-3771-8
DOI 10.1007/978-90-481-3771-8
Springer Dordrecht Heidelberg London New York

Library of Congress Control Number: 2009944068

Cover design: eStudio Calamar S.L.

Printed on acid-free paper

Springer is part of Springer Science+Business Media (www.springer.com)

Preface

For centuries it has been the desire of human beings to find better and better materials to achieve a variety of benefits for the mankind. This has been in vogue from the stone-age. The recent revolution discovering Smart Materials and Functionally Graded Materials is one of those attempts. These materials are inherently multi functional and they opened up possibilities which could not be imagined in the past. Materials can take a particular type of energy stimulus as input and generate an output belonging to a different type of energy. Typical example has been materials on which a mechanical force can produce electrical output. We are in twentyfirst century where in materials Research will totally concentrate on these new concepts and exploit them for variety of applications. Structural Health Monitoring leading to safety of operations is the primary application the aerospace, mechanical, nuclear and civil engineers will be expecting from this research. The Micro- and Nano-scale sensors and actuators form the basis for this technology and Departments of Science and Technology in all the countries are investing heavily on this highly potential field.

In India, the scientific group in this area is active for the past decade. They formed in 1999 a professional society named as Institute of Smart Structures & Systems (ISSS) which has been holding National and International seminars and also helping the government departments to fund research and development of relevance to this novel materials and structures. National Program on Smart Materials (NPSM) and the second phase program NPMASS are the consequences of these efforts. The current symposium is part of this effort and the Editors of this publication are also scientists involved in these developments.

This symposium covered novel theories, mathematical models, analyses, manufacturing methods and application of computational and experimental methods. This publication is thematically divided into seven sections with a total of 35 papers authored by International Scientific Community. The coverage is quite comprehensive and it is heartening that half of these papers are from R&D in Indian organizations. We trust this volume will provide to researchers and applied technologists on the potential fruitful directions.

The Editors express their sincere thanks to International Union of Theoretical and Applied Mechanics (IUTAM) for sponsoring this symposium on "Multi-Functional Materials Structures & Systems" (IUTAM-MMSS08) at the Indian Institute of

Science (IISc), Benguluru during 2008. The application was processed through Indian National Science Academy (INSA) and Prof. N.K. Gupta, one of the Vice-Presidents of INSA encouraged us to bid for it. Thanks to INSA and Prof. Gupta. The symposium was held between 10–12 December, 2008 at the Satish Dhawan Auditorium, IISc. Eightyfive delegates attended the symposium (with 22 of them coming from outside India) and the presentations brought out the core expertise available around the world in this newly emerging field. This book is the proceedings of this symposium. The papers were reviewed by International experts in this field and were revised before accepting for publication. Special thanks to all the reviewers who spent considerable amount of time on the papers in spite of their busy schedule.

From the time the symposium was approved by IUTAM, Prof. D.H. van Campen, Secretary General and Prof. J. Engelbrecht, Treasurer have been in constant touch with us and we were sending updates to them continuously. They were always prompt in replying to our enquiries. IUTAM provided funds to enable the organizers to support some of the participants towards Registration fee and accommodation and transport at Benguluru. There was a change in Office bearers of IUTAM at the end of the year 2008. Prof. Frederic Dias and Prof. Niels Olhoff, who are the Secretary General and Treasurer of the new committee, helped us to finalize the accounts and closure report to IUTAM. Thanks to all Office Bearers of IUTAM.

We have also received generous financial support from Aeronautics R&D Board (ARDB), Government of India; Institute of Smart Structures & Systems (ISSS) and the Indian Institute of Science. We would like to express thanks to all of them.

The Assigned Scientific Committee consisted of experts in the area from around the world. Prof. J. Dual from Switzerland was the IUTAM representative. Many of the Committee members presented papers at the symposium. Their acceptance to be on the committee has been a great inspiration and we thank them all for their help and suggestions from time to time. The International Steering Committee chaired by Dr. V.K. Aatre, one of the Editors of this volume, was deeply supportive for the conduct of the symposium. One of the members from outside India, Dr. F.G. Buchholz of the University of Paderborn, Germany attended the symposium and chaired two sessions.

Prof. B.N. Raghunandan, Chairman, Department of Aerospace Engineering and Prof. P.J. Paul, Convener, Joint Advanced Technology Program at IISc and some of our senior colleagues including Profs. C.R.L. Murthy, Ranjan Ganguli, Kartik Venkataraman extended their full support for the symposium and thanks are due to all of them. Research and Masters' students helped us in many ways and thanks to all of them for their help. Mr. Shivanand Bhavikatti helped us in organizing the manuscript of the proceedings and this is acknowledged with thanks.

December 2009
Benguluru

S. Gopalakrishnan
B. Dattaguru
V.K. Aatre

Contents

Section I New Materials A: Functionally Graded Materials & Shape Memory Alloys

Section II New Materials B: Nano Materials and Composites

Section III Multifunctional Material Systems

Section IV Smart Sensors, Structural Health Monitoring

Section V Applications

Section VI Computational Methods- I

Contributors

Sondipon Adhikari School of Engineering, Swansea University, Singleton Park, Swansea SA2 8PP, UK **(Paper 19)**

A. Arockiarajan Indian Institute of Technology, Chennai, India **(Paper 10)**

Charles E. Bakis Department of Aerospace Engineering, PENN State University, University Park, PA, USA **(Paper 27)**

Balakumar Balachandran Department of Mechanical Engineering, University of Maryland, College Park, MD 20742, USA **(Paper 33)**

V. Balmurugan Department of Mechanical Engineering, Indian Institute of Technology, Chennai, India **(Paper 01)**

B. Banerjee Structures Lab, Department of Civil Engineering, Indian Institute of Science, Benguluru – 560012, India **(Paper 28)**

S. Banerjee Centre for Advanced Composite Materials, Department of Mechanical Engineering, The University of Auckland, Auckland Mail Centre 1142, New Zealand **(Paper 22)**

Ayech Benejeddou Supméca – Paris, Structures, 3 rue Fernand Hainaut, 93407 Saint Ouen CEDEX, France **(Paper 31)**

D. Bhattacharyya Centre for Advanced Composite Materials and Department of Mechanical Engineering, University of Auckland, Private Bag 92019, Auckland, New Zealand **(Paper 12 & 22)**

Prashant Bhise Thapar University, Patiala 147004, India **(Paper 32)**

B. Boller Saarland University, Gemany **(Paper 18)**

Stéphane P.A. Bordas Department of Civil Engineering, University of Glasgow, Scotland, U.K **(Paper 30)**

Vidyashankar R. Buravalla General Motors (R&D), India Science Lab 3rd Floor, Creator Building, I T P B, Whitefield Road, Benguluru 560066 India **(Paper 04)**

P. Chellapandi Indira Gandhi Centre for Atomic Research, Kalpakkam-603 102, India **(Paper 17)**

Liang Cui School of Electrical, Electronic & Mechanical Engineering, University College Dublin, Dublin, Ireland **(Paper 02)**

B.M. Dawari Department of Civil Engineering, Indian Institute of Science, Bengaluru, India **(Paper 23)**

Sujay Deshmukh Department of Aerospace Engineering, Texas A&M University, USA **(Paper 11)**

Michael I. Friswell School of Engineering, Swansea University, Singleton Park, Swansea SA2 8PP, UK **(Paper 19)**

Hisao Fukunaga Department of Aerospace Engineering, Tohoku University, Aramaki-Aza-Aoba 6-6-01, Aoba-ku, Sendai 980-8579, Japan **(Paper 08)**

M. Ganapathi Bombardier India Center, Satyam Computer Services Ltd., Benguluru, India **(Paper 03)**

Ranjan Ganguli Department of Aerospace Engineering, Indian Institute of Science, Benguluru – 560012, India **(Paper 26)**

Michael D. Gilchrist School of Electrical, Electronic & Mechanical Engineering, University College Dublin, Dublin, Ireland **(Paper 02)**

Mahesh V. Hosur Center for Advanced Materials, Tuskegee University, Tuskegee, AL 36088, USA **(Paper 15)**

Ning Hu Department of Aerospace Engineering, Tohoku University, Aramaki-Aza-Aoba 6-6-01, Aoba-ku, Sendai 980-8579, Japan **(Paper 08)**

K. Jayabal Indian Institute of Technology, Chennai, India Institute of Mechanics, TU Dortmund, Germany **(Paper 10)**

K. Jayaraman Centre for Advanced Composite Materials, Department of Mechanical Engineering, The University of Auckland, Auckland Mail Centre 1142, New Zealand **(Paper 22)**

Shaik Jeelani Center for Advanced Materials, Tuskegee University, Tuskegee, AL 36088, USA **(Paper 15)**

Tarun Kant Department of Civil Engineering, IIT, Powai, Mumbai – 400078, India **(Paper 07)**

Rakesh K. Kapania Virginia Polytechnic Institute and State University, Blacksburg, VA, USA **(Paper 09)**

Yoshifumi Karube Department of Aerospace Engineering, Tohoku University, Aramaki-Aza-Aoba 6-6-01, Aoba-ku, Sendai 980-8579, Japan **(Paper 08)**

Ashish Khandelwal General Motors (R&D), India Science Lab 3rd Floor, Creator Building, I T P B, Whitefield Road, Benguluru 560066 India **(Paper 04)**

Stephen Kiernan School of Electrical, Electronic & Mechanical Engineering, University College Dublin, Dublin, Ireland **(Paper 02)**

B.H. Koh Departmental of Civil & Envi. Engineering and Mechanical Engineering & Material Science, Rice University, Houston, TX 77005, USA **(Paper 20)**

Ananth Kumar Department of Aerospace Engineering, Indian Institute of Science, Benguluru – 560012, India **(Paper 26)**

Zhiling Li Departmental of Civil & Environmental Engineering, Rice University, Houston, TX 77005, USA **(Paper 20)**

D. Roy Mahapatra Department of Aerospace Engineering, Indian Institute of Science, Benguluru 560012, India **(Paper 30)**

Prakash D. Mangalgiri General Motors (R&D), India Science Lab, 3rd Floor, Creator Building, I T P B, Whitefield Road, Benguluru 560066 India **(Paper 05)**

A. Menzel Indian Institute of Technology, Chennai, India **(Paper 10)**

Mira Mitra Department of Aerospace Engineering, IIT, Powai, Mumbai, India **(Paper 29)**

M.R. Mofakhami Saarland University, Germany **(Paper 18)**

Abhijit Mukherjee Thapar University, Patiala 147004, India **(Paper 32)**

Kaushik Mysore Purdue University, West Lafayette, IN 47907, USA **(Paper 34)**

Satish Nagarajaiah Departmental of Civil & Environmental Engineering and Mechanical Engineering & Material Science, Rice University, Houston, TX 77005, USA **(Paper 20)**

S. Narayanan Department of Mechanical Engineering, Indian Institute of Technology, Chennai, India **(Paper 01)**

Sundararajan Natarajan PhD Research Student, Theoretical, Applied and Computational Mechanics, Cardiff School of Engineering, U.K **(Paper 30)**

Zoubeida Ounaies Department of Aerospace Engineering, Texas A&M University, TX, USA **(Paper 11)**

P.C. Pandey Department of Civil Engineering, Indian Institute of Science, Bengaluru, India **(Paper 23)**

Prasad Potluri School of Materials, University of Manchester, Manchester M60 1QD, United Kingdom **(Paper 25)**

T. Prakash Satyam Computer Services Ltd., Benguluru, India **(Paper 03)**

Rudra Pratap Department of Mechanical Engineering, Indian Institute of Science, Benguluru, India **(Paper 21)**

Baldev Raj Indira Gandhi Centre for Atomic Research, Kalpakkam-603 102, India **(Paper17)**

Raghu Ram Department of Civil Engineering, IIT Bombay, Mumbai 400076, India **(Paper 32)**

Subramanian Ramakrishnan Department of Mechanical Engineering, University of Maryland College Park, MD 20742, USA (**Paper 33**)

Ananth Ramaswamy Department of Civil Engineering, Indian Institute of Science, Benguluru, India (**Paper 06**)

S. Rao Centre for Advanced Composite Materials, Department of Mechanical Engineering, The University of Auckland, Auckland Mail Centre 1142, New Zealand (**Paper 22**)

Harinadh Reddy Department of Civil Engineering, Indian Institute of Science, Benguluru, India (**Paper 06**)

J.N. Reddy Texas A&M University, College Station, TX 77843-3123, USA (**Paper 13**)

D. Roy Structures Lab, Department of Civil Engineering, Indian Institute of Science, Benguluru – 560012, India (**Paper 28**)

Samikkannu Raja Scientist, Structural Technologies Division, National Aerospace Laboratories, Benguluru – 560017, India (**Paper 24**)

V.L. Sateesh Department of Aerospace Engineering, Indian Institute of Technology, Kanpur, India (**Paper 16**)

S. Schmauder Institut für Materialprüfung, Werkstoffkunde und Festigkeitslehre (IMWF), Pfaffenwaldring 32 70569 Stuttgart, Germany (**Paper 35**)

Agam Sharan Department of Aerospace Engineering, IIT, Powai, Mumbai, India (**Paper 29**)

Sandeep Sharma Thapar University, Patiala 147004, India (**Paper 32**)

R.J. Shields Centre for Advanced Composite Materials and Department of Mechanical Engineering, University of Auckland, Private Bag 92019, Auckland, New Zealand (**Paper 12**)

S.M. Shiyekar Department of Civil Engineering, IIT, Powai, Mumbai – 400078, India (**Paper 07**)

A. Siddiq Institut für Materialprüfung, Werkstoffkunde und Festigkeitslehre (IMWF), Pfaffenwaldring 32 70569 Stuttgart, Germany (**Paper 35**)

M.K. Singha Indian Institute of Technology, Hauz Khas, New Delhi, India (**Paper 03**)

S.M. Sivakumar Indian Institute of Technology, Chennai, India (**Paper 10**)

Edward C. Smith Department of Aerospace Engineering, PENN State University, University Park, PA, USA (**Paper 27**)

Prashant Kumar Srivastava Department of Aerospace Engineering, Indian Institute of Science, Benguluru, India (**Paper 14**)

Ganesh Subbarayan Purdue University, West Lafayette, IN 47907, USA **(Paper 34)**

Mohammed Rabius Sunny Virginia Polytechnic Institute and State University, Blacksburg, VA, USA **(Paper 09)**

Amol G. Thakare General Motors (R&D), India Science Lab 3rd Floor, Creator Building, I T P B, Whitefield Road, Benguluru 560066, India **(Paper 05)**

Merlin Theodore Center for Advanced Materials, Tuskegee University, Tuskegee, AL 36088, USA **(Paper 15)**

C. Tiwari Department of Aerospace Engineering, PENN State University, University Park, PA, USA **(Paper 27)**

G.U. Unnikrishnan Texas A&M University, College Station, TX 77843-3123, USA **(Paper 13)**

V.U. Unnikrishnan Texas A&M University, College Station, TX 77843-3123, USA **(Paper 13)**

C.S. Upadhyay Department of Aerospace Engineering, Indian Institute of Technology, Kanpur, India **(Paper 16)**

C. Venkata Subbaiah Department of Civil Engineering, IIT, Powai, Mumbai – 400078, India **(Paper 07)**

C. Venkatesan Department of Aerospace Engineering, Indian Institute of Technology, Kanpur, India **(Paper 16)**

Kartik Venkatraman Department of Aerospace Engineering, Indian Institute of Science, Benguluru, India **(Paper 14)**

Robert J. Young School of Materials, University of Manchester, Manchester M60 1QD, United Kingdom **(Paper 25)**

Section I
New Materials A: Functionally Graded Materials & Shape Memory Alloys

Functionally Graded Shells with Distributed Piezoelectric Sensors and Actuators for Active Vibration Control

S. Narayanan and V. Balamurugan

Abstract A nine-noded piezolaminated shell element with functionally graded material (FGM) is formulated based on the volume fraction power law distribution. One side of the FGM is assumed to be metal rich while the other side to be ceramic rich. Active vibration control capability of FGM plates and shells with attached PZT layers as integrated sensors/actuators to suddenly applied pressure loads and random excitation is investigated using the LQR optimal control scheme for different volume fraction power law exponents.

Keywords Active vibration control · FGM · Functionally graded · Piezoelectric

1 Introduction

Advanced materials can be "Functionally Graded" to provide an ideal combination of characteristics desired. The laminated composite structures can be tailored to design advanced structures with high stiffness to weight ratios, high strength to weight ratios, and better thermal and transport properties. However, the sharp demarcation of the properties of each layer at the interface between the two adjacent layers of laminated composite structures causes large interlaminar shear stresses which eventually may give rise to initiation of delamination. Such detrimental effects can be circumvented if the properties are smoothly varied across the thickness direction. FGMs are microscopically inhomogeneous composites with the material properties graded in a continuous manner. FGMs find many important applications, e.g., super-heat resistance materials for thermal barrier coatings and furnace liners,

S. Narayanan (✉)
Department of Mechanical Engineering, Indian Institute of Technology Madras,
Chennai 600 036, India
e-mail: narayans@iitm.ac.in

V. Balamurugan
Centre for Engineering Analysis and Design, Combat Vehicles R & D Establishment,
DRDO, Chennai 600 054, India
e-mail: balamuruganv_cvrde@yahoo.com

B. Dattaguru et al. (eds.), *IUTAM Symposium on Multi-Functional Material Structures and Systems*, IUTAM Bookseries 19, DOI 10.1007/978-90-481-3771-8_1,

vehicle and personal body armor, electromagnetic sensors, and graded refractive index materials for optical applications. These materials are microscopically heterogeneous and are typically made from isotropic components, for example metals and ceramics. A finite element formulation based on the classical laminated plate theory is presented by He et al. [1] for the shape and vibration control of FGM plates with integrated piezoelectric sensors and actuators. Liew et al. [2] presented a finite element formulation based on FSDT, for the feedback control of FGM shells in the frequency domain with use of piezoelectric sensors and actuators. Woo et al. [3] studied the nonlinear free vibration behavior of FGM plates using closed form solution. Balamurugan and Narayanan [4] studied active vibration control of FGM plates/shells using PZT sensors/actuators. In this paper, the nine noded piezolaminated shear flexible, field consistent shell finite element model [5] is extended for the analysis of FGM plates and shells based on the volume fraction power law distribution [4]. Active vibration control of FGM plates and shells with attached PZT sensor/actuator layers is also considered using the LQR optimal control scheme.

2 Formulation

A doubly curved piezolaminated FGM shell is considered as shown in Fig. 1, with the curvilinear co-ordinates x, y along the inplane direction and z along the thickness direction. Thin PZT layers bonded on its top and bottom surfaces act as distributed sensors and actuators. The FGM shell is assumed to be made of two materials, and its properties are graded through the thickness direction according to a volume fraction power law distribution, so that one surface is ceramic rich and another surface is metal rich. The material properties can be expressed as,

$$P_{eff}(z) = P_1 V_1(z) + P_2 \left(1 - V_1(z)\right) \tag{1}$$

$$V_1 = \left(\frac{2z + h}{2h}\right)^n \tag{2}$$

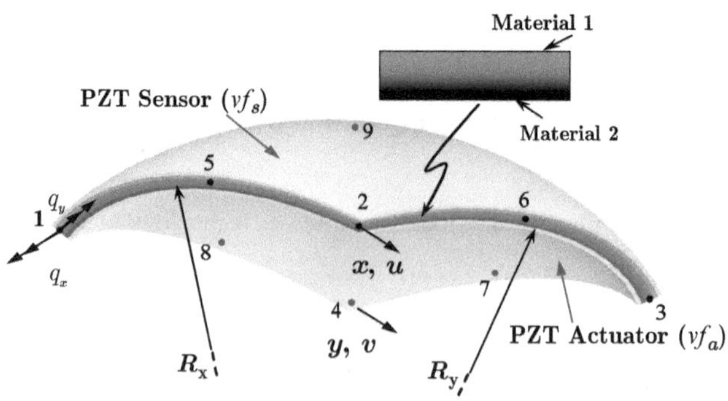

Fig. 1 Piezolaminated FGM shell

where P_{eff} is the effective material property of the FGM, and P_1 and P_2 are the properties of the materials 1 and 2 at the top and bottom surfaces, respectively and V_1 is the volume fraction of the constituent material 1 and n is the volume fraction exponent $(0 \leq n \leq \infty)$ and h is the thickness of the FGM shell. The material model is integrated in the nine-noded piezolaminated shell element [5].

The strain energy functional, U is given by,

$$U = \frac{1}{2}\int_{A_e}\left[\{\varepsilon_m\}^T[A]\{\varepsilon_m\} + \{\varepsilon_m\}^T[B]\{\varepsilon_b\} + \{\varepsilon_b\}^T[B]^T\{\varepsilon_m\} + \{\varepsilon_b\}^T[D]\{\varepsilon_b\}\right]dA$$
$$+\frac{1}{2}\int_{A_e}\{\gamma\}^T\left[\widetilde{A}\right]\{\gamma\}dA - \int_{V_p}\left(\{\varepsilon_m\}^T[e_p]^T\{E\} + z\{\varepsilon_b\}^T[e_p]^T\{E\} + \{\gamma\}^T[e_t]^T\{E\}\right)dV \tag{3}$$

where, $\{\varepsilon_m\}, \{\varepsilon_b\}, \{\gamma\}$ and $\{E\}$ are membrane stains, bending strains, shear strains and electric field. A_e and V_p are the area of the element and volume of the piezoelectric layer, respectively. The extensional stiffness coefficients, $[A_{ij}]$, bending-extensional stiffness coefficients, $[B_{ij}]$ and bending stiffness coefficients, $[D_{ij}]$, of the shell with $(i, j = 1,2,6)$ and transverse shear stiffness coefficients of the shell, $\left[\widetilde{A}_{rs}\right]$ $(r, s = 4,5)$, incorporating the FGM material model, are given by,

$$[[A_{ij}],[B_{ij}],[D_{ij}]] = \int_{-h/2}^{h/2}\left(\left(\left[Q_{ij}^1\right] - \left[Q_{ij}^2\right]\right)\left(\frac{2z+h}{2h}\right)^n(1,z,z^2) + \left[Q_{ij}^2\right](1,z,z^2)\right)dz \tag{4}$$

$$\left[\widetilde{A}_{rs}\right] = k_{\alpha r}k_{\alpha s}\int_{-h/2}^{h/2}\left(\left(\left[Q_{rs}^1\right] - \left[Q_{rs}^2\right]\right)\left(\frac{2z+h}{2h}\right)^n(1,z,z^2) + \left[Q_{rs}^2\right](1,z,z^2)\right)dz \tag{5}$$

where, $\left[Q^1\right]$ and $\left[Q^2\right]$ are the elastic properties of the materials 1 and 2. h is the thickness of the FGM shell. $k_{\alpha r}$, $k_{\alpha s}$ are shear correction factors $(= 5/6)$.

2.1 Strain Energy and Electrical Energy

Substituting for $\{\varepsilon_m\}, \{\varepsilon_b\}, \{\gamma\}$ and $\{E\}$ from equations in [5], in the above equation, U can be written as,

$$U = \frac{1}{2}\{q\}_e^T\int_{A_e}\begin{bmatrix}\wp_{\varepsilon_m} & \hat{N}\\ \wp_{\varepsilon_b} & \hat{N}\\ \wp_\gamma & \hat{N}\end{bmatrix}^T\begin{bmatrix}A & B & 0\\ B^T & D & 0\\ 0 & 0 & \tilde{A}\end{bmatrix}\begin{bmatrix}\wp_{\varepsilon_m} & \hat{N}\\ \wp_{\varepsilon_b} & \hat{N}\\ \wp_\gamma & \hat{N}\end{bmatrix}|J|d\xi d\eta\{q\}_e$$
$$-\{q\}_e^T\int_{V_p}\left[\left[\wp_{\varepsilon_m}\hat{N}\right]^T[e_p]^T + z\left[\wp_{\varepsilon_b}\hat{N}\right]^T[e_p]^T + \left[\wp_\gamma\hat{N}\right]^T[e_t]^T\right]\{B_\phi\}dz|J|d\xi d\eta\{\hat{\phi}\}_e$$
$$= \frac{1}{2}\{q\}_e{}^T[K_{uu}]_e\{q\}_e + \{q\}_e{}^T[K_{u\phi}]_e\{\hat{\phi}\}_e \tag{6}$$

where $|J|$ is the Jacobian of transformation between the global coordinates and the isoparametric coordinates. $[K_{uu}]_e$ is the element stiffness matrix and $[K_{u\phi}]_e$ is the element elastic–electric coupling stiffness matrix, which are given by,

$$[K_{uu}]_e = \int_{A_e} \begin{bmatrix} \wp_{\varepsilon m} & \hat{N} \\ \wp_{\varepsilon b} & \hat{N} \\ \wp_{\gamma} & \hat{N} \end{bmatrix}^T \begin{bmatrix} A & B & 0 \\ B^T & D & 0 \\ 0 & 0 & \tilde{A} \end{bmatrix} \begin{bmatrix} \wp_{\varepsilon m} & \hat{N} \\ \wp_{\varepsilon b} & \hat{N} \\ \wp_{\gamma} & \hat{N} \end{bmatrix} |J| d\xi\, d\eta\,; \tag{7}$$

$$[K_{u\phi}]_e = -\int_{V_p} \left[\left[\wp_{\varepsilon_m} \hat{N}\right]^T [e_p]^T + z\left[\wp_{\varepsilon_b} \hat{N}\right]^T [e_p]^T + \left[\wp_{\gamma} \hat{N}\right]^T [e_t]^T \right] \{B_\phi\} dz |J| d\xi d\eta \tag{8}$$

From the electric energy expression as described in [5], the electric stiffness matrix are obtained as,

$$\left[K_{\phi\phi}\right]_e = \int_{V_p} \left[B_\phi\right]^T [\epsilon] \left[B_\phi\right] dz |J| d\,\xi\, d\,\eta \tag{9}$$

where, $[\in]$ are permittivity constants (F/m).

2.2 Kinetic Energy

The element kinetic energy is given by,

$$\begin{aligned} T &= \frac{1}{2} \int_{A_e} \left[P(\dot{u}^{\circ 2} + \dot{v}^{\circ 2} + \dot{w}^{\circ 2}) + I(\dot{\theta}_x^2 + \dot{\theta}_y^2) \right] dA \\ &= \frac{1}{2} \{\dot{q}\}_e^T \int_{A_e} \hat{N}^T \left[\hat{M}_\rho\right] \hat{N} |J| d\,\xi\, d\,\eta \{\dot{q}\}_e = \frac{1}{2} \{\dot{q}\}_e^T [M_{uu}]_e \{\dot{q}\}_e \end{aligned} \tag{10}$$

where, A_e is the area of the element and $\left[\hat{M}_\rho\right] = \begin{bmatrix} P & 0 & 0 & 0 & 0 \\ 0 & P & 0 & 0 & 0 \\ 0 & 0 & P & 0 & 0 \\ 0 & 0 & 0 & I_P & 0 \\ 0 & 0 & 0 & 0 & I_P \end{bmatrix}$ and P & I_P given in [5], incorporating the FGM material model are replaced by,

$$\langle P, I_P \rangle = \int_{-h/2}^{h/2} \left((\rho^1 - \rho^2) \left(\frac{2z + h}{2h} \right)^n (1, z^2) + \rho^2 (1, z^2) \right) dz \tag{11}$$

where, ρ^1 and ρ^2 are density of the materials 1 and 2, respectively. $[M_{uu}]_e$ is the mass matrix given by,

$$[M_{uu}]_e = \int_A \hat{N}^T \left[\hat{M}_\rho\right] \hat{N} |J| d\,\xi\, d\,\eta \tag{12}$$

From the virtual work done by the surface force and applied electrical charge density as described in [5], the expressions for the applied elemental mechanical force and applied elemental electric charge are obtained.

2.3 *Piezoelastic Governing Equations of Motion*

Using the Hamilton's principle, the Langrange's equations of motion, for an element can be written as,

$$\begin{bmatrix} [M_{uu}]_e & [0] \\ [0] & [0] \end{bmatrix} \begin{Bmatrix} \{\ddot{q}\}_e \\ \{\ddot{\hat{\phi}}\}_e \end{Bmatrix} + \begin{bmatrix} [K_{uu}]_e & [K_{u\phi}]_e \\ [K_{\phi u}]_e & [K_{\phi\phi}]_e \end{bmatrix} \begin{Bmatrix} \{q\}_e \\ \{\hat{\phi}\}_e \end{Bmatrix} = \begin{Bmatrix} \{F_u\}_e \\ \{F_\phi\}_e \end{Bmatrix} \quad (13)$$

Equations (13) can be combined as,

$$[M_{uu}]_e\{\ddot{q}\}_e + \left[[K_{uu}]_e - [K_{u\phi}]_e[K_{\phi\phi}]_e^{-1}[K_{\phi u}]_e\right]\{q\}_e = \{F_u\}_e - [K_{u\phi}]_e[K_{\phi\phi}]_e^{-1}\{F_\phi\}_e \quad (14)$$

When second equation of (13) is applied to sensors, the sensed voltage is given by,

$$\{\hat{\phi}_s\}_e = -[K_{\phi\phi}]_{se}^{-1}\left([K_{\phi u}]_{se}\{q_s\}_e\right) \quad (15)$$

The global equations of motion is obtained by assembling elemental equations as,

$$[M_{uu}]\{\ddot{q}\} + [C_{uu}]\{\dot{q}\} + \left[[K_{uu}] - [K_{u\phi}][K_{\phi\phi}]^{-1}[K_{\phi u}]\right]\{q\} = \{F_u\} - [K_{u\phi}]\{\hat{\phi}_a\} \quad (16)$$

where, $[M_{uu}]$, $[K_{uu}]$, $\left([K_{u\phi}] = [K_{\phi u}]^T\right)$, $[K_{\phi\phi}]$ and $\{F_s\}$ are the corresponding global quantities, $\{\hat{\phi}_a\}$ is the actuator voltage vector. The governing global equations of motion given by (14) are converted into the modal state space form.

The active vibration control of FGM structures with PZT sensors/actuators are studied using the LQR optimal control [4, 5]. The feedback control system is designed to minimize a cost function or a performance index which is proportional to the required measure of the system's response. A state feedback rather than output feedback is adopted to enhance the control performance. The cost function used in this case is given by,

$$J = \int_0^\infty \left(\{\overline{y}\}^T[\overline{Q}]\{\overline{y}\} + \{\hat{\phi}_a\}^T[\overline{R}]\{\hat{\phi}_a\}\right) dt \quad (17)$$

where, $[\overline{Q}]$ and $[\overline{R}]$ are the semi-positive-definite and positive-definite weighting matrices on the outputs and control inputs, respectively.

3 Results and Discussion

3.1 Aluminum–Zirconia FGM Plate

The finite element formulation for the FGM structures is applied to the following free vibration and static problems, essentially for validation of the present finite element formulation and the active vibration control of FGM structures. As the first example, an aluminum–zirconia FGM simply supported plate, studied by Woo et al. [3] is considered. The size of the plate is 200 × 200 mm with a thickness of 10 mm and is as shown in Fig 2. The variation of zirconia through the plate thickness is as shown in Fig 3. The plate is assumed to be simply supported and modeled with the present finite element using a 4 × 4 mesh. The non-dimensional natural frequencies of vibration, for the volume fraction index of $n = 0, 0.5, 2$ and 1,000 are given in Table 2. The non-dimensional natural frequencies are given by, $\overline{\omega}_n = 2\pi\left(12(1-\nu_m^2)\rho_m a^4/(E_m h^2)\right)\omega_n$; where, E_m, ρ_m and ν_m are the Young's modulus, density and Poisson's ratio of the metal used in the FGM (Table 1), a and h

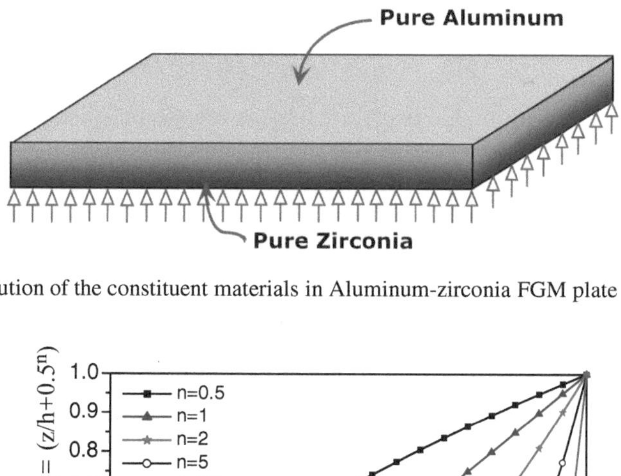

Fig. 2 Distribution of the constituent materials in Aluminum-zirconia FGM plate

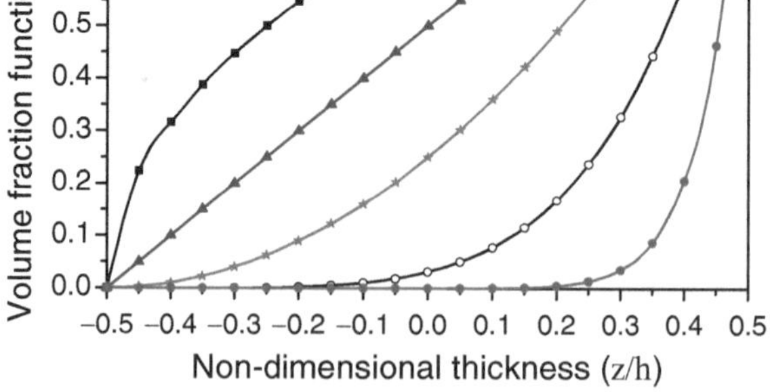

Fig. 3 Volume fraction function $V_l = (z/h + 0.5)^n$ with the non-dimensional thickness

Table 1 Material properties of aluminum–zirconia FGM

Material	Young's modulus (N/m^2)	Poisson's ratio	Density (kg/m^3)
Aluminum	70×10^9	0.3	2,707.0
Zirconia	151×10^9	0.3	3,000.0

Table 2 Non-dimensional natural frequencies $\overline{\omega}_n$ of aluminum–zirconia FGM plate

Volume fraction power law exponent	Woo et al. [3]	Present element					
	Mode 1	Mode 1	Mode 2	Mode 3	Mode 4	Mode 5	Mode 6
$n = 0$ (Zirconia)	27.50	27.31	67.92	67.92	107.4	137.68	137.68
$n = 0.5$	24.65	24.76	61.22	61.22	96.89	124.06	124.20
$n = 2$	22.80	23.04	56.46	56.46	89.37	114.14	114.44
$n = 1000$ (Aluminum)	19.65	19.61	48.76	48.76	77.11	98.84	98.84

are the length and thickness of the plate and ω_n is the natural frequency. It can be noted that $n = 0$ corresponds to pure zirconia and a large value of n, say $n = 1{,}000$ corresponds to pure aluminum. In Table 2, the first natural frequency for different values of n, obtained using a closed form solution given by Woo et al. [3] (Eq. (10)) are given. The first six natural frequencies and higher natural frequencies (nondimensional) for various values of n obtained using the finite element formulation presented in this paper are also given alongside in the table. It is observed that the first natural frequency for different volume fraction ratios, n, obtained by the finite element method matches closely with the first natural frequency obtained by the closed form expression.

3.2 *Active Vibration Control of Aluminum–Zirconia FGM Shell Panel with PZT Sensor and Actuator Patches*

A cantilevered FGM spherical shell panel made of aluminum (metal) and zirconia (ceramic) of size 400×200 mm shown in the Fig. 4, is considered. The thickness of the shell is 2 mm and is bonded with PZT sensor/actuator of size 100×100 mm having a thickness of 0.4 mm. This spherical panel is modeled with the present FGM shell element with 4×4 mesh size. It is subjected to suddenly applied radial pressure of 100 N/mm^2 and the resulting vibration is controlled using the LQR control with a controller switched on after 1 s. The first six natural frequencies of the above FGM shell for $n = 0, 0.5, 2, 15$ and 100 are given in Table 3.

The displacement of point P_1 of the FGM shell with LQR control, sensor voltage and actuator voltage for $n = 0.5$ and 15 are shown in Fig. 5 ($\overline{Q} = 10^9$ and for $\overline{R} = 1$). As n increases it can be noted the response also increases. This is so because, as n increases the volume fraction of aluminum increases and that of zirconia decreases resulting in less overall stiffness of the FGM, as the Young's modulus of zirconia is more than that of aluminum. It can also be noted that the control effectiveness remains unaffected with n and the control voltage increases.

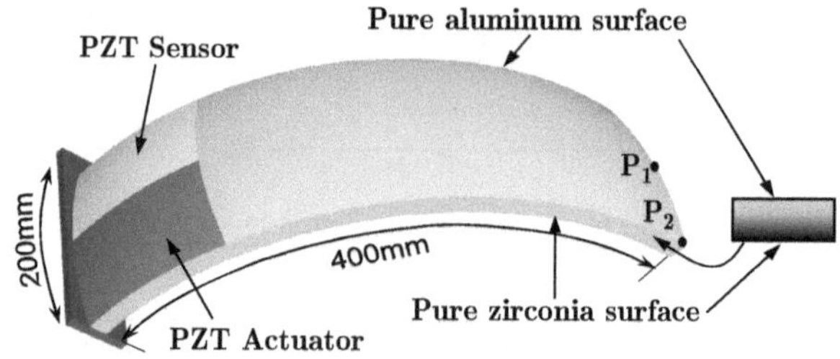

Fig. 4 Cantilevered aluminum–zirconia FGM spherical shell panel with PZT sensor and actuator subjected to the radial pressure

Table 3 Natural frequencies of the cantilevered aluminum–zirconia FGM spherical shell panel for various cases of volume fraction power law exponent n

	Natural frequencies (Hz)				
Mode no.	$n = 0.0$	$n = 0.5$	$n = 2.0$	$n = 15.0$	$n = 100.0$
1	20.8	18.9	17.5	16.4	15.9
2	66.8	60.3	55.7	52.4	50.2
3	114.8	104.1	95.8	89.5	86.0
4	210.4	189.9	175.1	164.5	157.3
5	297.9	269.3	247.5	230.5	220.6
6	406.7	366.6	337.6	316.0	301.7

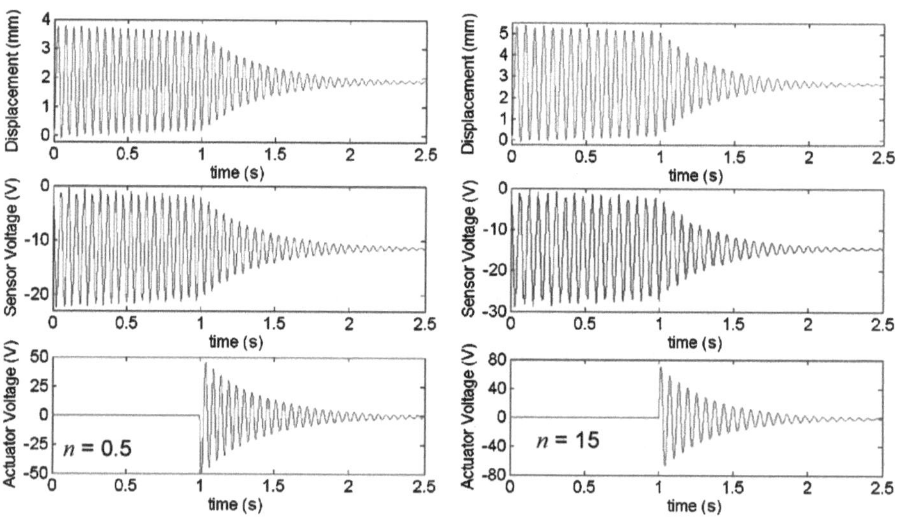

Fig. 5 Displacement at point P_1 of the FGM shell panels ($n = 0.5$ and $n = 15$) and the corresponding actuator voltages (for LQR control parameter, $\overline{Q} = 10^9$)

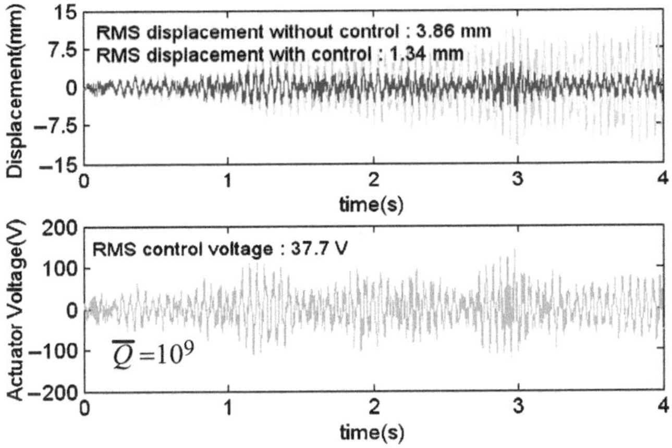

Fig. 6 Displacement of point 'P$_2$' of the FGM shell panel ($n = 0.5, 15$) without and with LQR control and the corresponding actuator voltage

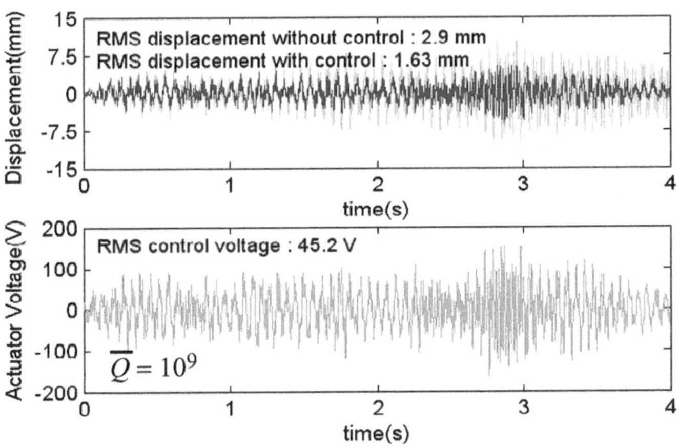

Fig. 7 Displacement of point 'P$_2$' of the FGM shell panel ($n = 0.5, 15$) without and with LQR control and the corresponding actuator voltage

LQR control of the functionally graded aluminum–zirconia cantilevered shell panel to white noise excitation with PSD of 2.5×10^{-3}N^2/(rad/s) acting at point P$_2$ with the PZT sensor and actuator located as shown in Fig. 4 is considered next. The displacement corresponding to point P$_2$ of the FGM shell panel for $n = 0.5$ and 15 without and with LQR control ($\overline{Q} = 10^9$; $\overline{R} = 1$) and the corresponding actuator voltages are shown in Figs. 6 and 7, respectively.

The rms displacement and control voltages indicated in the figures are also given in Table 4, indicating the effectiveness of the considered location of the PZTs and LQR control in controlling the random vibration of the FGM shell. The frequency response of the displacement at point P$_2$ for $n = 0.5$ and 15, due to harmonic

Table 4 RMS displacement responses and RMS control voltages

Volume fraction power law exponent	RMS response without control	LQR parameter $\overline{Q} = 10^9$		LQR parameter $\overline{Q} = 10^{10}$	
		RMS response with control	RMS control voltage	RMS response with control	RMS control voltage
$n = 0.5$	3.86	1.34	37.7	0.78	72.4
$n = 15$	2.9	1.63	45.2	0.98	90.6

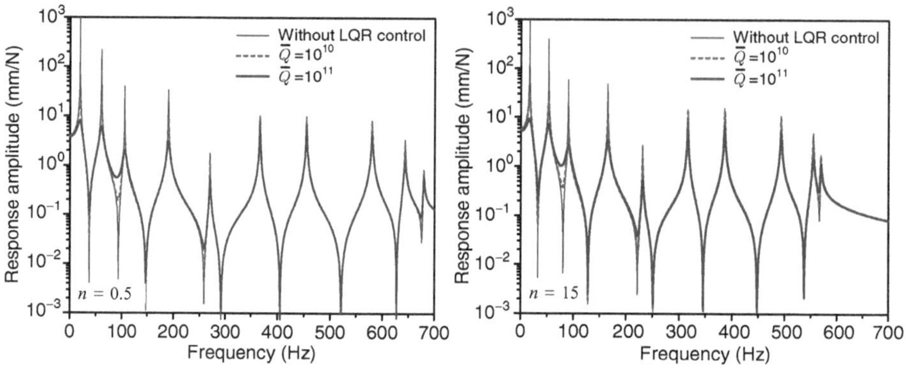

Fig. 8 The displacement at point P_2 of the FGM shell panel ($n = 0.5$ and $n = 15$) with LQR control in the frequency domain

excitation at P_2 is shown in Fig. 8. It can be noted that the location of the PZT sensor and actuator and the LQR control provide vibration damping for all the eight modes considered for estimating the response. The natural frequencies shift towards the left as n increases, which are due to the fact that as n increases, the volume fraction of aluminum in the FGM increases, resulting in the reduction of the overall stiffness.

4 Summary and Conclusions

In this paper, a finite element model for FGM plates and shells based on the nine noded piezoelectric shell element including piezoelectric bonded layers is presented. The FGM model is based on a volume fraction power law distribution. The finite element formulation is validated for static shape control and natural frequencies with results available from literature. The static shape control of the FGM plate with PZT actuators is investigated. Active vibration control to suddenly applied pressure load and random excitation of a cantilevered FGM shell with Aluminum (metal) and zirconia (ceramic) is studied using LQR control with the different values of weighting parameters $\overline{Q}$ with $\overline{R} = 1$. The effectiveness of the active vibration control is demonstrated both for suddenly applied pressure load and random excitation. In this case, the natural frequencies decrease with increase in '*n*' due to the decrease in the stiffness of the aluminum–zirconia FGM.

References

1. He, X.Q., T.Y. Ng, S. Sivashanker and K.M. Liew (2001) Active control of FGM plates with integrated piezoelectric sensors and actuators, *International Journal of Solids and Structures*, 38, 1641–1655.
2. Liew, K.M., X.Q. He and S. Kitipornchai (2004) Finite element method for the feedback control of FGM shells in the frequency domain via piezoelectric sensors and actuators, *Comput. Methods Appl. Mech. Engrg.*, 193, 257–273.
3. Woo, J., Meguida, S.A, and L.S. Ongb (2006) Nonlinear free vibration behavior of functionally graded plates,. *Journal of Sound & Vibration*, 289, 595–611.
4. Balamurugan, V. and S. Narayanan (2004), Active Control of FGM plates using distributed piezoelectric sensors/actuators, *XXI International Congress of Theoretical and Applied Mechanics (XXI ICTAM), Warsaw, Poland, August 15–21, 2004* (in CDROM).
5. Balamurugan, V. and S. Narayanan (2001), Shell finite element for smart piezoelectric composite plate/shell structures and its application to the study of active vibration control, *Finite Element in Analysis and Design*(37), 713–738.

A Numerical Investigation of the Dynamic Behaviour of Functionally Graded Foams

Stephen Kiernan, Liang Cui, and Michael D. Gilchrist

Abstract Two Finite Element models approximating the dynamic behaviour of functionally graded foam materials (FGFMs) have been developed under free weight drop impact and Kolsky wave propagation conditions. The FGFM is modeled by discretising the material into a large number of layers through the foam thickness. Each layer is described by a unique constitutive cellular response, which is derived from the initial relative density, ρ^*, unique to that layer. Large strain uniaxial compressive tests at strain rates of 0.001, 0.01 and 0.1/s were performed on expanded polystyrene (EPS) and ALPORAS® Aluminium (Al) foam and their $\sigma - \varepsilon$ response was used as input to a modified constitutive model from the literature. Simulations were then performed on both uniform and graded specimens. For both impact and wave propagation conditions it is found that under certain conditions an FGFM can outperform a uniform foam of equivalent density in terms of reducing peak accelerations imparted from an impact, or mitigating stress wave magnitudes through increased plastic deformation. These properties provide significant insight into the hypothesised behaviour of FGFMs and elucidate the potential for the future use in the design of next generation cushioning structures.

Keywords Aluminium foam · Energy absorption · Impact · Polymer foam · Stress wave propagation

1 Introduction

Cellular foams are widely used in energy absorbing applications where it is important to minimise the peak acceleration of the impacting body [1], e.g. packaging of fragile goods, protective headgear [2–6] and body garments. This is due to their low volume fraction of solid material and their complex microstructure which allows

---

S. Kiernan, L. Cui, and M.D. Gilchrist (✉)
School of Electrical, Electronic & Mechanical Engineering, University College Dublin, Dublin, Ireland
e-mail: stephen.kiernan@ucd.ie; liang.cui@ucd.ie; michael.gilchrist@ucd.ie

B. Dattaguru et al. (eds.), *IUTAM Symposium on Multi-Functional Material Structures and Systems*, IUTAM Bookseries 19, DOI 10.1007/978-90-481-3771-8_2,

large degrees of plastic crushing to occur at a fairly constant plateau stress value. This plastic crushing at a constant stress will continue until, depending on the initial density, a densification strain is reached when cell walls and struts impinge on one another and further crushing is of the material matrix itself rather than the foam cells. Understanding their dynamic stress–strain behaviour at finite deformations is therefore essential in order to predict their performance as cushioning materials.

The compressive static and impact loading response of polymeric foams has been previously characterised [7] using energy absorption and efficiency diagrams. They showed that, for a particular density, a foam is most efficient at absorbing the kinetic energy, KE, of an impact over a limited range of stress; if the imparted stress is too little the foam will not yield, too great and the foam will densify and become much like a solid. It is hypothesised that by means of a functionally graded foam, it may be possible to incorporate a continuously varying density in either one, two, or three dimensions to improve the energy absorbing efficiency over a wider range of stress. Work supporting this hypothesis has previously been presented by [8].

The Split Hopkinson Pressure Bar technique has proved to be extremely versatile in material characterization and has grown from its original configuration for compression testing to include tension, torsion and fracture testing [9]. It has been used to characterise the dynamic response of a multitude of materials such as soils [10], composites [11], and metals [12]. Foams [13] and virtual FGFMs [14] have also been studied using the SHPB technique. From 1-D elastic wave theory, $\sigma_s(t)$, $\varepsilon_s(t)$, and $\dot{\varepsilon}_s(t)$ of a SHPB specimen are obtained from the transmitted, (ε_{T}), and reflected, (ε_{R}), found from the induced strain pulses measured in the apparatus bars from the equations:

$$\dot{\varepsilon}_s(t) = -\frac{2C_0}{L_s}\varepsilon_{\mathrm{R}}(t) \tag{1}$$

$$\varepsilon_s(t) = -\frac{2C_0}{L_s}\int_0^t \varepsilon_{\mathrm{R}}(t)dt \tag{2}$$

$$\sigma_s(t) = E_b\frac{A_b}{A_s}\varepsilon_{\mathrm{T}}(t) \tag{3}$$

where $\dot{\varepsilon}_s(t)$, $\varepsilon_s(t)$, $\sigma_s(t)$ are strain rate, strain and stress respectively. C_0 is the wave speed within the bar, L_s is the length or thickness of the specimen, A_s and A_b are the cross-sectional area of the specimen and the bar, respectively, and E_b is Young's modulus of the bar.

Nature has employed Functionally Graded Materials (FGMs) to optimise weight to strength ratios in response to non-uniform load distributions (e.g., cortical/trabecular bone, bamboo) and ensure no peak stresses occur at any point in the structure [15]. This is an adaptive process, evolving over time, but increasingly designers are investigating methods of optimising structures through gradient material properties. The current study aims to design a virtual, novel FGFM which contains micro-scale cells varied continuously in a predefined manner for the purpose of improving its energy absorbing characteristics under physical impact and wave propagation conditions.

This paper first describes the methodology of obtaining the multiple $\sigma - \varepsilon$ curves required to describe the FGFM. Sections 2.1 and 2.2 then briefly detail the models used to approximate an FGFM under striker impact and Kolsky wave conditions respectively, along with the indicative results. Section 3 discusses these results with respect to improved energy management.[1] Finally, Section 4 presents the important conclusions from this work.

2 Methodology

Large strain uniaxial compressive tests, using a displacement controlled compression tower, at strain rates of 0.001, 0.01 and 0.1/s were performed on five EPS foams (50 × 50 × 50 mm) with densities of 15, 20, 25, 50, and 64 kg/m^3, and one ALPORAS® Al foam of density ≈ 250 kg/m^3 to determine their $\sigma - \varepsilon$ response–compressive curves are required as input to the ABAQUS [16] crushable foam model used in this work. Obtaining complete experimental $\sigma - \varepsilon$ curves for the many layers required in each model (up to 50 layers) was impractical so a FORTRAN program, incorporating an existing model from the literature that describes the uniaxial compressive behaviour of uniform foam [17], was written. Experimental data was used to calibrate the program's output, and based on the empirical observation that a foam's ρ^* largely dictates its compressive $\sigma - \varepsilon$ behavior, arbitrary $\sigma - \varepsilon$ curves could be generated through interpolation and extrapolation of ρ^* from the available experimental data. The curves generated by this program were validated against the $\sigma - \varepsilon$ curves of the EPS and ALPORAS® physical compression tests. By applying a unique $\sigma - \varepsilon$ definition to each layer using the ABAQUS crushable foam model, a quasi-graded (see Table 1) cellular material could then be virtually created. This

Table 1 Material gradients with density ranges used in striker impact simulations

Gradients	Density range[a] (kg/m^3) $\Delta\rho = 20$				
Uniform	44	54	64	84	104
Logarithmic	59.2–39.2	69.2–49.2	79.2–59.2	99.2–79.2	119.2–99.2
Square root	57.3–37.3	67.3–47.3	77.3–57.3	97.3–77.3	117.3–97.3
Linear	54.0–34.0	66.0–44.0	74.0–54.0	94.0–74.0	114.0 - 94.0
Quadratic	50.8–30.8	60.8–40.8	70.8–50.8	90.8–70.8	110.8–90.8
Cubic	49.2–29.2	59.2–39.2	69.2–49.2	89.2–69.2	109.2–89.2
Gradients	Density range[a] (kg/m^3) $\Delta\rho = 40$				
Uniform	44	54	64	84	104
Logarithmic	74.4–34.4	84.4–44.4	94.4–54.4	114.4–74.4	134.4–94.4
Square root	70.6–30.6	80.6–40.6	90.6–50.6	110.6–70.6	130.6–90.6
Linear	64.0–24.0	74.0–34.0	84.0–44.0	104.0–64.0	124.0 - 84.0
Quadratic	57.5–17.5	67.5–27.5	77.5–37.5	97.5–57.5	117.5–77.5
Cubic	54.2–14.2	64.2–24.2	74.2–34.2	94.2–54.2	114.2–74.2

[a]Densities are derived from EPS compression tests.

[1] Reducing peak g, increasing stress energy mitigation.

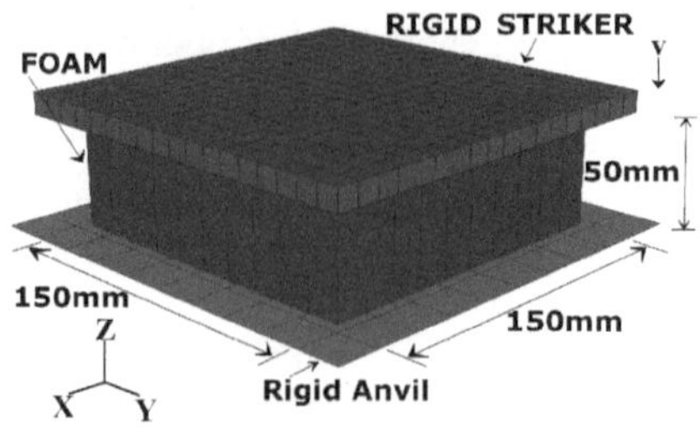

Fig. 1 The foam is composed of 50 layers, each defined with a unique cellular response. In addition to uniform foam, five gradients were investigated (Table 1), which all monotonically decrease from the striker to anvil face. Preliminary simulations showed that monotonically decreasing the density from the striker to anvil face resulted in a decrease in peak *g* values[2]. The anvil is encastre constrained and the striker may only move in the Z-direction. A single impact velocity of 5.425m/s was chosen

technique has been implemented on the liner of a generic equestrian helmet finite element model [18, 19]. All simulations were carried out using reduced integration, 3-D, 8-node hexahedral elements with hourglass control, as an explicit dynamic analysis and solved using an explicit central difference integration rule. Strain rate effects are not considered within the scope of this study.

2.1 *Striker Impacts*

Figure 1 illustrates the virtual test configuration and Table 1 shows the test parameters used in the physical impact simulations, and refers to a single striker mass and density range ($\Delta\rho$). Striker masses of 1, 2, 4, 6, 8, 10, 12 and 14 kg were also simulated, giving a total of 440 simulations to parametrically study the effect of the selected variables.

2.2 *Split Hopkinson Pressure Bar*

A three dimensional model comprised of projectile, incident bar, specimen and transmission bar was created to simulate one dimensional wave propagation through an FGFM specimen, the bar strains of which were validated against experiments carried out using a wide area SHPB apparatus. This ensures that valid boundary conditions are applied to the virtual FGFM specimens. The experimental transmission and incident bars were made from 30% glass filled nylon, to minimise impedance mismatch with the specimen, and were 1,000 mm long with a 50 mm diameter. These were approximated as linear elastic during modeling. Projectile lengths of 150 mm and 250 mm were simulated at the same impact velocities used during the validation experiments; 19 and 13 m/s respectively. A virtual specimen length of 300 mm is used to illustrate the differing response between a uniform and a

Table 2 Material gradients with density ranges used in SHPB simulations

	Density range[a] (kg/m^3)		
Gradients	$\Delta\rho = 100$	$\Delta\rho = 50$	$\Delta\rho = 20$
Uniform	250.0	250.0	250.0
Logarithmic	179.3–279.3	—	—
Square root	184.4–284.4	—	—
Linear	200.0–300.0	225–275	240–260
Quadratic	215.8–315.8	—	—
Cubic	223.7–323.7	—	—

[a]Densities are derived from ALPORAS® compression tests.

functionally graded foam. This methodology is sound because results are taken directly from the F.E. specimen rather than derived from the bar strains, thus negating the assumptions that Eqs. (1), (2) and (3) are derived from. During the validation experiments wave dispersion was observed in the incident and transmission bars due to their viscoelastic nature and so wave corrections using Fourier Theory and following the works of [20, 21] was carried out during post-processing to shift the incident and transmissions waves to the bar specimen interfaces. Table 2 shows the test parameters used in the simulations.

3 Discussion

3.1 Striker Impacts

The shape of Fig. 2 is indicative of an FGFMs impact response for the different average densities examined. For low kinetic energy impacts, the graded foam performs better[2] than the uniform foam (e.g., $\rho_{\text{average}} = 54$ kg/m^3, mass = 1 kg) and the convex gradients (e.g., quadratic) perform better than the concave gradients (e.g., square root). However, as the impacting mass (and therefore KE) is increased to 14 kg, an opposite trend is observed (see Fig. 2). The marked improvement of the FGFM over the uniform foam in reducing the peak acceleration of the lower energy impacts can be explained as follows. It has been stated previously that foam with a single density is most efficient at absorbing energy when it works within the plateau strain region, up to densification, as it absorbs most energy under large plastic strains with little corresponding increase in stress. From simulation it was found that for a uniform foam of 44 kg/m^3, the stress imparted at the time of peak acceleration was 198 kPa for a striker mass of 1 kg and was 581 kPa for a striker mass of 14 kg. From the experimental $\sigma - \varepsilon$ compression tests and FORTRAN program (Section 2) it can be deduced that 44 kg/m^3 EPS foam will yield at about 310 kPa and thus will not yield when struck with a striker of 1 kg at 5.425 m/s, but rather

[2] Greatly reduced peak accelerations relative to that of a uniform foam of equal ρ_{average}.

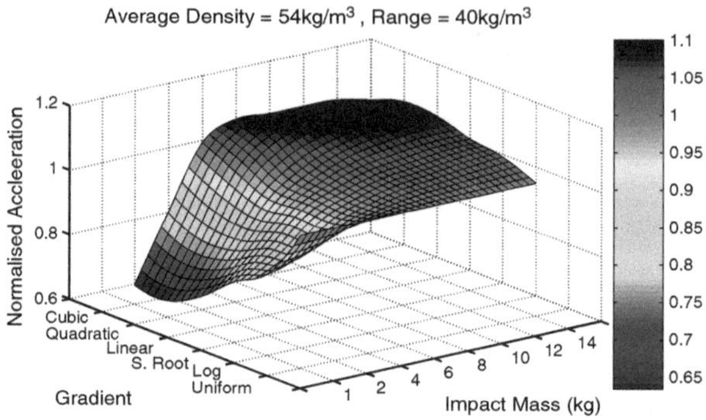

Fig. 2 The influence of multiple input parameters (in this case impact mass and material gradient) on a desired output, such as peak acceleration can be clearly seen through the use of design surface plots. Acceleration has been normalised against the peak acceleration of the striker impacting a uniform foam of equal ρ^*_{average}

will behave elastically with very little deformation, resulting in high peak accelerations. However, when struck with a striker of 14 kg at 5.425 m/s it will absorb the corresponding kinetic energy within the plateau stress region up to 0.6 strain. The FGFMs perform distinctly better than the uniform foam when absorbing the lower energies due to their spatially varying yield surface, a direct result of the density gradient. From Table 1, for example, the density of a cubically varying foam with $\rho_{\text{average}} = 44\ \text{kg/m}^3$ and $\Delta\rho = 40\ \text{kg/m}^3$ will vary from 54.2 to 14.2 kg/m^3. At 14.2 kg/m^3, local plastic deformation was found from simulations to initiate at about 100 kPa, deforming to almost 0.7 strain, and approximately 20% by volume (14.2–28 kg/m^3) of the graded foam will yield plastically at a stress of 198 kPa. This is in stark contrast to the equivalent uniform foam, which exhibits no yielding at this stress level. As the kinetic energy of the striker is increased the advantage gained by a varying yield surface diminishes rapidly. Low yielding regions of the FGFM are no longer effective and local deformation beyond their densification strains occurs while mitigating only a small fraction of the total energy. Results show that a uniform 44 kg/m^3 foam experiences 0.54 strain at the incident surface and 0.52 strain at the distal surface when impacted by a 14 kg striker at 5.425 m/s. In contrast, the quadratically varying FGFM deforms locally to only 0.2 strain at the incident surface and yet there is 0.98 strain at the distal face. Intuitively, and from previous work [7], it is more advantageous for a foam's entire volume to deform up to, but not beyond, its densification strain if it is to act most effectively as a cushioning structure.

3.2 *Split Hopkinson Pressure Bar*

Figures 3(a) and (b) shows strain-time traces of the strain wave recorded in the virtual transmission bar for a uniform foam of density 250 kg/m^3 and five gradiented

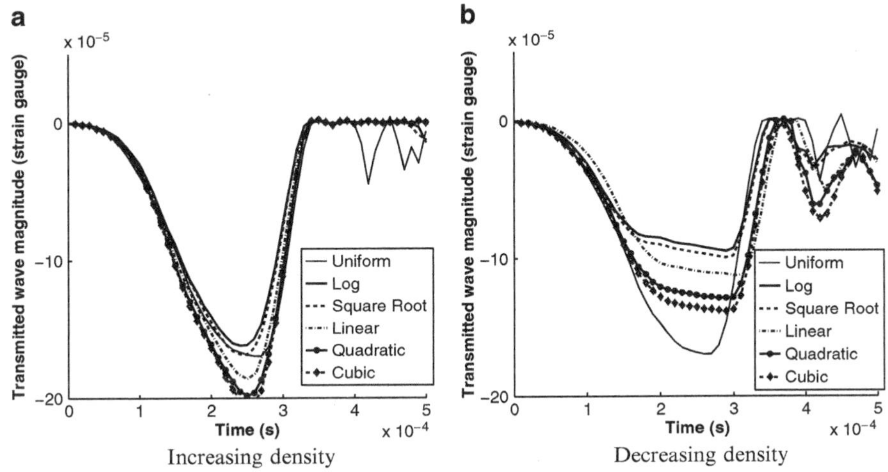

Fig. 3 Comparison of transmitted wave as measured from the strain gauge for the uniform specimen and the FGFM specimens ($\Delta\rho$=100 kg/m^3)

foams for the cases of increasing and decreasing density in the direction of wave propagation respectively. Although Fig. 3(a) shows logarithmic and square root functions to be slightly more efficient at reducing the amount of stress transmitted, Fig. 3(b) highlights the importance of the orientation of the material gradient. The decrease in transmitted strain (and consequently stress) achieved by employing a gradient foam indicates that the specific stress absorbed (Pa $\cdot$ m$^{-3}\cdot$ kg^{-1}) by the gradient foam, from the incident wave, is increased. The main mechanism by which a foam absorbs energy or attenuates stress is through plastic deformation. As the stress wave propagates along the z-direction through a uniform foam, it will plastically deform the foam until its magnitude σ_w becomes lower than the foam's yield stress σ_y. Once this occurs, the remainder of the wave will propagate elastically with little additional energy dissipated since no more plastic deformation will occur. It can be deduced that if σ_y were to decrease in the z-direction there would be a time delay in σ_w attenuating to the value of σ_y, allowing greater amounts of energy to be absorbed plastically. This is exactly what an FGFM is designed to achieve: the yield stress σ_y of a graded foam diminishes in the z-direction as the density decreases. Ideally, in order to maximise the amount of plastic deformation in a graded foam, the yield stress σ_y at any point z should be equal to the stress wave magnitude σ_w at z, while the change in yield stress between any two points z and $z + \delta z$ should be greater than $\delta\sigma_w$ over the distance δz. Symbolically:

$$\sigma_y \mid_z = \sigma_w \mid_z, \quad \forall z \tag{4}$$

$$\frac{\partial \sigma_y}{\partial z} \geq \frac{\partial \sigma_w}{\partial z} \tag{5}$$

4 Conclusions

It has been shown that a FGFM can, under certain conditions, outperform traditional foams when employed as cushioning structures.

- It is shown that an FGFM can exhibit superior energy absorption over equivalent uniform foams under low energy impacts, and that convex perform better than concave gradients. This advantage is negated when the impact energy becomes significantly high such that low density regions of the graded foam become ineffective at bearing the higher load and densify after absorbing only a small fraction of the total energy. What constitutes a 'high energy impact' is somewhat difficult to define but will depend on $\rho_{average}$, the density gradient, and the matrix composition of the foam.
- Five density gradients were examined and compared, although others could certainly have been chosen. However, determining rigorously the ideal gradient for cushioning applications would require the solving of an optimisation problem.
- The variation in a cellular FGFM would make it extremely difficult to dynamically test and obtain valid results using a traditional SHPB apparatus, due to the limiting assumptions that are made for Eqs. (1)–(3). Variables that a FGM would introduce, such as a spatially varying yield stress, varying density and/or varying strain rate sensitivities, would almost certainly invalidate any efforts in ensuring stress–strain uniformity, in which case only an average stress–strain result could be obtained. This problem is overcome by virtually testing such materials since Eqs. (1)–(3) are not necessary to calculate the constitutive outputs.
- FGFMs are capable of reducing the duration of high accelerations during an impact event. This property could have wide implications in the head protection industry as many head injury criteria (e.g., HIC [22]) rely on acceleration durations as indicators of the likelihood for a person suffering significant head trauma. In this respect, protective headgear, e.g., safety helmets, employing FGFMs as the liner constituent may be advantageous to the wearer in reducing the risk of brain injury after a fall.
- Traditionally, many helmet certification standards (e.g., [23]) require a helmet to keep the acceleration of a headform dropped from a single drop height below some certain target level – achieving this is quite simple. However, recent helmet standards (e.g., [24]) demand that helmets be effective at multiple drop heights, thus simulating both high and low energy impacts. This can be more difficult to achieve with current helmet liner technologies. FGFMs have been shown to exhibit significant advantages under low energy impact conditions while still performing nearly as well as their uniform counterpart under high energy conditions. These foams, carefully manufactured, may be one possible answer to the more stringent requirements of emerging helmet standards.

Acknowledgement This work has been supported by the COST Action 357 initiative. (http://www.cost357.org)

References

1. Hilyard, N.C., Djiauw, L.K., (1971) Observations on the Impact Behaviour of Polyurethane Foams; I. The Polymer Matrix. Journal of Cellular Plastics. 7:33–42
2. Mills, N.J., Gilchrist, A., (1991) The Effectiveness of Foams in Bicycle and Motorcycle Helmets. Accid. Anal. and Prev. 23:153–163
3. Di Landro, L., Sala, G., Olivieri, D. (2002) Deformation mechanisms and energy absorption of polystyrene foams for protective helmets. Polymer Testing 21:217–228
4. Gilchrist, M.D., O'Donoghue, D., (2000) Simulation of the development of frontal head impact injury. Computational Mechanics, 26(3):229-235
5. Gilchrist, M.D., (2003) Modelling and accident reconstruction of head impact injuries. Key Engineering Materials, 245-246:417-430
6. O'Riordain, K., Thomas, P.M., Phillips, J.P., Gilchrist, M.D., (2003) Reconstruction of real world head injury accidents resulting from falls using multibody dynamics. Clinical Biomechanics, 18:590-600
7. Avalle, M., Belingardi, G., Montanini. R., (2001) Characterization of polymeric structural foams under compressive impact by means of energy absorption diagram. International Journal of Impact Engineering 25:455–472
8. Cui, L., Kiernan, S., Gilchrist, M.D., (2009) Designing the energy absorption capacity of functionally graded foam materials. Materials Science and Engineering A 507:215-225
9. Al-Mousawi, M. M., Reid, S. R., Deans, W. F., (1997) The use of the split Hopkinson pressure bar techniques in high strain rate materials testing. Proc Instn Mech Engrs Part C 211:273–292
10. Bragov, A.M., Lomunov, A.K., Sergeichev, I.V., Tsembelis, K., Proud, W.G., (2008) Determination of physicomechanical properties of soft soils from medium to high strain rates. International Journal of Impact Engineering 35:967 - 976
11. Ninan, L., Tsai, J., Sun, C.T. (2001) Use of split Hopkinson pressure bar for testing off-axis composites. International Journal of Impact Engineering 25:291–313
12. Lee, O. S., Kim, M.S., (2004) Dynamic material property characterization by using split Hopkinson pressure bar (SHPB) technique. Nuclear Engineering and Design 226:119–125
13. Chen, W., Song, B., Frew, D. J., Forrestal, M. J., (2003) Dynamic Small Strain Measurements of a Metal Specimen with a Split Hopkinson Pressure Bar. Society for Experimental Mechanics 43:20–23
14. Kiernan, S., Cui, L., Gilchrist, M.D., (2009) Propagation of a stress wave through a virtual functionally graded foam. Int. J. Non-Linear Mechanics 44:456-468
15. Miyamoto, Y, Kaysser, W.A., Rabin, B.H., Kawasaki, A., Ford, R.G., (1999) Functionally Graded Materials: Design, Processing and Applications. Materials Technology Series, Kluwer Academic Publishers
16. ABAQUS, ABAQUS Analysis User's Manual, Version 6.7, in: ©ABAQUS, Inc, 2007
17. Schraad, M. W., Harlow, F. H., (2006) A stochastic constitutive model for disordered cellular materials: Finite-strain uni-axial compression. International Journal of Solids and Structures 43:3542–3468
18. Forero Rueda, M.A., Cui, L., Gilchrist, M.D., (2009) Optimisation of energy absorbing liner for equestrian helmets. Part I: Layered foam liner. Journal of Materials & Design 30:3405–3413
19. Cui, L., Forero Rueda, M.A., Gilchrist, M.D., (2009) Optimisation of energy absorbing liner for equestrian helmets. Part II: Functionally graded foam liner. Journal of Materials & Design 30:3414–3419
20. Zhao, H., Gary, G., (1995) A Three Dimensional Analytical Solution Of The Longitudinal Wave Propagation In An Infinite Linear Viscoelastic Cylindrical Bar. Application To Experimental Techniques. Journal of the Mechanics and Physics of Solids 43:1335–1348
21. Blanc, R. H., (1993) Transient Wave Propagation Methods for Determining the Viscoelastic Properties of Solids. Journal of Applied Mechanics 60:763–768

22. Newman, J. A., (1998) Kinematics of head injury - An overview, in: Frontiers in Head and Neck Trauma: Clinical and Biomechanical, IOS Press Inc., Burke, Virginia
23. BSI, EN 1384: 1996 Specification for helmets for equestrian activities, in: London, British Standards Distribution, 1996
24. BSI, EN 14572:2005: High performance helmets for equestrian activities, in: London, British Standards Institution, 2005

Nonlinear Stability of Functionally Graded Plates Subjected to Aero-thermo-mechanical Loads

M. Ganapathi, T. Prakash, and M.K. Singha

Abstract The nonlinear stability behaviors of functionally graded material plates subjected to aero-thermo-mechanical loads are studied. The material properties are graded in the thickness direction according to power-law distribution. A finite element method is employed with a eight-noded C^0 shear flexible quadrilateral plate element. The influences of material gradient index, thickness of plate, boundary condition, and aspect ratio on the nonlinear characteristics of functionally graded plates are investigated.

Keywords FEM · FGM · Flutter · Inplane load · Neutral axis · Postbuckling · Power law · Supersonic · Thermal load

1 Introduction

Functionally graded materials (FGM) are inhomogeneous and the material properties vary smoothly in one or more directions, and obtained by changing the volume fraction of the constituent materials [1]. The result is a microstructure that produces continuously varying thermal and mechanical properties at the macroscopic or continuum level. In practice, the use of these materials in aerospace industries has necessitated to understand the stability behaviors of functionally graded structures subjected to aero-thermo-mechanical loads. The notable contributions pertaining to the buckling analysis of rectangular FGM plates under in-plane load are available in the literature [2–4], but they deal with eigen-buckling analysis, which may not reveal the actual behavior. A nonlinear analysis based on exact position of

M. Ganapathi (✉)
Bombardier India Center, Satyam Computer Services Ltd., Bengaluru, India
e-mail: ganapathi.manickam@aero.bombardier.com

T. Prakash and M.K. Singha
Indian Institute of Technology Delhi, Haus Khas, New Delhi, India
e-mail: thiruprakash@rediffmail.com, maloy@am.iitd.ac.in

B. Dattaguru et al. (eds.), *IUTAM Symposium on Multi-Functional Material Structures and Systems*, IUTAM Bookseries 19, DOI 10.1007/978-90-481-3771-8_3,

neutral surface is needed to understand the actual stability characteristics. Also, the thermal stability problems have been attempted using eigen approach by many authors [5–7]. However a nonlinear thermal postbuckling analysis is needed to bring out the extension-bending coupling of FGM plates. Apart from that, FGM is a strong candidate for aerospace applications and hence studying its flutter characteristics is of great interest. Though there were many investigations done on isotropic [8] and composite members, the work on FGM plate is limited [9, 10]. In this paper, the nonlinear stability behaviors of FGM plates subjected to aero-thermo-mechanical loads are studied comprehensively. The influences of material gradient index, thickness of plate, boundary condition, and aspect ratio on the nonlinear characteristics of functionally graded plates are investigated using a finite element formulation.

2 Formulation

A functionally graded material (FGM) plate of thickness h made by mixing metal and ceramic is studied here. Two different datum planes are considered for the measurement of z, namely, z_{ms}, and z_{ns} measured from the middle surface, and the neutral surface of the plate, respectively (Fig. 1).

The volume-fraction of ceramic (V_c) is expressed as $V_c = \left(\frac{2z_{ms}+h}{2h}\right)^k = \left(\frac{2z_{ns}+h+2C}{2h}\right)^k$ where k is the volume fraction exponent ($k \geq 0$) and C is the distance of neutral surface from the mid-surface. The volume fraction of metal can be written as $V_m(z) = 1 - V_c(z)$. The effective Young's modulus E is expressed as $E(Z) = E_c V_c(z) + E_m V_m(z)$ where E_c and E_m are the Young's modulus of the ceramic and metal, respectively. The neutral surface distance, C is determined from the mid-surface as

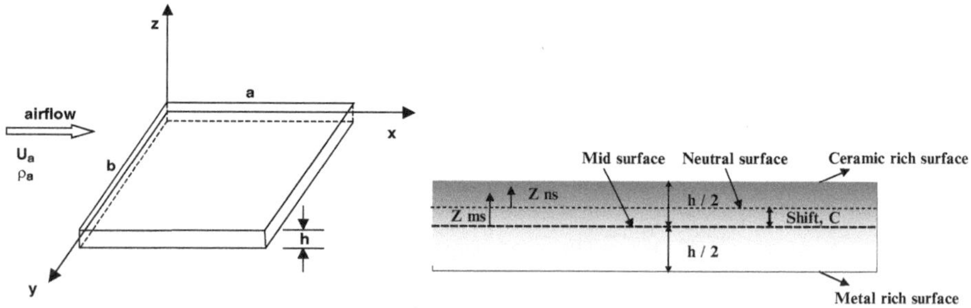

Fig. 1 Configuration and coordinate system of FGM plate

$$C = \frac{\int_{-h/2}^{h/2} E(z_{ms}) z_{ms} dz_{ms}}{\int_{-h/2}^{h/2} E(z_{ms}) dz_{ms}} \tag{1}$$

First order shear deformation theory based on exact position of neutral surface is developed here. The displacements u, v, w at a point (x, y, z_{ns}) in the plate are expressed as functions of neutral-plane displacements u_0, v_0, w and independent rotations θ_x and θ_y as

$$\begin{aligned} u(x,y,z) &= u_0(x,y) + z_{ns}\theta_x(x,y) \\ v(x,y,z) &= v_0(x,y) + z_{ns}\theta_y(x,y) \\ w(x,y,z) &= w_0(x,y) \end{aligned} \tag{2}$$

von Karman's assumptions for moderately large deformation allows Green's strains to be written in terms of the displacements given in Eq. (2) for a plate as,

$$\{\varepsilon\} = \begin{Bmatrix} \varepsilon_m^L \\ 0 \end{Bmatrix} + \begin{Bmatrix} z_{ns}\varepsilon_b \\ \varepsilon_s \end{Bmatrix} + \begin{Bmatrix} \varepsilon_m^{NL} \\ 0 \end{Bmatrix} \tag{3}$$

The in-plane strains $\{\varepsilon_m^L\}$, bending strains $\{\varepsilon_b\}$, shear strains $\{\varepsilon_s\}$ and the nonlinear components of in-plane strains $\{\varepsilon_m^{NL}\}$ in Eq. (3) are written as

$$\{\varepsilon_m^L\} = \begin{Bmatrix} u_{o,x} \\ v_{o,y} \\ u_{o,y} + v_{o,x} \end{Bmatrix}, \quad \{\varepsilon_b\} = \begin{Bmatrix} \theta_{x,x} \\ \theta_{y,y} \\ \theta_{x,y} + \theta_{y,x} \end{Bmatrix},$$

$$\{\varepsilon_s\} = \begin{Bmatrix} \theta_x + w_{o,x} \\ \theta_y + w_{o,y} \end{Bmatrix}, \quad \{\varepsilon_m^{NL}\} = \begin{Bmatrix} (1/2)w_{o,x}^2 \\ (1/2)w_{o,y}^2 \\ w_{o,x}\, w_{o,y} \end{Bmatrix} \tag{4}$$

In the present formulation, the membrane stress resultants $\{N\}$ and the bending stress resultants $\{M\}$ are expressed as $\{N\} = \{N_{xx}\, N_{yy}\, N_{xy}\}^T = [A_{ij}]^{ns}\{\varepsilon_m\}$ and $\{M\} = \{M_{xx}\, M_{yy}\, M_{xy}\}^T = [D_{ij}]^{ns}\{\varepsilon_b\}$. The corresponding relations in conventional mid-surface-based formulation are given by $\{N\} = \{N_{xx}\, N_{yy}\, N_{xy}\}^T = [A_{ij}]\{\varepsilon_m\} + [B_{ij}]\{\varepsilon_b\}$ and $\{M\} = \{M_{xx}\, M_{yy}\, M_{xy}\}^T = [B_{ij}]\{\varepsilon_m\} + [D_{ij}]\{\varepsilon_b\}$. Here, the matrices $[A_{ij}]^{ns}$ and $[D_{ij}]^{ns}$ are extensional, and bending stiffness matrices for neutral-surface based formulation, whereas, $[A_{ij}]$, $[B_{ij}]$ and $[D_{ij}]$ $(i, j = 1, 2, 6)$ are extensional, extensional-bending coupling and bending stiffness matrices for mid-surface based formulation. These stiffness coefficients are defined by $[A_{ij}, D_{ij}]^{ns} = \int_{-h/2-C}^{h/2-C} [\overline{Q}_{ij}](1, z_{ns}^2) dz_{ns}$ and

$[A_{ij}, B_{ij}, D_{ij}] = \int_{-h/2}^{h/2} [\overline{Q}_{ij}](1, z_{ms}, z_{ms}^2)dz_{ms}$. It may be noted that the extension-bending coupling matrix does not exist in the neutral surface based formulation. In the mid-surface based formulation the offset of stress resultants from mid-surface is incorporated through the extension-bending coupling matrix $[B_{ij}]$. The transverse shear force $\{Q\}$ representing the quantities $\{Q_{xz}, Q_{yz}\}$ is related to the transverse shear strains $\{\varepsilon_s\}$ through the constitutive relations as $\{Q\} = [E_{ij}]\{\varepsilon_s\}$ i, $j = 4, 5$ where $E_{ij} = \int_{-h/2-C}^{h/2-C} [\bar{Q}_{ij}]\kappa_i \kappa_j dz_{ns}$. Here κ_i is the transverse shear correction factor taken as $\sqrt{5/6}$ for non-uniform shear strain distribution through the plate thickness. $\overline{Q}_{ij}$ is the standard reduced stiffness matrix.

The strain energy of the plate can be expressed in terms of the field variables $u_0, v_0, w_o, \theta_x, \theta_y$ and their derivatives. The potential energy due to external in-plane mechanical forces, $(N_{xx}^o, N_{yy}^o, N_{xy}^o)$ acting along the mid-surface can be evaluated in terms of field variables $u_0, v_0, w_o, \theta_x, \theta_y$. The potential energy due to thermal stress resultants $(N_{11}^{th}, N_{22}^{th}, N_{12}^{th})$ can also be expressed in the similar way. The work done by the applied non-conservative aero-load can be evaluated in terms of free stream air density, velocity and Mach number using Piston theory within the 2D static approximation [11]. Following standard procedure (minimization of total potential energy) the nonlinear finite element equations may be written as [12]

$$[M]\{\ddot{\delta}\} + ([K] + [K_{NL}] + \lambda_T [K_{GT}] + \lambda_M [K_{GM}] + \lambda_f [\overline{A}])\{\delta\} = \{F\} \quad (5)$$

where [M] is the consistent mass matrix; $\{\ddot{\delta}\}$ is the acceleration vector, $[K]$ is the linear stiffness matrix and $[K_{NL}]$ is the nonlinear stiffness matrix. $[K_{GT}]$ and $[K_{GM}]$ are the geometric stiffness matrices due to unit thermal and mechanical loads; $[\overline{A}]$ is the aero-dynamic force. Here $\lambda_T, \lambda_M, \lambda_f$ are the temperature rise $\Delta T(= T_c - T_m)$, mechanical load and the aerodynamic pressure $\left(= \frac{\rho_a U_a^2}{\sqrt{M_\infty^2 - 1}}\right)$ respectively. $\{\delta\}$ is the displacement vector. In the present work, an eight-noded C^0 continuous shear flexible plate bending element with five degrees of freedom per node [11] is modified with respect to neutral-surface to study the nonlinear behavior of FGM plates.

3 Results and Discussion

With the formulation developed above, three different problems are attempted. They are explained below.

3.1 Mechanical Post-Buckling of Si_3N_4/SUS304 FGM Plates

For the case of pure ceramic or metallic plate under compressive in-plane load, the neutral surface coincides with the mid-surface and hence the Euler type of bifurcation buckling occurs. The critical buckling load can be found from solving the eigenvalue problem:

$$[K + K_{NL} + \lambda K_G]\{\delta\} = \{0\} \tag{6}$$

In case of FGM plates, the neutral surface shifts from the mid-surface and hence the applied in-plane loads do not coincide with the stress resultants. In such cases, the nonlinear equilibrium Eq. (5) is written as

$$[K + K_{NL}]\{\delta\} = \{F\} \tag{7}$$

where $\{F\}$ is the force vector consisting of in-plane compressive load applied along the mid-surface and the moment developed due to the offset in the neutral surface. Eq. (7) is solved by Newton-Raphson technique.

The nonlinear behavior of simply supported thin square Si_3N_4/SUS304 [13] FGM plates (a/h = 100, a/b = 1) under uni-axial compression is investigated in Fig. 2 for various values of material gradient index. Here, the variation of out-of-plane displacement (w/h) with respect to non-dimensional compressive load $(\lambda^* = a^2\lambda_{cr}/\pi^2 D_c.\ D_c = E_c h^3/12(1 - \nu^2))$ is presented. The isotropic plates exhibit bifurcation buckling and the corresponding postbuckling paths for pure

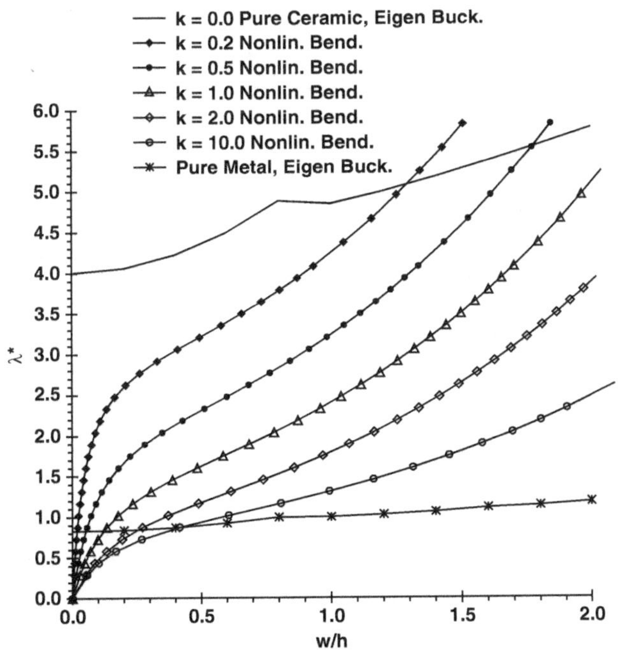

Fig. 2 Postbuckling path of simply supported thin FGM square plate under uniaxial compression (a/h = 100)

ceramic and pure metal cases are traced by the eigenvalue analysis. For FGM plates the non-linear bending behavior is obtained through Newton-Raphson technique and the corresponding load-displacement curves evaluated are shown in the figure. It is observed that, with the increase in material gradient index value (k), the resistance of the plate reduces, i.e., for a specified in-plane load the out-of-plane displacement increases, and this is because of the stiffness reduction due to the more metal inclusion in FGM plate. It is to be noted here that as the in-plane compressive load is applied at the mid-surface, the FGM plates show the nonlinear bending behavior. If the in-plane load is assumed to be applied at the neutral surface for some reasons, the plate would exhibit bifurcation buckling behavior.

3.2 *Thermal Post-Buckling of Aluminum–Alumina FGM Plates*

For the case of pure ceramic or metallic plate under uniform temperature rise, only membrane forces are generated. The critical buckling temperature at which Euler type of buckling occurs is found from solving the following eigenvalue problem.

$$[K] + [K_{NL}] + [K_{G0}] + \lambda_T\,[K_{GT}] = \{0\} \tag{8}$$

where λ_T is the temperature rise ($\lambda_T = \Delta T = T_c - T_m$) and the lowest eigenvalue is the critical buckling temperature. In case of FGM plates under linear or nonlinear temperature distribution through the thickness, bending moments develop together with the membrane forces, which are not included in the geometric stiffness matrix. Hence, the force vector on the right hand side exists and the eigenvalue type of buckling will not occur. In such case, the nonlinear equilibrium equation is solved by Newton-Raphson technique to get the temperature-displacement curves.

$$[K] + [K_{NL}] + [K_{G0}] + \lambda_T\,[K_{GT}] = \{F\} \tag{9}$$

The nonlinear bending behavior of simply supported thin square aluminum/alumina [14] FGM plates (a/h = 100, a/b = 1, T_m = 305 K) under nonlinear through-thickness temperature distribution is investigated and shown in Fig. 3. The non-linear governing equation is solved by Newton-Raphson technique to include the effects of extension-bending coupling and non-symmetric-through-thickness thermal load (thermal moment $\{M_T\}$). It is observed that, due to strong bending moment $\{M_T\}$ developed in FGM plate, the plate starts bending towards the upper side of the plate wherein ceramic content is high. For comparison purpose, the study is also made using eigenvalue approach by neglecting the effect of bending moment $\{M_T\}$, and the corresponding postbuckling paths are also highlighted in the Figure. It is observed that the postbuckling curves are symmetric about the Y-axis for pure ceramic and pure metal cases, whereas, it looses its symmetry and is shifted towards ceramic rich upper side for FGM plates. This is attributed to the extension-bending coupling and a shift in the neutral surface towards the high stiff ceramic side of the plate. The difference between solid curves (nonlinear analysis) and

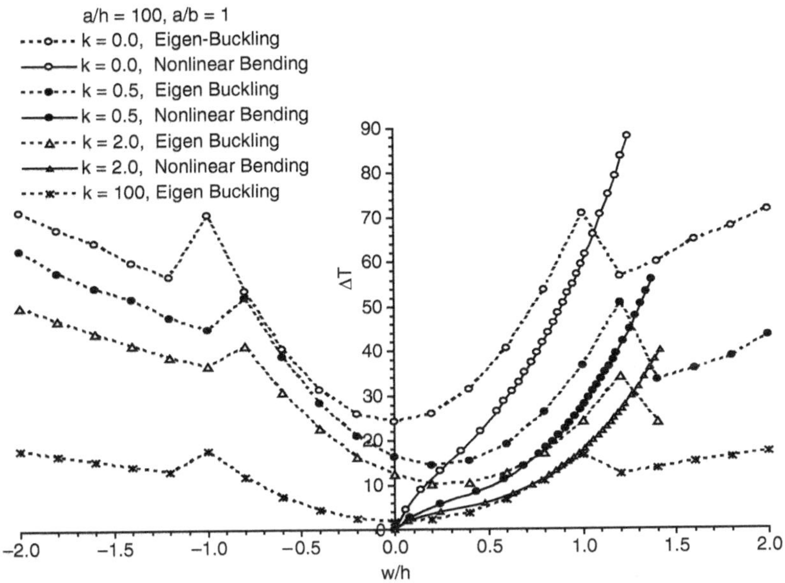

Fig. 3 Effect of gradient index, k on thermal postbuckling paths of simply supported thin aluminum/alumina FGM square plate under nonlinear temperature rise through thickness (a/h = 100, a/b = 1, T_m = 305 K)

dotted curves (eigenvalue analysis) clearly indicate the effect of thermal bending moment $\{M_T\}$.

3.3 *Flutter Characteristics of FGM Plates*

To analyze the supersonic panel flutter behavior of functionally graded plates, the equation (Eq. 5) is deduced to the following.

$$[M]\{\ddot{\delta}\} + \left[[K] + [K_G] + \lambda_f\left[\overline{A}\right]\right]\{\delta\} = \{0\} \tag{10}$$

Substituting the harmonic motion $\{\delta\} = \{\delta_0\}\, e^{i\,\omega\, t}$, the governing equation leads to

$$[[\overline{K}] - \overline{k}[I]]\{\delta_0\} = 0 \tag{11}$$

where $\left[\overline{K}\right] = \left[[K] + [K_G] + \lambda_f\left[\overline{A}\right]\right]^{-1}[M]$ and $\overline{k}$ is equal to $1/\omega^2$. $[I]$ is the identity matrix. The problem is now reduced to that of finding out the eigenvalues. As λ_f increases monotonically from zero, two of the eigenvalues will approach each other and coalesce to ω_{cr} at $\lambda_f = \lambda_{fcr}$.

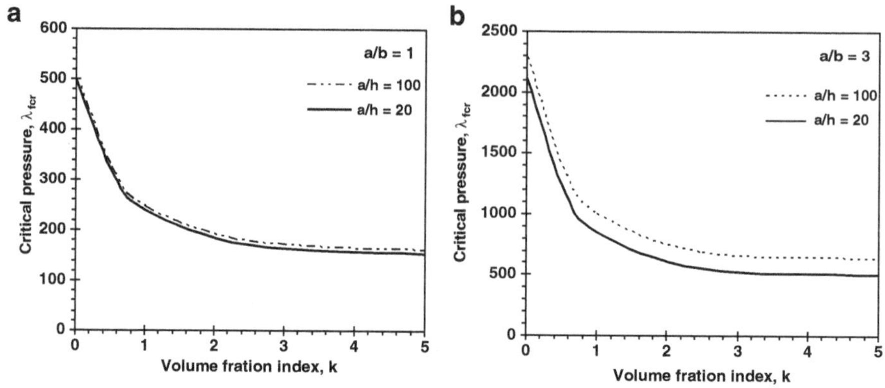

Fig. 4 Effect of thickness on the critical aerodynamic pressure of simply supported aluminum-alumina FGM plate: (**a**) a/b = 1; (**b**) a/b = 3

The flutter characteristics of square and rectangular plates made of aluminum-alumina [14] is investigated and presented in Fig. 4. It is inferred from this figure that the critical non-dimensional aerodynamic pressure decreases with the increase in volume fraction index k. However, the rate of decrease in the flutter speed is high for the low value of k. This is due to the fact that the stiffness is high for the ceramic panel and minimum for the metallic panel, and it degrades gradually with the increase in k. The effect of side-to-thickness ratio on critical non-dimensional aerodynamic pressure of functionally graded plates is examined here assuming two values for a/h(= 20 and 100). It can be noticed in Fig. 4 that the increase in thickness results in the reduction of critical non-dimensional flutter speed or dynamic pressure and the rate of decrease in the value of flutter speed increases with the increase in the aspect ratio and volume fraction index k.

References

1. Koizumi M. The concept of FGM. Ceramic Transaction: Functionally Graded Material. 1993; 34: 3–10.
2. Birman V. Buckling of functionally graded hybrid composite plates. Pro-ceedings of the 10th Conference on Engineering Mechanics, Boulder, Colo-rado, 1995; 2: 1199–1202.
3. Feldman E, and Aboudi J. Buckling analysis of functionally graded plates subjected to uniaxial loading. Composite Structures, 1997; 38: 29–36.
4. Javaheri R, and Eslami MR. Buckling of functionally graded plates under in-plane compressive loading. ZAMM, 2002; 82: 277–283.
5. Javaheri R, Eslami MR. Thermal buckling of functionally graded plates. AIAA Journal 2002;40: 162–169.
6. Wu Lanhe. Thermal buckling of a simply supported moderately thick rectangular FGM plate. Composite Structures 2004;64:211–218.
7. Najafizadeh MM, Eslami MR. First-order-theory-based thermoelastic stability of functionally graded circular plates. AIAA Journal 2002;40:1444–1450.

8. Dowell EH. Nonlinear oscillations of a fluttering plate. AIAA J 1966;4(7): 1267–1276.
9. Sohn KJ, Kim JH. Structural stability of functionally graded panels subjected to aero-thermal loads. Compos Struct 2008;82(3):317–325.
10. Sohn KJ, Kim JH. Supersonic flutter of stepwise functionally graded panels. Proceedings of fourth international conference of advanced materials and processing, Hamilton, New Zealand, 10–13 December; 2006.
11. Ganapathi M, Touratier M. Supersonic flutter analysis of thermally stressed laminated composite flat panels. Compos Struct 1996;34:241–248.
12. Zienkiewicz OC, Taylor RL. "The finite element method." McGraw-Hill, Singapore, 1989.
13. Jae-Sang Park, Ji-Hwan Kim, Thermal postbuckling and vibration analyses of functionally graded plates. Journal of Sound and Vibration, 289 (2006) 77–93.
14. Prakash T, Singha MK, and Ganapathi M. Thermal Postbuckling Analysis of FGM Skew Plates. Engineering Structures, 2008; 30: 22–32.

A Generalized Three Species Model for Shape Memory Alloys

Ashish Khandelwal and Vidyashankar R. Buravalla

Abstract In this work, an existing three-species model for describing Shape Memory Alloy (SMA) behavior is generalized to incorporate more relevant effects like asymmetric hysteresis, two-way memory and different forms of hardening. Using the notion of multiple yield surfaces for describing different transformation paths within the existing thermodynamic framework, necessary and sufficient conditions are derived for describing phase transformation. A few case studies are provided to illustrate the efficacy of the proposed generalized model.

Keywords Dissipation potential · Multiple yield surface plasticity · Phase transformation

1 Introduction

Recently, Nallathambi et al. [1] have proposed a 1-D model for SMAs incorporating three internal variables to denote the three phases that are involved. A consistent thermodynamic based formalism amenable for hierarchical development is followed to develop a model analogous to classical rate independent multiple yield surface plasticity. In a one dimensional context, considering different transformation paths between the three different variants viz., austenite (A), plus-martensite (M^+) and minus-martensite (M^-), six transformation conditions are derived as yield surfaces. The model, under some assumptions, describes these transformations and a phase-diagram is constructed to illustrate different transformation scenarios. The striking feature of their approach is its ability to accommodate additional variables for introducing other complex effects such as residual deformations, asymmetry in hysteresis, texture effects and material degradation. Nallathambi et al. [1] show that their model, developed using some simplistic assumptions, captures the essential SMA response. While several similarities between this model and those by Lubliner

A. Khandelwal and V.R. Buravalla (✉)
General Motors (R&D), India Science Lab 3rd Floor, Creator Building, I T P B, Whitefield Road, Bengaluru 560066 India
e-mail: ashish.khandelwal@gm.com; vidyashankar.buravalla@gm.com

B. Dattaguru et al. (eds.), *IUTAM Symposium on Multi-Functional Material Structures and Systems*, IUTAM Bookseries 19, DOI 10.1007/978-90-481-3771-8_4,

and Auricchio [2] and Qidwai and Lagoudas [3] exist, there are characteristic differences. Important effects like, two-way memory, asymmetric hysteresis and functional degradation also need to be included in the model developed by Nallathambi et al. [1], thus requiring generalization. One of the important aspects involved is in handling simultaneous evolution of two species into a single product species. In the model developed by Nallathambi et al. [1], the kinetics is written in an independent fashion, albeit, satisfying the conservation of mass constraint. Hence, as is shown here, the kinetics as proposed yields discontinuous or non-uniform evolution of the product phase. This is reflected in the stress–strain–temperature response also. Upon generalization of the existing model to incorporate asymmetry and different possible simultaneous evolutions, the yield surfaces derived therein involve eighteen material parameters, not all of which are independent. Case studies using the existing model show that some of parameters are not independent and depending on the load path and the possible transformations, suitable constraint or consistency conditions are necessary.

2 Thermodynamic Framework of the Model

The inherent phase transformation involved in 1-D SMA response is captured using three phases, and hence the minimum number of species in the material should be at least three. Following the thermodynamic formalism proposed by Nallathambi et al. [1] based on three volume fractions, viz, austenite (α), plus-martensite (ξ^+) and minus-martensite (ξ^-), the following form of dissipation inequality in terms of the driving forces (f) is obtained.

$$(f^+ - f^\alpha)\,\dot{\xi}^+ + (f^- - f^\alpha)\,\dot{\xi}^- \geq 0. \tag{1}$$

Upon, relaxing the symmetry assumption, Eq. (1) takes the following form for α to M^+ and its reverse transformation,

$$|f^+ - f^\alpha| = k_{1f} \quad \text{for} \quad \dot{\xi}^+ > 0 \quad \text{and} \quad |f^+ - f^\alpha| = k_{1r} \quad \text{for} \quad \dot{\xi}^+ < 0, \tag{2}$$

where k_{1f} and k_{1r} are the maximum magnitude of driving force. This relaxation captures wider variety of SMA behavior. Similar conditions can be written for other transformations. Extending this further, the following expression for the forward and reverse driving forces (D) in terms of derivatives of free energy (G) are obtained as:

$$D^{\alpha+} = \sigma H_f^+ + \Delta G^{\alpha+} - \frac{\partial G^0}{\partial \xi^+} + \frac{\partial G^i}{\partial \alpha}, \tag{3}$$

$$D^{+\alpha} = \sigma H_r^+ + \Delta G^{\alpha+} - \frac{\partial G^0}{\partial \xi^+} + \frac{\partial G^i}{\partial \alpha}. \tag{4}$$

Similarly other forward and reverse driving forces can be stipulated. The normalized driving forces with respect to transformation strain (H) and the equivalent stress forms are obtained as:

$$\Gamma_1 = \frac{D^{\alpha+}}{H_+^f} = \sigma + \beta_1^{\alpha\xi^+} - \beta_{1f}^{\varepsilon^t} + \beta_{1f}^{\alpha}, \qquad \Gamma_2 = \frac{D^{+\alpha}}{H_+^r} = -\sigma + \beta_1^{\xi^+\alpha} - \beta_{1r}^{\varepsilon^t} + \beta_{1r}^{\alpha},$$

$$\Gamma_3 = \frac{D^{\alpha-}}{H_-^f} = -\sigma + \beta_2^{\alpha\xi^-} - \beta_{2f}^{\varepsilon^t} + \beta_{2f}^{\alpha}, \qquad \Gamma_4 = \frac{D^{-\alpha}}{H_-^r} = \sigma + \beta_2^{\xi^-\alpha} - \beta_{2r}^{\varepsilon^t} + \beta_{2r}^{\alpha},$$

$$\Gamma_5 = \frac{D^{-+}}{H_\pm^f} = 2\left(\sigma - \beta_{3f}^{\varepsilon^t}\right), \qquad \Gamma_6 = \frac{D^{+-}}{H_\pm^r} = -2\left(\sigma - \beta_{3r}^{\varepsilon^t}\right).$$

It can be recognized that the hardening parameter $\beta_j^{s\to p}$ [1] corresponds to the bulk or chemical free energy. Using Clausius–Clapeyron relation, following hardening function for the first order phase transformation is obtained:

$$\beta_j^{s\to p} = m\left(\theta - \theta_0\right), \tag{5}$$

where θ_0 is the equilibrium temperature and 'm' represents the slope. It may be noted that the transformation or orientation hardening parameter (β^{ε^t}) will vary from zero to maximum during forward transformation and *vice versa* during the reverse transformation. The interaction hardening parameter (β^{α}) will have the opposite nature. The corresponding thermomechanical yield surface equations (F_i) are:

$$F_1 = \sigma + m\left(\theta - \theta_0\right) - \beta_{1f}^{\varepsilon^t} + \beta_{1f}^{\alpha} + k_{1f}, \qquad F_4 = \sigma + m\left(\theta - \theta_0\right) - \beta_{2r}^{\varepsilon^t} + \beta_{2r}^{\alpha} - k_{2r},$$

$$F_2 = -\sigma + m\left(\theta - \theta_0\right) - \beta_{1r}^{\varepsilon^t} + \beta_{1r}^{\alpha} - k_{1r}, \qquad F_5 = 2\left(\sigma - \beta_{3f}^{\varepsilon^t}\right) - k_{3f},$$

$$F_3 = -\sigma + m\left(\theta - \theta_0\right) - \beta_{2f}^{\varepsilon^t} + \beta_{2f}^{\alpha} - k_{2f}, \qquad F_6 = -2\left(\sigma - \beta_{3r}^{\varepsilon^t}\right) - k_{3r}.$$

Assuming linear hardening (linear kinetics), in general, 18 parameters are required for model description. Using thermodynamic principles and relevant phenomenology, this number can be reduced. The transformation condition, χ_i, depends on the magnitude of yield function, F_i, feasibility, χ_i^c, defined by consistency condition and the source species availability, χ_i^{SA}, given by:

$$\chi_i^{tr} = \frac{1 + sgn\left(F_i\right)}{1 + \left|sgn\left(F_i\right)\right|}, \qquad \chi_i^c = sgn\left((-1)^i \Delta_i\right), \qquad \chi_i^{SA} = sgn(s).$$

For transformation,

$$\chi_i = \chi_i^{tr}\chi_i^{c}\chi_i^{SA} = 1. \tag{6}$$

It may be noted that consistency parameter Δ_i is given by:

$$\Delta_i = \left[\frac{\partial \Gamma_i}{\partial \sigma}\dot{\sigma} + \frac{\partial \Gamma_i}{\partial \theta}\dot{\theta}\right] \begin{cases} > 0 & \text{for forward transformation,} \\ < 0 & \text{for reverse transformation.} \end{cases}$$

Robust implementation of the present model is given in the following section.

[1] Superscripts 's' and 'p' represent the source and product phases respectively.

3 Integration Algorithm for Time Discrete Model

A predictor–corrector type of algorithm is used to integrate the time discrete model. The rate form of yield surface equation is:

$$\dot{F}_i = \frac{\partial F_i}{\partial \sigma}\dot{\sigma} + \frac{\partial F_i}{\partial \theta}\dot{\theta} + \frac{\partial F_i}{\partial \xi^+}\dot{\xi}^+ + \frac{\partial F_i}{\partial \dot{\xi}^-}\dot{\xi}^- + \frac{\partial F_i}{\partial \alpha}\dot{\alpha}. \tag{7}$$

The generalized incremental evolution equation is then obtained as:

$$A_{ij}x_j = b_i, \quad i = 1..6 \quad \text{and} \quad j = 1..2, \tag{8}$$

where,

$$A_{i,1} = \chi_i\left[-EH_f^+\frac{\partial F_i}{\partial \sigma} - H_f^+\frac{\partial E}{\partial \xi^+} + \frac{\partial F_i}{\partial \xi^+} - \frac{\partial F_i}{\partial \alpha}\right],$$

$$A_{i,2} = \chi_i\left[EH_f^+\frac{\partial F_i}{\partial \sigma} + H_f^+\frac{\partial E}{\partial \xi^-} + \frac{\partial F_i}{\partial \xi^-} - \frac{\partial F_i}{\partial \alpha}\right],$$

$$b_i = -E\dot{\varepsilon}\frac{\partial F_i}{\partial \sigma} - \frac{\partial F_i}{\partial \theta}\dot{\theta} - \frac{\partial F_i}{\partial \alpha}. \tag{9}$$

Using the following finite difference approximation for x

$$x = \begin{bmatrix}\dot{\xi}^+\\ \dot{\xi}^-\end{bmatrix} \approx \begin{bmatrix}\Delta\xi^+\\ \Delta\xi^-\end{bmatrix}, \tag{10}$$

the solution of Eq. (8) updates the value of the volume fractions (ξ^+, ξ^-). It may be noted that in Eq. (9), E is the modulus and ε is the total strain. In the next section, relevant constraint conditions important for tailoring the model to any specific SMA are derived.

4 Constraint on the Hardening Functions

In order to have this model applicable to a wide variety of SMAs, it is essential to understand how to obtain the necessary parameters from experimental results. In SMAs important physical parameters of interest include the transition temperatures, typically obtained by DSC. In the case of stress free thermal cycling, as in DSC, equal fractions of M^+ and M^- form from austenite. During cooling, at a specific temperature, the magnitude of yield surfaces F_1 and F_3 become zero, defined as the martensite start temperature, M_s. On further cooling, at a specific temperature, interaction hardening parameter becomes minimum, defining M_f. Similarly, during heating, using F_2 and F_4 one can obtain the austenite start (A_s) and finish (A_f)

temperatures. It may be noted that there will be two definitions of these temperatures corresponding to $\alpha \Leftrightarrow M^+$ and $\alpha \Leftrightarrow M^-$ transformation. Their equivalence at zero stress is used for constraining the hardening function. Using the yield surface F_1 and F_2, i.e., $\alpha \Leftrightarrow M^+$, and similarly, for the yield surface F_3 and F_4, i.e., $\alpha \Leftrightarrow M^-$, following can be obtained:

$$M_s^+ = \theta_o + \frac{\beta_{1f}^{\alpha} - k_{1f} + \sigma}{m}, \qquad M_s^- = \theta_o + \frac{\beta_{2f}^{\alpha} - k_{2f} - \sigma}{m}, \tag{11}$$

$$M_f^+ = \theta_o + \frac{\beta_{1f}^{\varepsilon^t} - k_{1f} + \sigma}{m}, \qquad M_f^- = \theta_o + \frac{\beta_{2f}^{\varepsilon^t} - k_{2f} - \sigma}{m}, \tag{12}$$

$$A_s^+ = \theta_o + \frac{k_{1r} - \beta_{1r}^{\varepsilon^t} + \sigma}{m}, \qquad A_s^- = \theta_o + \frac{k_{2r} - \beta_{2r}^{\varepsilon^t} - \sigma}{m}, \tag{13}$$

$$A_f^+ = \theta_o + \frac{\beta_{1r}^{\alpha} + k_{1r} + \sigma}{m}, \qquad A_f^- = \theta_o + \frac{\beta_{2r}^{\alpha} + k_{2r} - \sigma}{m}. \tag{14}$$

Using the equivalence of transition temperatures (like $M_s^+ = M_s^-$), four constraint conditions are obtained. Further, satisfaction of II law of thermodynamics (non-negative hysteresis area) yields following.

$$F_1|_{finish} \geq F_2|_{start}, F_3|_{finish} \geq F_4|_{start}, \qquad F_5|_{finish} \geq F_6|_{start} \quad \forall[\sigma, \theta]. \tag{15}$$

Additional constraint conditions can be obtained from SMA response to the following two loading cases.

1. Loading path ℓ_1: Under zero stress, pure austenite is cooled to $\theta < M_f$; then mechanically loaded till complete detwinning $\xi^+ = 1$.
2. Loading path ℓ_2: At a positive constant stress level, material is cooled from temperature above $A_f(\sigma)$, to a temperature below $M_f(\sigma)$.

For ℓ_2, as seen from Eq. (11), the formation of M^+ will be favoured over the formation of M^-. So, at a specific stress, only M^+ is obtained. The magnitude of stress at the finish of transformation should satisfy the finish limit of F_5 (similarly, under compressive stress for F_6). This can be stated as:

$$F_3 < F_1 \quad \forall\, [\sigma, \theta] \in \ell_2. \tag{16}$$

The above inequality, in the limit gives critical stress which provides the finish stress for F_3, i.e.,

$$F_3|_{finish}\left(\sigma_f^{cr}\right) = 0. \tag{17}$$

It may be noted that the nature of constraints depends on the specific thermomechanical characteristics. In the following section, the phase diagram for a typical NiTi SMA with Type-1 phase diagram [4] is derived as a special case of this general model.

5 Case Study: Symmetric Dissipation Potential for Forward and Reverse Transformation

In the case of symmetric dissipation potential, enthalpy of transformation and transformation strain are same during forward and reverse transformation. Assuming a linear hardening rule for both transformation and interaction hardening, nine parameters are required. The yield surfaces under this condition (identical to those obtained by Nallathambi et al. [1]) are:

$$F_1 = \sigma - m(\theta - \theta_0) - h_1\varepsilon^t + h_2\alpha - k_1, \qquad F_4 = \sigma + m(\theta - \theta_0) - h_1\varepsilon^t + h_2\alpha - k_2,$$
$$F_2 = -\sigma + m(\theta - \theta_0) + h_1\varepsilon^t - h_2\alpha - k_1, \qquad F_5 = \sigma - h_1\varepsilon^t - k_3,$$
$$F_3 = -\sigma - m(\theta - \theta_0) + h_1\varepsilon^t + h_2\alpha - k_2, \qquad F_6 = -\sigma + h_1\varepsilon^t - k_3.$$

It may be noted that the symmetry assumption satisfies the previously discussed constraint conditions by taking $k_1 = k_2$ and $k_3 = 0$. This further reduces the number of parameters to 7. In this case there is an assumption of Young's modulus being constant. By using Eqs. (11–14) the characteristic transformation temperature are given as:

$$M_s = \theta_o + \frac{h_2 - k_1 + \sigma}{m}, \qquad M_f = \theta_o + \frac{h_1 H - k_1 + \sigma}{m}, \tag{18}$$

$$A_s = \theta_o + \frac{k_1 - h_1 H + \sigma}{m}, \qquad A_f = \theta_o + \frac{h_2 + k_1 + \sigma}{m}. \tag{19}$$

Using the material parameters given in Table 1, the resulting characteristic phase diagram is shown in Fig. 1. This phase diagram is for positive stress and a

Table 1 Material properties of a typical NiTi

Elastic moduli		Transformation parameters	
E_a = 67 GPa	θ_o = 294.73°K	h_1 = 500 MPa	H = 0.067
E_m = 26.3 GPa	m = 7 MPa/°K	h_2 = 150 MPa	k_1 = 150 MPa

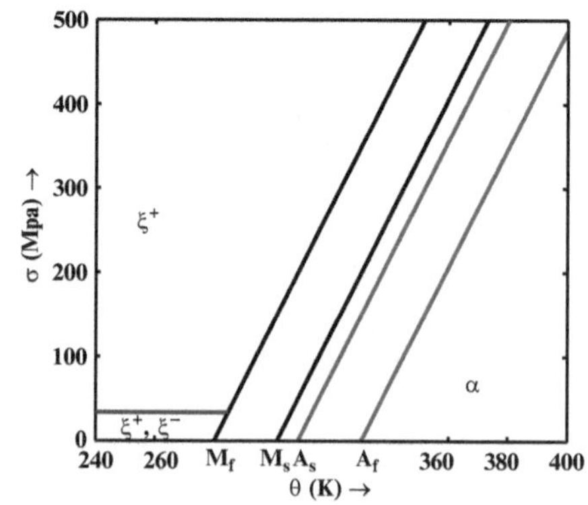

Fig. 1 Phase diagram for symmetric dissipation potential

symmetric counterpart exists for negative stress. The capability of the present model in modelling a wide variety of thermomechanical hysteresis is elucidated with a few numerical simulations.

6 Role of Hardening Functions

In this section, the ability of the present model to incorporate other intricate phenomenon in SMAs is shown using two important cases.

6.1 *Effect of Interaction Hardening*

In this study, simulation of mechanical loading and unloading at constant temperature ($\theta = 370K > A_f$) is investigated. In Fig. 2 the response of two different cases with different hardening functions for reverse transformation is presented. From Fig. 2 at the end of mechanical unloading, for asymmetric case there is a residual strain in the material. This behavior is seen in SMAs used for actuation, due to certain types of processing, like those with remnant cold work. Also, this study suggests that making hardening as a function of temperature helps in modelling an SMA with different asymmetric and nonlinear hardening.

6.2 *Effect of Constraint Condition on Hardening Parameters*

The other important aspect discussed here, is the constraint condition on the transformation hardening function. The importance of this is elicited by a simulation for the case with symmetric potential and the load path as shown in Fig. 3.

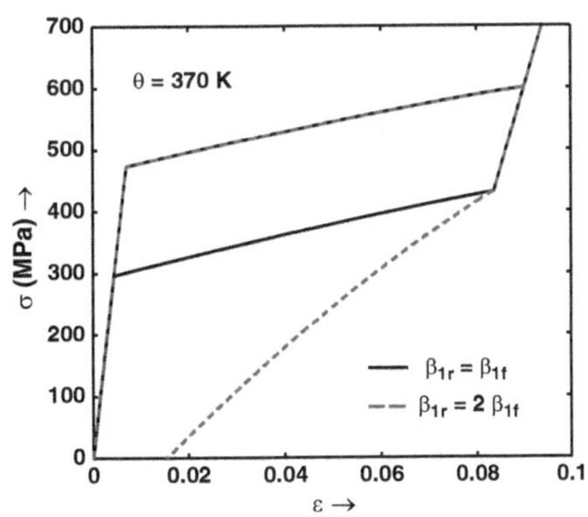

Fig. 2 Stress–strain response for different interaction hardening

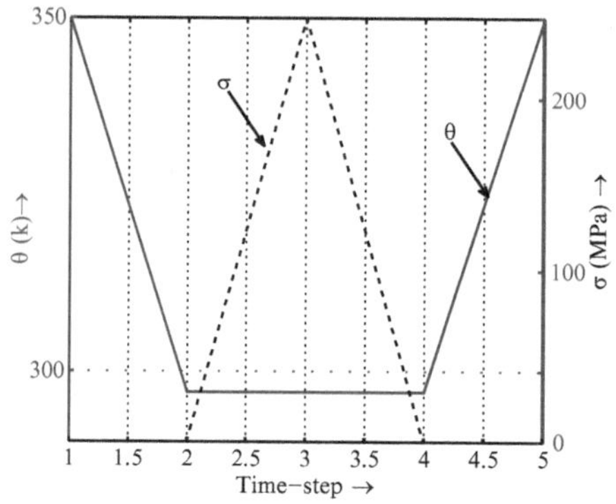

Fig. 3 Thermomechanical load path

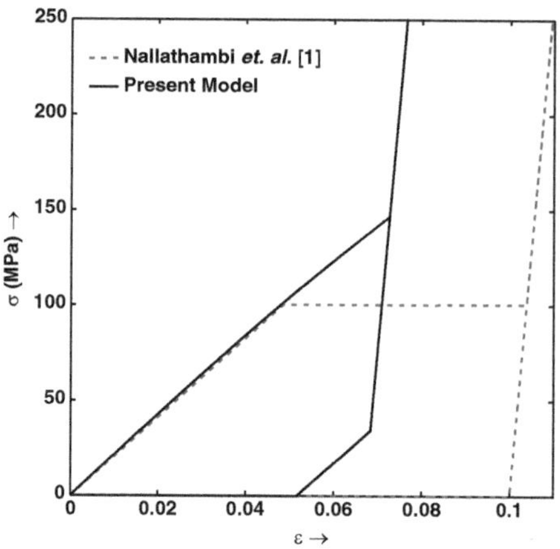

Fig. 4 Stress–strain response for the load path shown in Fig. 3

The stress–strain results for cases with (present model) and without necessary constraints Nallathambi et al. [1] are plotted in Fig. 4. Undesirable discontinuities present in Nallathambi et al. [1] are not present in the proposed model. Also, another point to be noted is that a suitable penalty on the yield surfaces F_1 and F_2 can incorporate the two-way strain.

7 Summary Remarks

An existing three-species model for SMAs developed by Nallathambi et al. is generalized to incorporate many other relevant effects seen in SMAs. Appropriate consistency conditions for material parameter functions are derived and a numerical algorithm is provided for its robust implementation. Case studies show the efficacy of this model to predict different complex thermomechanical phenomena. It also shows that the framework proposed by Nallathambi et al. [1] is hierarchical and amenable for generalization.

Acknowledgement The authors are grateful to Prof. Srinivasan M. Sivakumar for bringing this approach to their cognizance and for several useful technical discussions.

References

1. Nallathambi A, Doraiswamy S, Chandrasekar A S, and Srinivasan S M. A three species model for shape memory alloys. *To appear in Int. J. structural changes in solids*
2. Lubliner J and Auricchio F 1996. Generalized plasticity and shape memory alloys *Int. J. Sol. Struct.* 33, 991–1003
3. Qidwai M A and Lagoudas D C 2000. Numerical implementation of a shape memory alloy thermomechanical constitutive model using return mapping algorithms. *Int. J. Num. Meth. Engg.* 47 1123–1168
4. Duerig W T, Melton N K, Stockel D, and Wayman C M 1990. Engineering aspects of shape memory alloys Butterworth-Heinemann London Boston MA

Use of SMA Constitutive Model in Finite Element Analysis of Wire-Based Actuators

Prakash D. Mangalgiri and Amol G. Thakare

Abstract Suitable SMA material constitutive models are required to include the SMA behavior in the Finite Element (FE) analysis of structures with SMA based active elements. The complex behavior of SMA with path dependency, phase transformations and hysteresis preclude straight-forward incorporation into an FEM code and special schemes need to be devised to address such issues, especially under combined thermo-mechanical loading. This paper describes one such approach. A phenomenological material model to describe behavior of SMA under an arbitrary loading path in stress (strain) – temperature regime developed earlier by the research group at ISL is examined and then used to develop a special material routine to be used in a general purpose FEM analysis code. The user material routine (UMAT) facility in ABAQUS is used for this exercise. Further, calculations are carried out for a simple actuator configuration to demonstrate its working in the FE environment.

Keywords Shape memory alloys · FEM · Smart actuator · User material routine

1 Introduction

Shape memory alloys are the class of materials which are able to undergo crystallographic phase transformations brought about by application of stress, change in temperature or both leading to large deformations. The SMAs present two main characteristic properties viz., superelasticity and shape memory effect. Superelasticity represents the recovery of large deformations during mechanical loading-unloading at high temperatures. In shape memory effect, material recovers residual deformation developed due to loading-unloading at low temperature through thermal cycle. These behaviors are the consequences of martensitic phase

P.D. Mangalgiri (✉) and A.G. Thakare
General Motors (R&D), India Science Lab, 3rd Floor, Creator Building, I T P B, Whitefield Road, Bengaluru 560066, India
e-mail: prakash.mangalgiri@gm.com; amol.g.thakare@gm.com

B. Dattaguru et al. (eds.), *IUTAM Symposium on Multi-Functional Material Structures and Systems*, IUTAM Bookseries 19, DOI 10.1007/978-90-481-3771-8_5,

transformations. Due to these unique properties, SMAs are playing a key role in the development of actuators required to build smart systems. In such systems, the SMA element (often as a wire) forms a "structural" part of the device or the system. While numerical methods, such as Finite Element Methods (FEM), are well developed to handle structural analysis of conventional structures, they do not yet satisfactorily address structures where one or more structural elements are built out of SMA. The complex behavior of SMA with path dependency, phase transformations and hysteresis preclude straight-forward incorporation into a conventional FEM code and special schemes need to be devised to address such issues, especially under combined thermo-mechanical loading. This paper describes one such approach focused on Shape Memory Alloys.

A number of constitutive models for SMA have been proposed in the literature. Recently, Buravalla and Khandelwal [1, 2] have given a good account of various models and proposed a rigorous and robust SMA constitutive law and transformation kinetics using a phenomenological approach of Brinson [3]. This will form the basis of the present development for FE implementation. Existing implementations of 1-D SMA models in FE analysis of structures focus on specific problems and usually carry out SMA analysis decoupled from the main structural analysis [4, 5]. Such investigations treat the SMA wire elements in a lumped manner and couple the SMA to other structure through kinematic constraints. Also, the loading is only simple monotonic loading and thus there is no manifestation of issues such as path dependency and hysteresis under non-monotonic loadings. Further, the current commercial general purpose (GP) FE software such as ANSYS, LS-DYNA do recognize the need for providing the material models for materials like SMA. However, material models currently available in such GP FE packages are rather simplistic and do not capture many aspects of the SMA behavior.

This paper is an attempt to address the two issues mentioned above: one that the FE analysis needs to capture the complexity of SMA behavior and the other that such material model needs to be implemented in GP FE. The present paper illustrates the implementation of Buravalla and Khandelwal model [1, 2] as UMAT in the general purpose FE software, ABAQUS/Standard [6]. Simulations are carried out for a simple actuator configuration to demonstrate the validity and robustness of the implementation. Issues such as partial phase transformations, cyclic loading, etc. are addressed. Finally, the use of the technique is demonstrated by carrying out a few parametric studies on the actuator design parameters.

2 Constitutive Model for SMA

Figure 1 depicts a characteristic phase diagram for an SMA on stress (σ)-temperature (T) plane. The shaded regions represent the transformation zones and M_s, M_f, A_s, A_f are the start and finish temperatures at zero stress for the martensitic and the austenitic transformations, respectively. The slopes of transformation boundary line for martensite and austenite transformation are given by C_M and C_A, respectively.

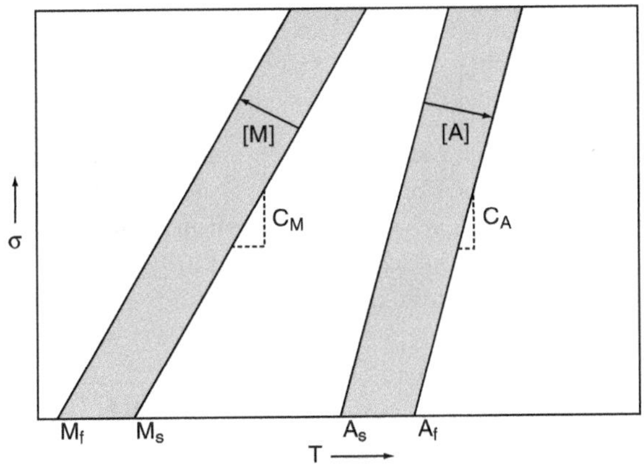

Fig. 1 Typical phase diagram for a shape memory alloy

Following Buravalla and Khandelwal [1], constitutive relations for such an SMA can be written in the following form.

$$\sigma = D(\xi)\varepsilon + \Omega(\xi)\xi + \Theta(\xi)\Delta T \tag{1}$$

where, σ is the second Piola-Kirchoff stress, ε is the Green strain, ξ is martensite fraction, varying from 0 (fully austenite) to 1 (fully martensite) and T is the temperature. D, Ω and Θ are defined as:

$$D = D_a + \xi\,(D_m - D_a)\,, \quad \Omega = -\varepsilon_l D \quad \text{and} \quad \Theta = -D\alpha \tag{2}$$

where, ε_l is the maximum residual strain, D_a and D_m represent the elastic moduli of austenite and martensite phase, respectively, and α is the thermal coefficient of expansion. More details of the constitutive model can be seen in Buravalla and Khandelwal [1].

The martensitic phase fraction ξ is to be determined from the transformation kinetics. Because of the nonlinear nature of the problem needing incremental approach, the rate form of kinetics of Buravalla and Khandelwal [2] is adopted. However, their distance based memory parameter is not used in the current formulation to keep the implementation simple in this initial attempt.

For $T \geq M_f$ and $C_M(T - M_s) \leq \sigma \leq C_M(T - M_f)$

$$\xi^{AS} = \chi^{AS}\,(1-\xi)\,\frac{\pi}{C_M(M_s - M_f)}\cot\left\{\frac{\pi}{2C_M\,(M_s - M_f)}\,[\sigma - C_M\,(T - M_f)]\right\}(\sigma - C_M T) \tag{3}$$

$$\text{where} \quad \chi^{AS} = \begin{cases} 1 \text{ when } \begin{cases} (\dot{\sigma} - C_M\,\dot{T}) > 0 \\ C_M(T - M_s) \leq \sigma \leq C_M(T - M_f) \end{cases} \\ 0 \text{ otherwise} \end{cases}$$

For $T \geq A_s$ and $C_A(T - A_f) \leq \sigma \leq C_A(T - A_s)$

$$\xi^{SA} = \chi^{SA}\xi\frac{\pi}{C_A\,(A_f - A_s)}\left\{\cot\left[\frac{\pi}{2C_A\,(A_s - A_f)}\,[\sigma - C_A\,(T - A_s)]\right]\right\}^{-1}(\sigma - C_A T) \tag{4}$$

$$\text{where} \quad \chi^{SA} = \begin{cases} 1 & \text{when} \begin{cases} (\dot{\sigma} - C_A \dot{T}) < 0 \\ C_A(T - A_f) \le \sigma \le C_A(T - A_s) \end{cases} \\ 0 & \text{otherwise} \end{cases}$$

Here, the scalar parameters χ^{AS} and χ^{SA} represent the phase transformation activation conditions. Further, $\dot{\xi} = \dot{\xi}^{AS} + \dot{\xi}^{SA}$ and $\dot{\xi} + \dot{\xi}_A = 0$ for the conservation of mass where, ξ_A is volumetric fraction of austenite. More detailed discussion about this evolution kinetics can be found in [2].

The time-discrete evolution equations can be written in the residual form as

$$R^{AS} = \Delta\xi^{AS} - \chi^{AS}(1-\xi)\frac{\pi}{C_M(M_s - M_f)}(\Delta\sigma - C_M\Delta T) \cdot \cot\left\{\frac{\pi}{2C_M(M_s - M_f)}\left[\sigma - C_M\left(T - M_f\right)\right]\right\} = 0 \tag{5a}$$

$$R^{SA} = \Delta\xi^{SA} - \chi^{SA}\xi\frac{\pi}{C_A\left(A_f - A_s\right)}(\Delta\sigma - C_A\Delta T) \cdot \left\{\cot\left[\frac{\pi}{2C_A\left(A_s - A_f\right)}\left[\sigma - C_A\left(T - A_s\right)\right]\right]\right\}^{-1} = 0 \tag{5b}$$

Further, the above residue equations are linearized in terms of ξ^{AS}, ξ^{SA}, ε and T to give

$$\mathrm{R}\begin{Bmatrix} d\Delta\xi^{AS} \\ d\Delta\xi^{SA} \end{Bmatrix} = -\begin{Bmatrix} \frac{\partial R^{AS}}{\partial \varepsilon}d\varepsilon + \frac{\partial R^{AS}}{\partial T}dT \\ \frac{\partial R^{SA}}{\partial \varepsilon}d\varepsilon + \frac{\partial R^{SA}}{\partial T}dT \end{Bmatrix}, \quad \text{where } \mathrm{R} = \begin{bmatrix} \frac{\partial R^{AS}}{\partial \Delta\xi^{AS}} & \frac{\partial R^{AS}}{\partial \Delta\xi^{SA}} \\ \frac{\partial R^{SA}}{\partial \Delta\xi^{AS}} & \frac{\partial R^{SA}}{\partial \Delta\xi^{SA}} \end{bmatrix} \tag{6}$$

These equations can then be solved for $d\xi^{AS}$ and $d\xi^{SA}$. Further, for the calculation of material tangent stiffness modulus (C) and thermal modulus (H), the constitutive equation (Eq. (1)) is linearized in terms of strain (ε) and temperature (T). The parameters C and H are required for the implementation of user material model in ABAQUS/Standard as described below.

3 FE Implementation

The UMAT facility of the FE package ABAQUS/Standard is utilized for implementation of the above mentioned constitutive model and evolution kinetics for SMA. The FE solver (ABAQUS) passes the displacement and temperature increments to UMAT and requires updated values of stress, material tangent stiffness and temperature moduli to be returned from UMAT. The nonlinearity of the constitutive model and the evolution law necessitates iterations while solving for stress. A step by step description of the algorithm is presented below.

1. At the start of a time step t_{n+1}, the given quantities include increments in strain ($\Delta\varepsilon$) and temperature (ΔT) and values of stress (σ_n) and martensite fraction (ξ_n) from the previous time step t_n.
2. For the first iteration, set $\xi^i_{n+1} = \xi_n$.
3. Prediction of stress (σ^i_{n+1}) is performed based on $\Delta\varepsilon_{n+1}$, ΔT_{n+1} and ξ^i_{n+1}. For the first iteration, material is treated as elastic with no phase transformation.
4. Compute increment in martensite ($\Delta\xi^i_{n+1}$) from the time-discrete form of evolution kinetics, Eq. (6) and update $\xi^i_{n+1} = \xi^i_{n+1} + \Delta\xi^i_{n+1}$.
5. Compute σ^i_{n+1} using the constitutive equation, Eqs. (1), (2).
6. Check for convergence. If $|R^{AS}|$ and $|R^{SA}| \leq Tol$, a prescribed tolerance, then proceed to step 8. Else, increment the iteration counter and go to step 4.
7. Update algorithmic tangent stiffness (C_{n+1}) and thermal modulus (H_{n+1}).
8. Return values of ξ_{n+1}, σ_{n+1} and C_{n+1}, H_{n+1} and go to the next time step.

The optimal *Tol* value found from convergence study is 0.0001. The same is used for all the numerical studies presented herein. The UMAT accounts only for the thermo-mechanical behavior of the SMA material and the rest of the finite element part is completely handled by ABAQUS in the usual way.

4 Results and Discussion

The material parameters used in this study are summarized in Table 1.

4.1 Validation

The UMAT and the incremental scheme of computations is first validated by studying the behavior of an SMA wire through ABAQUS (B22 element) and comparing the results obtained by treating the entire SMA wire as a lumped parameter model (LPM). The SMA wire is considered to be under axial load so that it is under constant stress. Several thermo-mechanical loading cases for phase transformations are considered such as (i) stress excursion at constant temperature in austenitic range, (ii) temperature excursion at constant stress, (iii) temperature excursions followed by stress excursions, and (iv) stress excursions followed by temperature excursions.

Table 1 SMA Material properties used in the analysis

Property	Martensite	Austenite
Modulus	$D_m = 26.3$ GPa	$D_a = 67.0$ GPa
Theta (Θ)	0.55 MPa/°C	–
Transformation temperatures	$M_s = 5.9$°C, $M_f = -2.85$°C	$A_s = 34.5$°C, $A_f = 49.0$°C
Stress–temperature slopes	$C_M = 8.0$ MPa/°C	$C_A = 13.8$ MPa/°C
Transformation strain	$\varepsilon_l = 0.067$	–

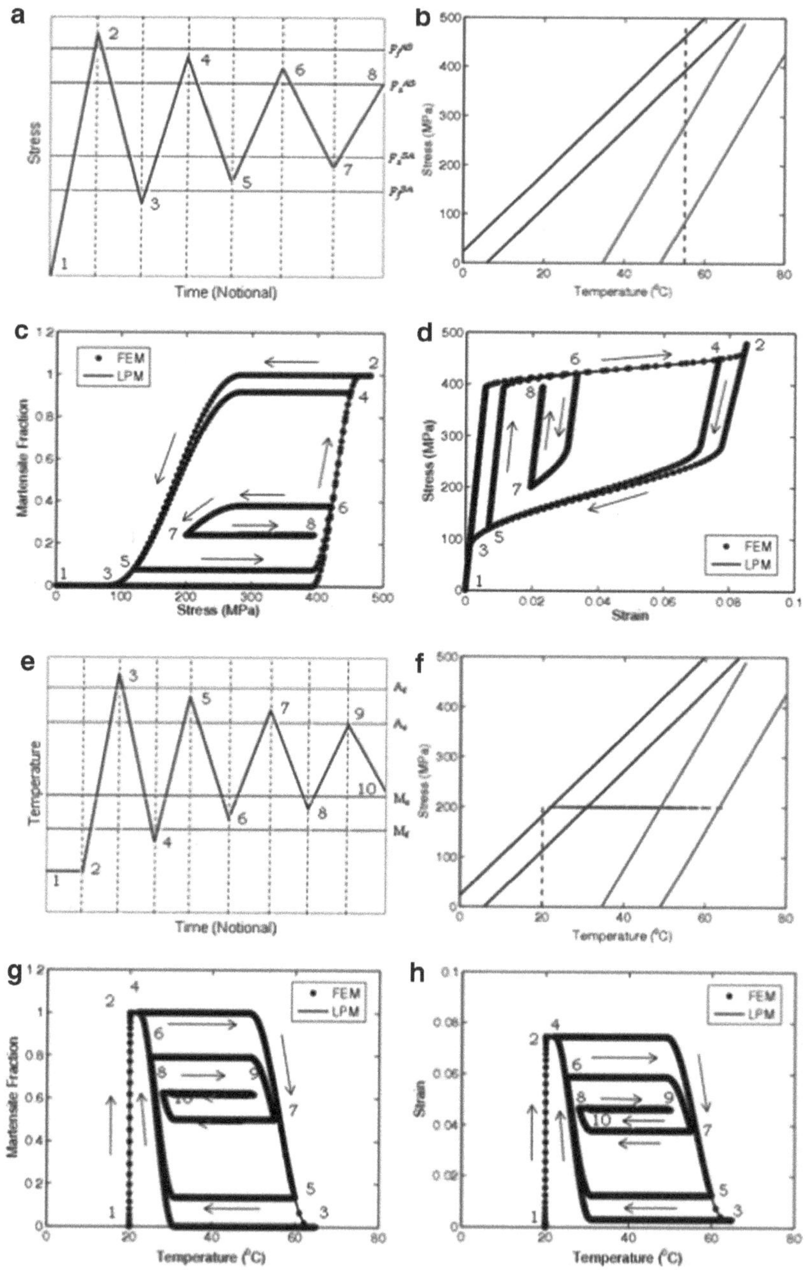

Fig. 2 (**a**) Schematic of stress loading history, (**b**) stress–temperature phase diagram, (**c**) evolution of martensite with stress, and (**d**) stress–strain response for cyclic stress loading at constant temperature of 55°C. (**e**) Schematic of temperature loading history, (**f**) stress–temperature phase diagram, (**g**) evolution of martensite with temperature, and (**h**) strain–temperature response for cyclic thermal loading at constant stress level of 200 MPa

In each case the FE model of the wire had ten elements of equal length and the results were then compared with those of the lumped parameter model. The results were identical. As an example, Fig. 2 depicts the results for the case of cyclic loading at constant temperature (Figs. 2(a, b, c and d)) and constant stress (Figs. 2(e, f, g and h)). Note that the transformations including partial transformations are captured well.

Figure 3(a) shows the schematic of a cantilever beam actuator where an SMA wire of diameter $d = 1$ mm is attached to a steel beam (2 mm width and 1 mm depth in cross section) offset by a distance $a = 2$ mm. The material properties used

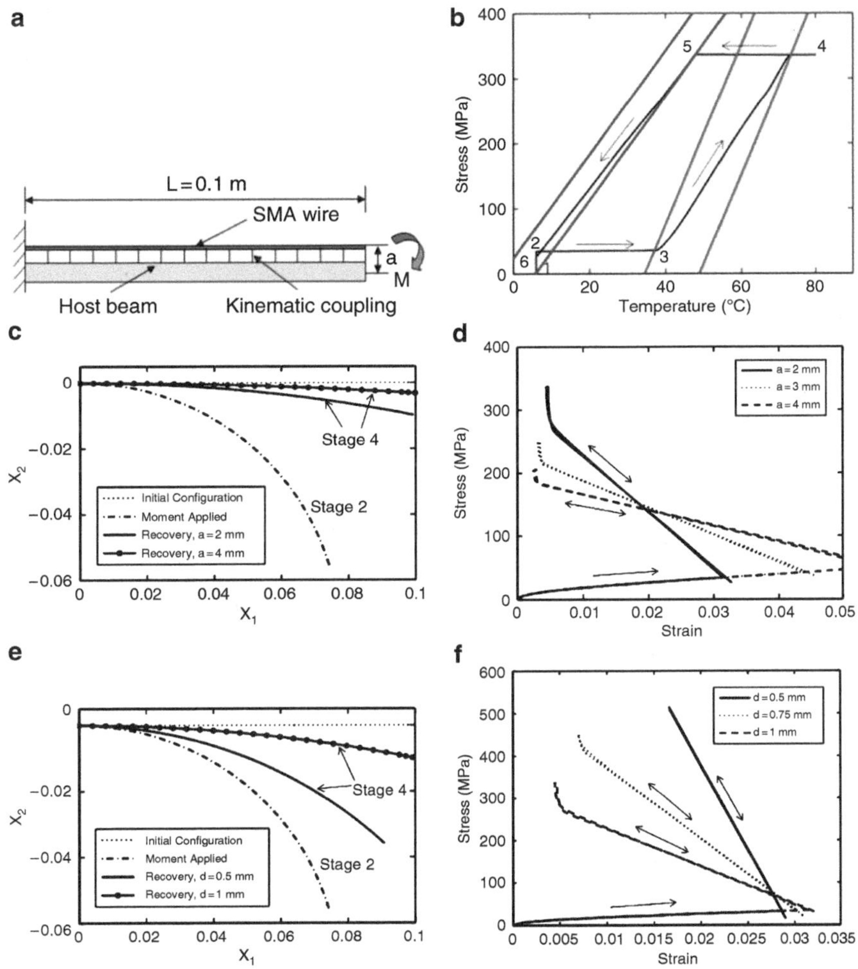

Fig. 3 (**a**) Schematic of beam actuator and (**b**) stress–temperature history of SMA wire. (**c**) Shapes of the host beam at various loading stages corresponding to offset distance a= 2 and 4 mm. (**d**) Stress–strain response of the SMA wire for various values of offset distance. (**e**) Shapes of the actuator at various loading stages corresponding to SMA wire diameter d= 1 and 0.5 mm. (**f**) Stress–strain response of the SMA wire for various values of SMA wire diameter

for host beam are, $E = 210$ GPa and $\upsilon = 0.3$. The kinematic constraints facility in ABAQUS is used for prescribing the connection between SMA wire and the host beam.

Initially, SMA wire is assumed to be at 6°C (just above M_s) and in austenite phase ($\xi_0 = 0$). Numerical results are obtained for various values of the offset distance (a) and the wire diameter (d), which are the two of the main design parameters. Moment is applied at the free end causing bending of the beam and stressing of SMA wire leading to formation of martensite. The SMA wire is then heated to temperature of 80°C followed by cooling to initial temperature of 6°C.

The complete stress-temperature history for SMA wire is shown in Fig. 3(b) for a= 2 mm and d= 1 mm. Figures 3(c, d) and 3(e, f) show host beam deformations and corresponding stress-strain response of SMA wire for various values of a and d, respectively considering equal amount of host beam bending. In Figs. 3(c and e), the shape of host beam in bent configuration represents stage 2 in Fig. 3(b) while shape recovery after heating represents stage 4 in Fig. 3(b). It can be seen that the structural analysis can be performed in an unified way. The results for the variations in the values of a and d show the influence of these design parameters on performance. For example, it can be seen that for better recovery one needs to use larger offset distance and larger diameter wire. Of course, these may have other implications (such as thermal response and power required), but those issues are not within the scope of this paper.

5 Concluding Remarks

The recently developed SMA constitutive model and the phase transformation kinetics are implemented as a UMAT to be used with a general purpose FE package. Numerical simulations are carried out to validate the implementation. Finally, as an illustration, behavior of an SMA wire based actuator is analyzed.

Acknowledgement Authors are grateful to Vidyashankar Buravalla and Ashish Khandelwal for many useful discussions.

References

1. Buravalla V R and Khandelwal A, 2007. Differential and integrated form consistency in 1-D phenomenological models for shape memory alloy constitutive behavior *Int J Solids & Struct.* 44, 4369–4381
2. Buravalla V R and Khandelwal A, 2008. Phenomenological Modeling of Shape Memory Alloys, Smart Devices: Modeling of Material Systems: An International Workshop, IIT Madras, Chennai (India), 10–12 Jan 2008, Sivakumar S M, Buravalla V R, Srinivasa A R, Eds., AIP Conf. Proc. 1029, 104–123

3. Brinson L C, 1993. One dimensional constitutive behavior of shape memory alloys: thermo-mechanical derivation with non-constant material functions *J. Intell. Mater. Syst. Struct.* 4 229–242
4. Wang G and Shahinpoor M, 1996. Design, prototyping and computer simulations of a novel large bending actuator made with a shape memory alloy contractile wire *Smart Mater. Struct.* 6, 214–221
5. Shu S G, Lagoudas D C, Hughes D and Wen J T, 1997. Modeling of a flexible beam actuated by shape memory alloy wires *Smart Mater. Struct.* 6, 265–277
6. *ABAQUS/Standard User's Manual*, Version 6.5.5, Hibbit, Karlsson & Sorensen, Inc., Pawtucket, Rhode Island, USA

Time Dependent Deformations in Concrete: A Multi-scale Approach

Ananth Ramaswamy and Harinadha Reddy

Abstract Estimation of creep and shrinkage are critical in order to compute loss of prestress with time in order to compute leak tightness and assess safety margins available in containment structures of nuclear power plants. Short-term creep and shrinkage experiments have been conducted using in-house test facilities developed specifically for the present research program on 35 and 45 MPa normal concrete and 25 MPa heavy density concrete. The extensive experimental program for creep, has cylinders subject to sustained levels of load typically for several days duration (till negligible strain increase with time is observed in the creep specimen), to provide the total creep strain versus time curves for the two normal density concrete grades and one heavy density concrete grade at different load levels, different ages at loading, and at different relative humidity's. Shrinkage studies on prism specimen for concrete of the same mix grades are also being studied. In the first instance, creep and shrinkage prediction models reported in the literature has been used to predict the creep and shrinkage levels in subsequent experimental data with acceptable accuracy.

While macro-scale short experiments and analytical model development to estimate time dependent deformation under sustained loads over long term, accounting for the composite rheology through the influence of parameters such as the characteristic strength, age of concrete at loading, relative humidity, temperature, mix proportion (cement: fine aggregate: coarse aggregate: water) and volume to surface ratio and the associated uncertainties in these variables form one part of the study, it is widely believed that strength, early age rheology, creep and shrinkage are affected by the material properties at the nano-scale that are not well established. In order to understand and improve cement and concrete properties, investigation of the nanostructure of the composite and how it relates to the local mechanical properties

A. Ramaswamy (✉)
Department of Civil Engineering, Indian Institute of Science, Benguluru, India
e-mail: ananth@civil.iisc.ernet.in

H. Reddy
Graduate Student, Civil Engg. IISc
e-mail: harinadh@civil.iisc.ernet.in

B. Dattaguru et al. (eds.), *IUTAM Symposium on Multi-Functional Material Structures and Systems*, IUTAM Bookseries 19, DOI 10.1007/978-90-481-3771-8_6,

is being undertaken. While results of creep and shrinkage obtained at macro-scale and their predictions through rheological modeling are satisfactory, the nano and micro indenting experimental and analytical studies are presently underway. Computational mechanics based models for creep and shrinkage in concrete must necessarily account for numerous parameters that impact their short and long term response. A Kelvin type model with several elements representing the influence of various factors that impact the behaviour is under development. The immediate short term deformation (elastic response), effects of relative humidity and temperature, volume to surface ratio, water cement ratio and aggregate cement ratio, load levels and age of concrete at loading are parameters accounted for in this model. Inputs to this model, such as the pore structure and mechanical properties at micro/nano scale have been taken from scanning electron microscopy and micro/nano-indenting of the sample specimen.

Keywords Creep · Shrinkage · Normal and heavy density concrete · Nano-indenting · Scanning electron microscopy

1 Introduction

Probably the most uncertain and most difficult aspect of the design of reinforced and prestressed concrete structures is the prediction of time-dependent behavior. However, realistic prediction of concrete creep and shrinkage is of crucial importance for durability and long-time serviceability of concrete structures. For some structures the long-term performance from the safety view point prediction of time dependent deformations are critical. Creep and shrinkage cause increases in deflection and curvature, cracking of concrete, loss of prestress, redistribution of stresses and leakages. Predicting delayed strains in concrete proves to be critical to a large number of pre-stressed concrete structures, such as containment vessels of nuclear power plants. These delayed strains include *autogeneous shrinkage*, as deformation related to the water consumption during the hydration reaction in early-age concrete; *drying shrinkage*, as deformation related to the moisture diffusion from the inner core to the outside of the concrete member; *basic creep strains*, as time-dependent deformation of a loaded specimen, without drying; and *drying creep strains*, as additional deformation, which occurs in drying in a loaded specimen (Pickett effect). Concrete is a multiphase system where the voids of the skeleton are filled partly with liquid water and partly with a gas mixture composed of dry air (incondensable constituent) and water vapor (condensable gas) and in case of higher than standard temperatures several phenomenon, such as heat conduction, vapor diffusion and liquid water flow should be considered. Liquid water flow is due to pressure gradients, or capillary effects caused by the meniscus curvature inside pores. Factors effecting creep in concrete include the relative humidity, temperature, cement and aggregate content, water cement ratio, specimen geometry and the load levels that are sustained. The present study reports on an extensive test program on creep and shrinkage in normal and heavy density concretes together with validation using the B3 model [1], that accounts for uncertainties associated with the contributing parameters in the long term. A multi-physics model based on the thermal, chemical, hygral and mechanical processes is under development.

2 Experimental Program

Tables 1, 2 and 3 provides details of the constituents used in the mix for the 35, 45 MPa normal concrete and 25 MPa heavy density concrete, respectively. The sample specimens (three each of 150 mm diameter ×300 mm cylinder) were placed in the creep rig, and the loads were applied using a jack up to the prescribed level and locked-into the frame. Constant load conditions were maintained using a load cell positioned in the system and strains measured using demec gages. The factors studied in the present program include relative humidity (50%, 60%, 70%), age of concrete at loading (7 days, 14 days, 1 month, 3 month, 6 month, 1 year and 2 year) and magnitude of constant uniaxial load applied (20% F_{ck}, 30% F_{ck}, 40% F_{ck}, 50% F_{ck}, 60% F_{ck}, 70% F_{ck}). Figures 1, 2, and 3 show variations in creep measured in the tests for the different mixes. As expected, creep levels are lower when load is applied at a later age and at higher relative humidity.

The shrinkage specimens were prismatic having a 100 by 100 mm cross-section and 500 mm length. Six specimen in each mix were kept in an environmental controlled room at a constant relative humidity and temperature level. Demec gauges

Table 1 Mass (kg) of different constituents per cubic meter of 35 MPa normal density mix

Cement	350	Coarse aggregate (10 mm)	330.9
Water/ice	158	Fine aggregate (natural)	470
Coarse aggregate (40 mm)	330.9	Fine aggregate (crushed)	320
Coarse aggregate (20 mm)	441.2	Admixture (%)	1.3

Table 2 Mass (kg) of different constituents per cubic meter of 45 MPa normal density mix

Cement	385	Fine aggregate (natural)	492
Water/ice	153	Fine aggregate (crushed)	334
Coarse aggregate (20 mm)	612.6	Admixture (%)	1.28
Coarse aggregate (10 mm)	408.4		

Table 3 Mass (kg) of different constituents per cubic meter of 25 MPa heavy density mix

Cement	360	Fine aggregate (iron sand)	1,340
Water/ice	172.8	Fine aggregate (crushed)	82
Coarse aggregate (20 mm)	1,092.0	Admixture (%)	1.2
Coarse aggregate (10 mm)	728.0		

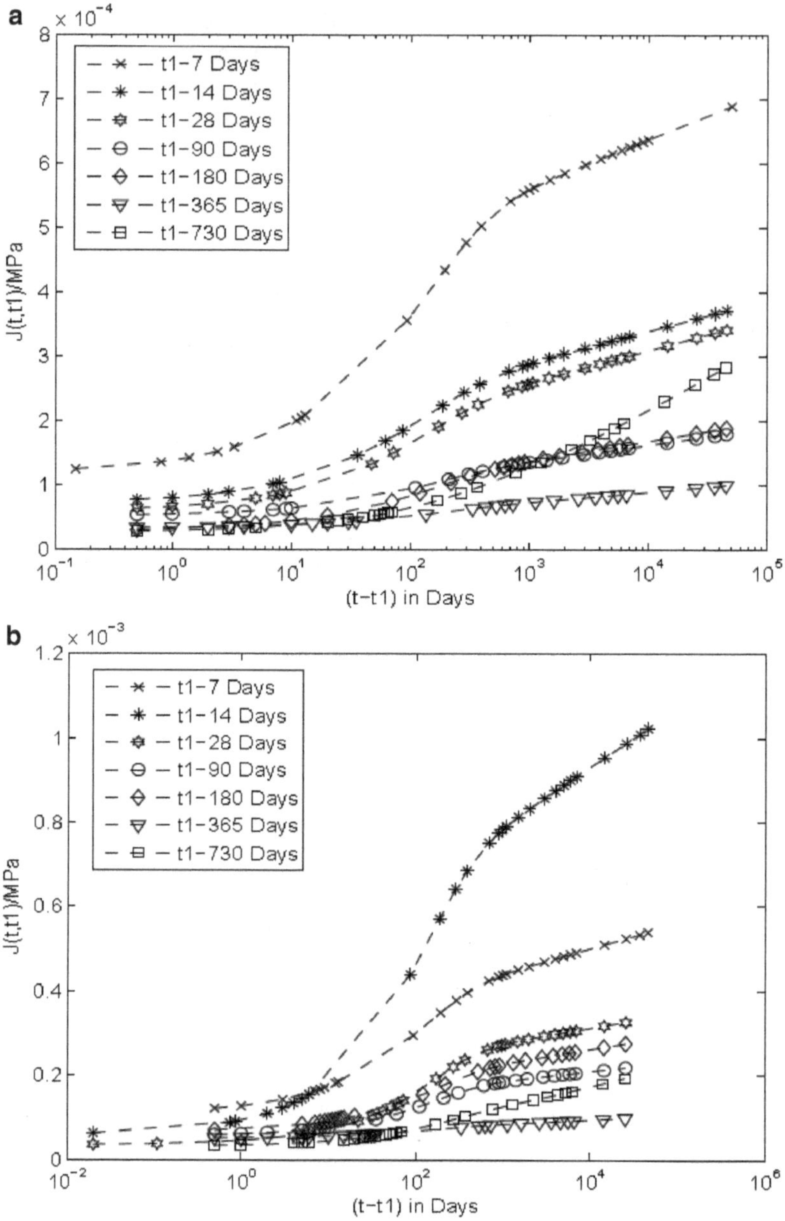

Fig. 1 Creep in Normal concrete at different ages of loading (**a**) 45 MPa concrete at 60% relative humidity (**b**) 35 MPa concrete at 50% relative humidity

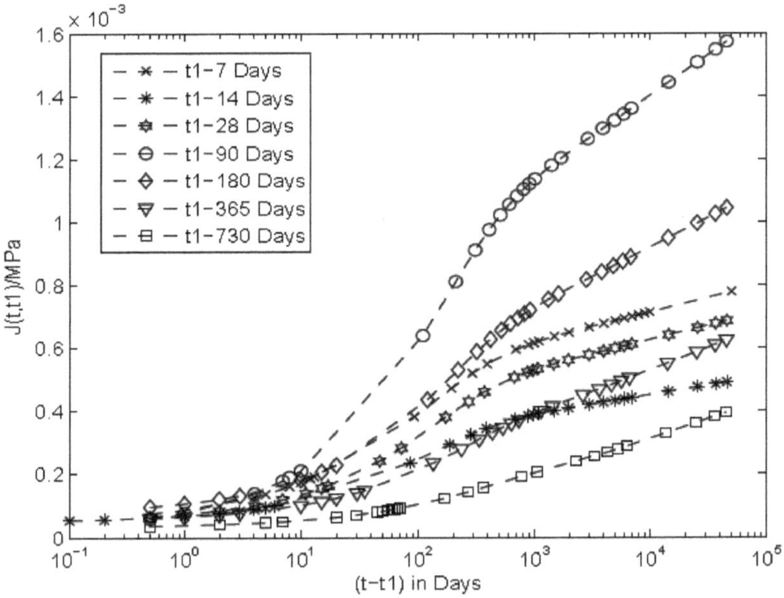

Fig. 2 Creep in 25 MPa heavy density concrete at 70% relative humidity at different ages of loading

were mounted on each of the side faces and the strains were measured regularly. The weight of the prism specimens were also recorded along with measured strains in order to estimate the water loss. The water loss data is necessary in order to estimate the limiting shrinkage strain. Figures 4-6 show plots of the shrinkage strains versus time for each of the mix proportions. The shrinkage strain is higher for concrete with higher strength cured at lower humidity.

3 Analysis and Discussion

The B3 model [1] has been used to predict the creep in concrete for the different mix grades. Essentially the model accounts for the instantaneous deformations, the creep under insulated conditions of no moisture movement and the deformations under drying process.

Figures 1(a and b) and 2 shows creep micro-strain versus time strain prediction for 35 and 45 MPa normal concrete and 25 MPa heavy density concrete at different age at loading at different relative humidity's. It may be seen that if we consider only the first four data and use B3 model [1] to predict the next four data, the predicted and the experimental data are in good agreement. Figures 3(a and b) and 4 shows the shrinkage strains predicted using the B3 model [1–3], in the different mix specimen with time. Both creep and shrinkage predictions account for the parameter

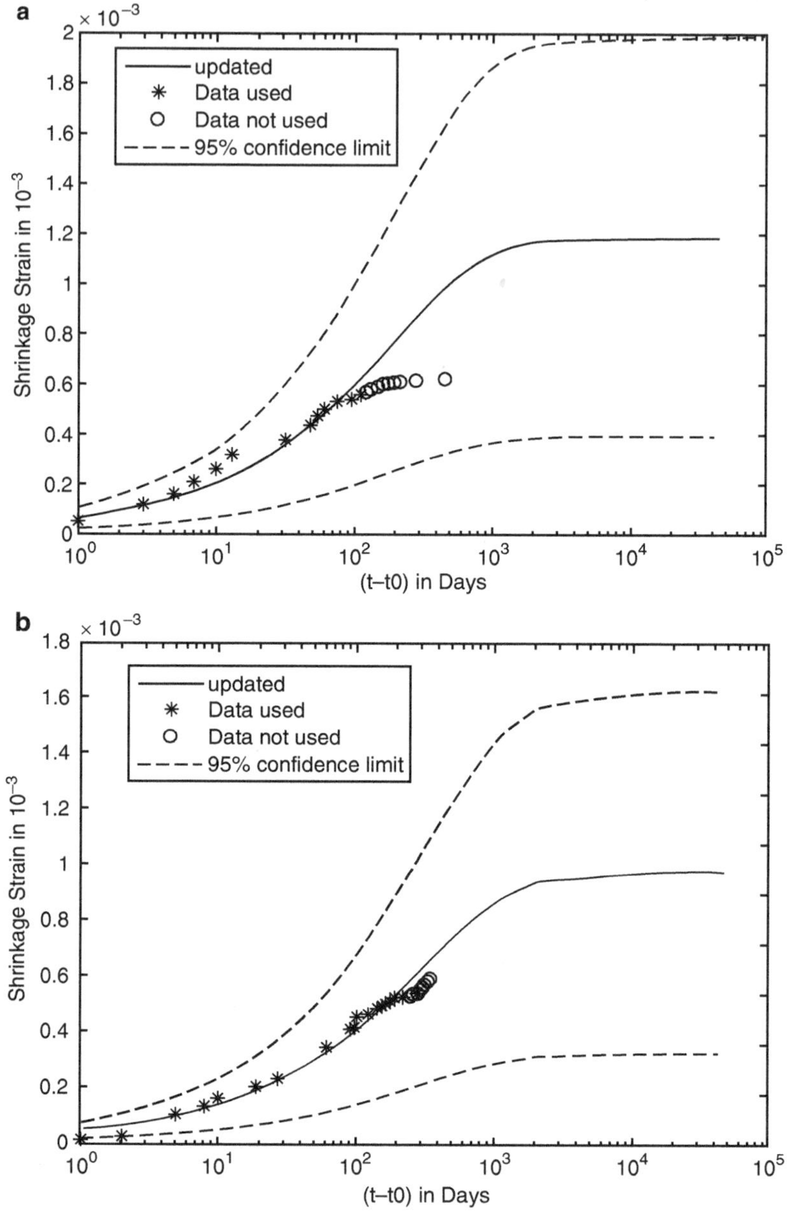

Fig. 3 Shrinkage (**a**) 45 MPa normal density concrete at 60% relative humidity, (**b**) 35 MPa normal density concrete at 50% relative humidity

uncertainties using Bayesian updating discussed in the cited papers. While these procedures were developed for normal concrete, the present study has extended its application to heavy density concrete.

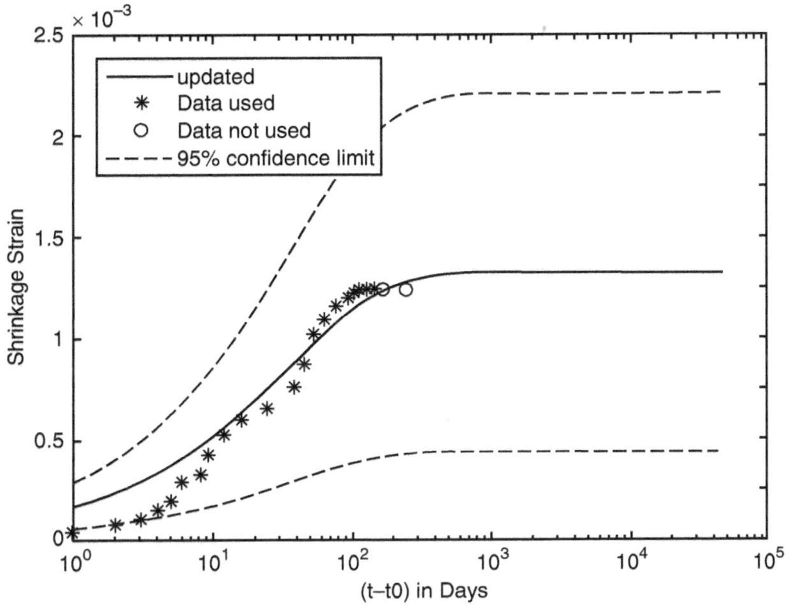

Fig. 4 Shrinkage in 25 MPa heavy density concrete at 70% relative humidity

A mechanistic model that accounts for the thermo-hygro-chemo-mechanical coupling in the form of Kelvin elements that depend on parameters such as load level and age of concrete at loading, temperature, relative humidity, chemical composition of the mix, water cement ratio and rate of hydration is under development. The approach proposed by Oh and Cha [4] for moisture and temperature distribution as a function of the rate of hydration provides the basis for the present study in developing a model for creep and shrinkage prediction. The study employs the diffusion formulation for the determination of temperature and moisture distribution. Heat balance is expressed with the transient Fourier equation, for heat conduction with internal sources (an internal sink in the moisture field) expressed as follows:

$$\frac{\partial \phi}{\partial t} = \nabla .\left(D_{\phi} \nabla \phi\right) \tag{1}$$

$$(C\rho)_T = K\nabla^2 T + q_T \tag{2}$$

Where ϕ is the Humidity, D_{ϕ} is the Diffusion coefficient, C is the Specific heat capacity, $\rho =$ Density, K is the Constitutive matrix, T is the Temperature and q_T is the Internal rate of energy production caused by cement hydration expressed as follows:

$$\mathrm{q(t)} = \max \mathrm{Q}(\partial\alpha/\partial \mathrm{t}) \tag{3}$$

where Q max is the maximum heat of hydration of young concrete and α is the degree of reaction of the cement and is a function of the maturity age and other

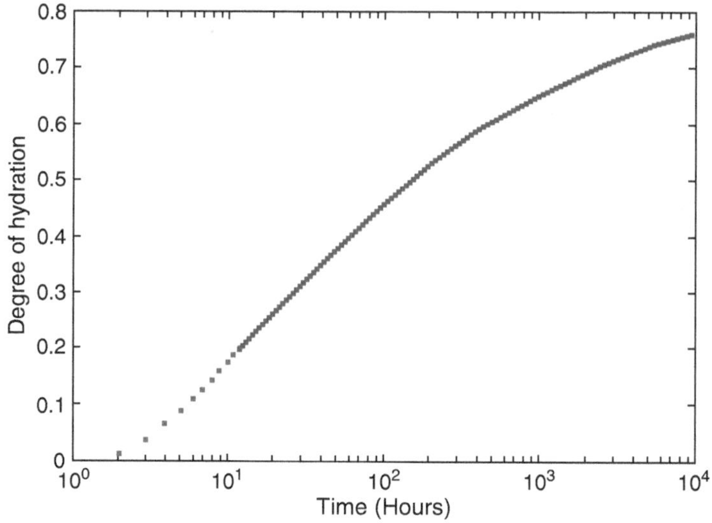

Fig. 5 Rate of hydration of 45 MPa concrete having water cement ratio of 0.45

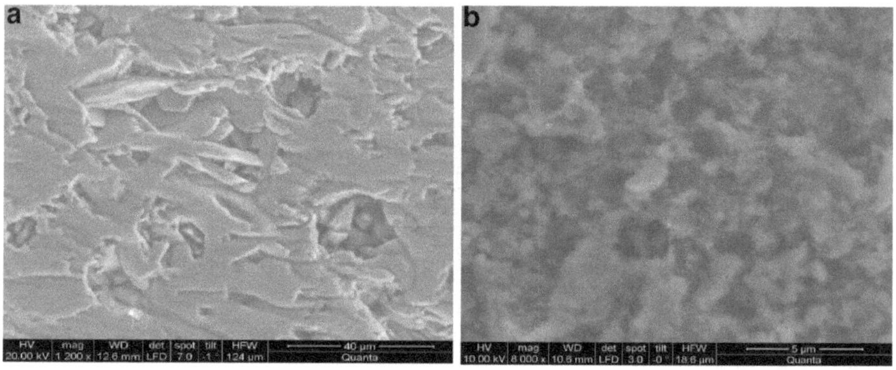

Fig. 6 Scanning electron microscope image of (**a**) 25 MPa heavy density concrete at 1 year showing partially hydrated long ettringite crystals and (**b**) 45 MPa normal density concrete at 1 year showing a flower like hydrated mono sulfate hydrate particle

empirical constants detailed in Oh and Cha [4]. Figure 5 shows the rate of hydration for a 45 MPa concrete having a water cement ratio of 0.45. In addition, the pore structure of the composite, the magnitude of the mechanical load and the age of the concrete at the load application provide the coupled framework of the formulation. The pore structure of the concrete is a function of its constituents in the mix (cement, sand, coarse aggregate, water, super plasticizers and additives) and their particle size gradations that control the hydration process. Figures 6(a and b) are scanning electron microscopic images that illustrate the differences between two different mixes at the same age. Figure 7 shows the composition analysis for 45 MPa concrete. Carbon, oxygen, silicon and calcium are the predominant elements in the mix.

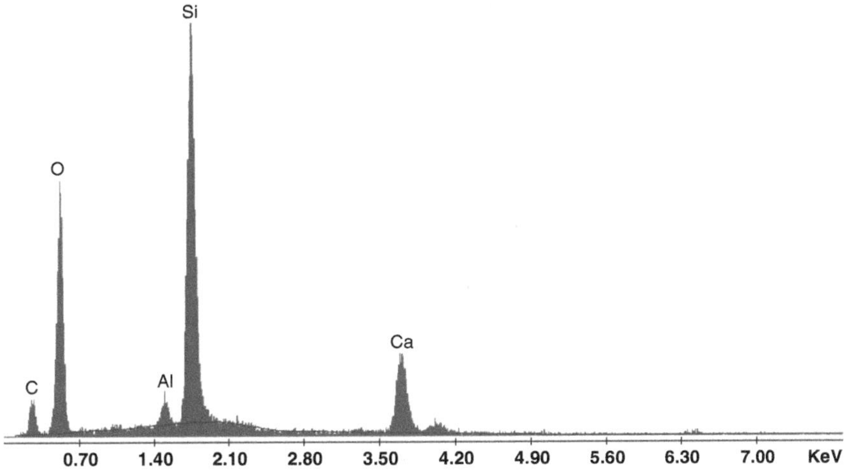

Fig. 7 Composition analysis for 45 MPa mortar mix indicating elemental peaks at different electron volt levels (Kev)

Early age rheology, creep and shrinkage are affected by the material properties at the nano-scale that are not well established. Recent studies by Corr and Shah [5] and Hughes and Trtik [6] have used nano indenting to study the properties of normal concrete and cement mortar while and Lu et al. [7] has employed the procedure for studying polymer composites. In order to understand and improve cement and concrete properties, investigation of the nanostructure of the composite and how it relates to the local mechanical properties has been attempted. As concrete is a heterogeneous composite, micro and nano indenting applied at different points of the concrete have been attempted to obtain the hardness and compliance of the material at a very local scale. Scanning electron microscopy and indenting is expected to offering insights in selecting some parameters in the coupled multi-physics model, such as the average pore size and its distribution, average hardness as an aggregation of the aggregate and mortar hardness. Figure 8 shows the compliance obtained from micro indenting applied in 45 MPa concrete in a 1 year cured (at 70% relative humidity) sample.

4 Closing Remarks

From the studies that have been undertaken so far it is seen that as expected creep strains reduce with increase in age of loading and for higher levels of relative humidity. B3 Model with calibration using test data from short term measurements is able to predict the Creep and shrinkage strains satisfactorily for long term. A mechanistic model that accounts for the influence of load level and age of concrete at loading, temperature, relative humidity, chemical composition and hydration rate, is under development.

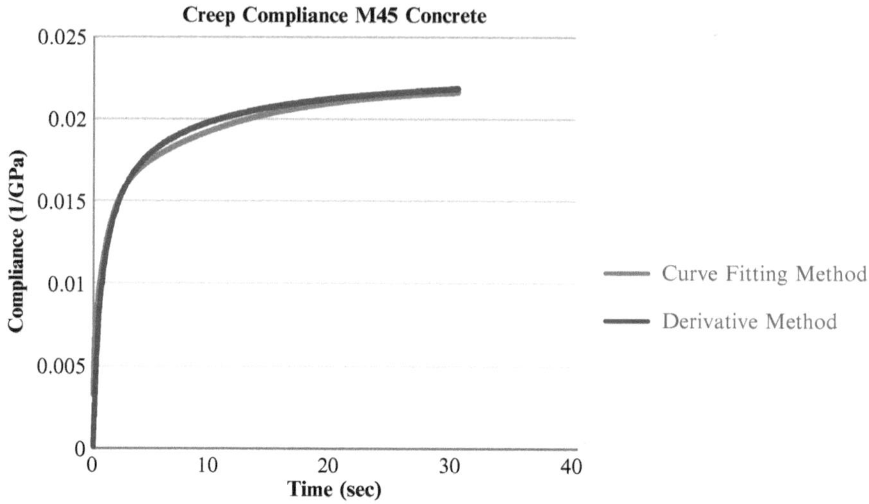

Fig. 8 Creep compliance in 45 MPa concrete

Acknowledgements Financial support provided by Board for Research in Nuclear Sciences (BRNS) (Sanction No.2003/36/28/BRNS/2039 dated 17-2-2004) is gratefully acknowledged.

References

1. Bazant, Z.P., and Baweja, S. (2000). "Creep and Shrinkage Prediction Model for Analysis and Design of Concrete Structures: Model B3," *As submitted for:, Adam Neville Symposium: Creep and Shrinkage-Structural Design Effects*, ACI SP-194, A.Al-Manaseer, ed., Am. Concrete Institute, Farmington Hills, Michigan, 2000, 1–83.
2. Bazant, Z.P., and Chern, J.C. (1984). "Bayesian Statistical Prediction of Con crete Creep and shrinkage",*ACI JOURNAL*, 81, July–August, 1984.
3. Bazant, Z.P., Kim, J.K., Wittmann, F.H., and Alou, F. (1987). "Statistical Extrapolation of Shrinkage Test Data- Part II Bayesian Updating," ACI Mat terials Journal, 84, 83–91.
4. Oh, B.H., and Cha, S.W. (2003) "Nonlinear Analysis of Temperature and Mois- ture Distributions in Early Age Concrete Structures Based on Degree of Hydra tion", ACI Materials Journal, 361–370.
5. Corr, D. and Shah, S.P. (2005) "Concrete material science at the nan-scale, Application of nanotechnology in concrete design", Proc. of Int. conference at University of Dundee, Scotland, U.K., Thomas Telford, July, pp.1–12.
6. Hughes, J.J. and Trtik, P. (2004) "Micro-mechanical properties of cement paste measured by depth sensing nano-indentation: A preliminary correlation of physical properties with phase type", Materials characterization, 53, 223–231.
7. Lu, H. Wang, B, Ma, J, Huang, G., and Viswanathan, H. (2003) "Measurement of Creep Compliance of Solid Polymers by Nano-indentation", Mechanic of Time-Dependent Materials, 7, 189–207.

Higher Order Theories of Functionally Graded Beams and Plates

Tarun Kant, S.M. Shiyekar, and C. Venkata Subbaiah

Abstract Estimation of displacements and stresses in functionally graded (FG) beams and plates are carried out using higher order shear and normal deformation theories (HOSNTs). Taylor's series expansion is adopted to approximate primary displacements in the thickness direction. This leads to the reduction of three-dimensional (3D) problem to two-dimensional (2D) problem, which ultimately reduces the complexity, normally associated with 3D analysis. The expansions of primary displacements are considered in such a way that it accounts the effects of transverse shear and transverse normal stresses and strains. Governing equations of equilibrium are derived using the virtual work principle. Present results are validated with exact and other available solutions. Analytical formulation and solutions to the static analysis of simply supported functionally graded beams and plates based on higher order theories are presented in this paper.

Keywords Beams · FGM · Functionally graded material · Higher order theories · Plates · Power law

1 Introduction

Functionally Graded Material (FGM) is a special combination of ceramics and metals designed to endure high thermal environment and to provide stiffness to the host structure. Volume fraction and material properties of such materials are continuous and smooth along the direction under consideration. Abrupt change in the elastic properties and thermal coefficients in the composite laminates results in high interlaminar stresses leading to delamination. FGM is a better alternative to the composite laminates because of a gradual change in their elastic properties, particularly

T. Kant (✉), S.M. Shiyekar, and C.V. Subbaiah
Professor of Civil Engineering, Indian Institute of Technology Bombay Powai, Mumbai-400 076, India
e-mail: tkant@civil.iitb.ac.in; tkant@iitb.ac.in; spendhari@zentech-usa.com

B. Dattaguru et al. (eds.), *IUTAM Symposium on Multi-Functional Material Structures and Systems*, IUTAM Bookseries 19, DOI 10.1007/978-90-481-3771-8_7,

along the thickness direction. Koizumi [7] proposed the concept of FGM, which was initiated by material scientists in the Sendai area of Japan [8]. Wetherhold et al. [23] created special FGM beam by varying volume fraction through the thickness of symmetric laminates to control thermal deformations. Aboudi et al. [1] studied FGMs under thermal gradients and later Aboudi et al. [2] presented HOSTs for analysis of FGM plates. Praveen and Reddy [15] contributed to the study of non-linear thermoelastic static and dynamic analyses of FG plates. Sankar [21] presented elasticity solution for FG beams subjected to transverse loads. Croce and Venini [3] discussed the finite element (FE) model based on Reissner-Mindlin [10, 20] first order shear deformation plate theory (FOST) and provided thermo-mechanical analysis of FG plates. Nguyen et al. [12] presented the analysis of FG plates with FOST and carried out various parametric studies on the shear correction factor. Pan [13] presented exact solution of single and bi-layer simply supported FG plates under surface traction with five different exponential factors considered for the variation of stiffness properties in the thickness direction. Ramirez et al. [16] developed a discrete layer theory (DLT) in combination with Ritz method for FG plate and compared their results with exact [13] solution. Ray and Sachade [17] presented exact solution for the analysis of FG plates subjected to electric loading. Reddy [18], Reddy and Cheng [19], and Vel and Batra [22] presented elasticity solutions for FG plates. Matsunaga [9] presented a high order shear deformation theory (HOST) for the analysis of FG plate with simple power law blending scheme. Initially Kant [4] presented a refined HOST and derived equilibrium equations based on energy principles. Later Pandya and Kant [14] presented FE solution based on C^o continuity. Kant and Manjunatha [5] extended the HOST for unsymmetric laminates. Further Kant and Swaminathan [6] presented analytical solution for sandwiches and laminated composites. In this paper, a higher order shear and normal deformation theory is formulated for the analysis of FG beams and plates. Results are compared with elasticity and other solutions.

2 Theoretical Formulation

Rectangular FGM plate simply (diaphragm) supported on all four edges is considered here for the analysis. The plate is having a and b lengths along x and y axes respectively, while the thickness dimension h is along z-axis. The positive axes, positive displacements and origin for z-axis are measured from the mid-plane of the plate, as shown in Fig. 1. Top surface of the plate is subjected to mechanical surface traction. The 3D elasticity problem is approximated as a 2D plate problem by assuming displacement field as: *Model HOST12* [6]:

$$\begin{aligned}
u(x,y,z) &= u_0(x,y) + z\theta_x(x,y) + z^2 u_0^*(x,y) + z^3\theta_x^*(x,y)\\
v(x,y,z) &= v_0(x,y) + z\theta_y(x,y) + z^2 v_0^*(x,y) + z^3\theta_y^*(x,y)\\
w(x,y,z) &= w_0(x,y) + z\theta_z(x,y) + z^2 w_0^*(x,y) + z^3\theta_z^*(x,y)
\end{aligned} \tag{1}$$

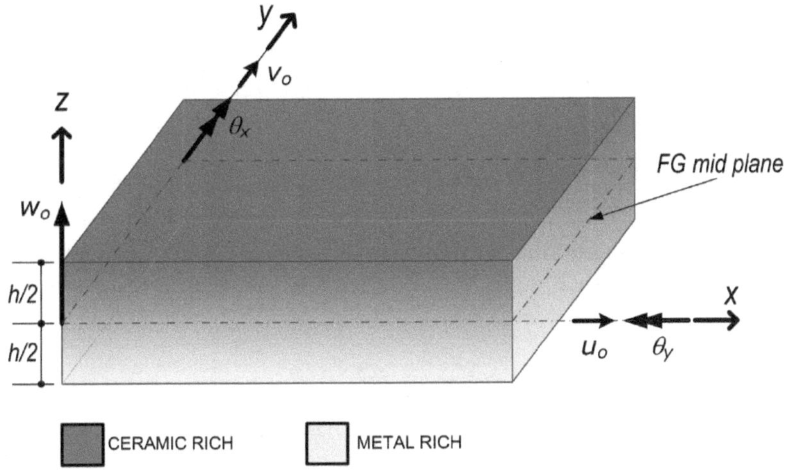

Fig. 1 Geometry of FG plate simply (diaphragm) supported on all edges and positive set of displacements and directions

Similarly, to approximate the 2D plane stress elasticity problem to a 1D beam problem, the displacement components u(x,z) and w(x,z) at any point in the beam are expanded in Taylor's series in terms of thickness coordinate. The various higher order displacement fields and the Timoshenko displacement field considered in the formulation are:

Timoshenko beam theory (Timo)

$$\begin{aligned} u(x,z) &= u_o(x) + z\theta_x(x) \\ w(x,z) &= w_o(x) \end{aligned} \tag{2}$$

Higher order beam theory 1 (Host5)

$$\begin{aligned} u(x,z) &= u_o(x) + z\theta_x(x) + z^2u_o^*(x) + z^3\theta_x^*(x) \\ w(x,z) &= w_o(x) \end{aligned} \tag{3}$$

Higher order beam theory 2 (Host6)

$$\begin{aligned} u(x,z) &= u_o(x) + z\theta_x(x) + z^2u_o^*(x) + z^3\theta_x^*(x) \\ w(x,z) &= w_o(x) + z\theta_z(x) \end{aligned} \tag{4}$$

Higher order beam theory 3 (Host7)

$$\begin{aligned} u(x,z) &= u_o(x) + z\theta_x(x) + z^2u_o^*(x) + z^3\theta_x^*(x) \\ w(x,z) &= w_o(x) + z\theta_z(x) + z^2w_o^*(x) \end{aligned} \tag{5}$$

Higher order beam theory 4 (Host8)

$$u(x,z) = u_o(x) + z\theta_x(x) + z^2 u_o^*(x) + z^3\theta_x^*(x)$$
$$w(x,z) = w_o(x) + z\theta_z(x) + z^2 w_o^*(x) + z^3\theta_z^*(x) \quad (6)$$

where the parameters u_0, v_0 are the in-plane displacements and w_0 is the transverse displacement at any point (x, y) on the middle plane of the plate and θ_x and θ_y are the rotations of the normal to the mid-plane about y and x axes respectively. Parameters in HOST, such as $u_0^*, \theta_x^*, v_0^*, \theta_y^*, w_0^*, \theta_z^*$ are the corresponding higher order terms in the Taylor series expansion.

2.1 Effective Moduli of Two-phase Composites

The Mori–Tanaka [11] scheme for estimating the effective moduli is applicable to regions of the graded microstructure that have a well-defined continuous matrix and a discontinuous particulate phase. It is assumed that the matrix phase, denoted by the subscript *1*, is reinforced by spherical particles of a particulate phase, denoted by the subscript 2. In this notation, K_1, m_1 and V_1 denote respectively the bulk modulus, the shear modulus and the volume fraction of the matrix phase; K_2, m_2 and V_2 denote the corresponding material properties and the volume fraction of the particulate phase. The effective local bulk modulus K and the shear modulus μ are obtained by the Mori–Tanaka scheme for a random distribution of isotropic particles in an isotropic matrix and are given by [22].

$$\frac{K-K_1}{K_2-K_1} = \frac{V_2}{\left(1+(1-V_2)\dfrac{K_2-K_1}{K_1+\frac{4}{3}\mu_1}\right)}, \quad \frac{\mu-\mu_1}{\mu_2-\mu_1} = \frac{V_2}{\left(1+(1-V_2)\dfrac{\mu_2-\mu_1}{\mu_1+f_1}\right)},$$
$$f_1 = \frac{\mu_1(9K_1+8\mu_1)}{6(K_1+2\mu_1)}. \quad (7)$$

The volume fraction of the ceramic phase is given by the power-law function, $V_c = V_c^- + (V_c^+ - V_c^-)\left(\frac{1}{2}+\frac{z}{h}\right)^p$. Here V_c^+ *and* V_c^- are respectively the volume fractions of the ceramic phase on the top and bottom surfaces of the plate, and p is a parameter that describes the volume fraction profile through the thickness. Matsunaga [9] used simple power law scheme in which, variation of Young's modulus E is given by $E(z) = E_m + E_{cm}V_c$, where E_m and E_c are Young's moduli of metal and ceramic and $E_{cm} = E_c - E_m$. In this scheme, the Poisson's ratio is kept constant.

2.2 Equations of Equilibrium

Equations of equilibrium are derived from principle of minimum potential energy.

$$\delta u_0 : \frac{\partial N_x}{\partial x} + \frac{\partial N_{xy}}{\partial y} = 0 \qquad \delta u_0^* : \frac{\partial N_x^*}{\partial x} + \frac{\partial N_{xy}^*}{\partial y} - 2S_x = 0$$

$$\delta v_0 : \frac{\partial N_y}{\partial y} + \frac{\partial N_{xy}}{\partial x} = 0 \qquad \delta v_0^* : \frac{\partial N_y^*}{\partial y} + \frac{\partial N_{xy}^*}{\partial x} - 2S_y = 0$$

$$\delta w_0 : \frac{\partial Q_x}{\partial x} + \frac{\partial Q_y}{\partial y} + (q_z^+) = 0 \qquad \delta w_0^* : \frac{\partial Q_x^*}{\partial x} + \frac{\partial Q_y^*}{\partial y} - 2M_z^* + \frac{h^2}{4}(q_z^+) = 0$$

$$\delta \theta_x : \frac{\partial M_x}{\partial x} + \frac{\partial M_{xy}}{\partial y} - Q_x = 0 \qquad \delta \theta_x^* : \frac{\partial M_x^*}{\partial x} + \frac{\partial M_{xy}^*}{\partial y} - 3Q_x^* = 0$$

$$\delta \theta_y : \frac{\partial M_y}{\partial y} + \frac{\partial M_{xy}}{\partial x} - Q_y = 0 \qquad \delta \theta_y^* : \frac{\partial M_y^*}{\partial y} + \frac{\partial M_{xy}^*}{\partial x} - 3Q_y^* = 0$$

$$\delta \theta_z : \frac{\partial S_x}{\partial x} + \frac{\partial S_y}{\partial y} - N_z + \frac{h}{2}(q_z^+) = 0 \quad \delta \theta_z^* : \frac{\partial S_x}{\partial x} + \frac{\partial S_y}{\partial y} - 3N_z^* + \frac{h^3}{8}(q_z^+) = 0 \tag{8}$$

Navier's solution procedure is used to obtain the solution of displacement variables, satisfying the simply supported boundary conditions.

3 Numerical Results and Discussions

In this section, numerical results are presented for simply supported FG beams and plates.

Example 1. The distribution of stresses and displacements of a simply (diaphragm) supported, on all the four edges, FG plate subjected to doubly sinusoidal mechanical loading is presented in this example. The bottom surface of FG plate is metal rich where as top surface is of ceramic rich. In this example, metal matrix phase is considered. The material properties of metal and ceramics are given as, $K_m = 227.24$ GPa, $\mu_m = 65.55$ GPa, for Monel (Metal) and $K_c = 125.83$ GPa, $\mu_c = 58.077$ GPa for Zirconia (Ceramic) [15].

Figure 2 shows the variation of stresses and displacements across the thickness of FG plate. All the variations demonstrate the results very close to exact results.

Example 2. A power law bounded FG simply supported plate is analyzed by using simple power law. In this scheme it is possible to integrate the stress quantities for the p more than *1* as Poisson's ratio is kept constant across the thickness. The material properties of metal and ceramics are given as, $E_m = 70$ GPa, $\nu_m = 0.3$, for Aluminum (Metal) and $E_c = 380$ GPa, $\nu_c = 0.3$ for Alumina (Ceramic) [9]. Figure 3 demonstrates variations of stresses and displacements across the thickness of the FG plate subjected to mechanical loading for aspect ratio a/h = 10 and volume fraction indices, $p = 0, 1, 2, 4$ and 10.

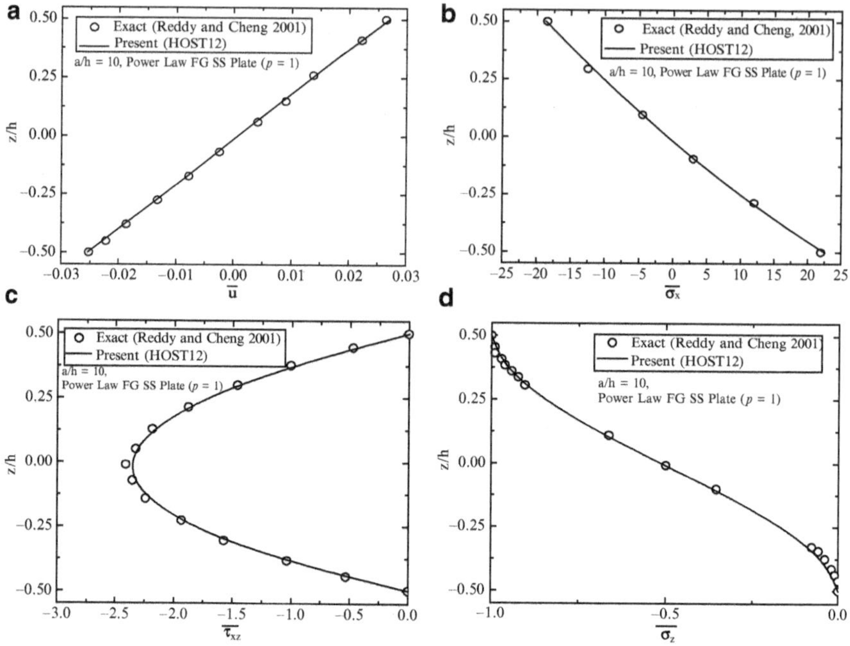

Fig. 2 Variations of displacement and stress across the thickness of the Monel/Zirconia FG plate

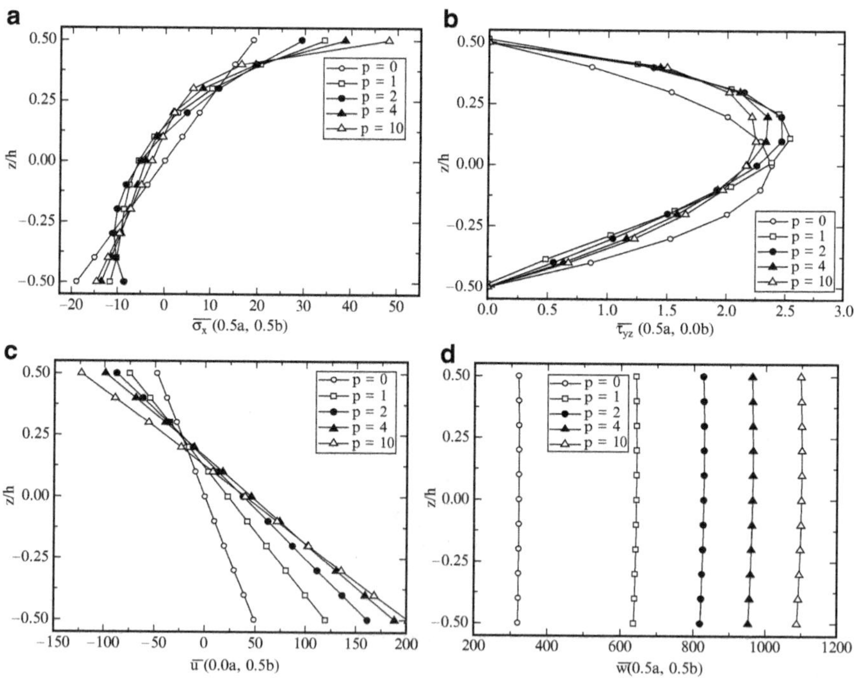

Fig. 3 Variations of stresses and displacements across the thickness of the Aluminum/Alumina FG plate

Accuracy of various higher order theories for the stress analysis of FG beams is established through numerical experimentation. The following sets of data are used. Variation of Young's modulus of the FGM can be defined as:

$$E(z) = \bar{E}_o e^{\lambda z}, \quad \bar{E}_o = \frac{E_o}{1 - \nu^2}.$$

If E_o and E_h are the Young's moduli at bottom and top of the of the FG beam respectively, then the gradation factor is defined by $\lambda = \ln(E_h/E_o)$.

Material set 1 [15, 18]
Eo = 70 GPa (Aluminum), Eh = 151 GPa (Zirconia), $y = 0.3$

Material set 2 [21]
Eo = 1 GPa, Eh/Eo = 5, 10, 20, 30 and 40, $v = 0.3$

Example 3. A simply supported functionally graded beam under sinusoidal transverse load is considered. Material set 1 is used. Figure 4 shows the through thickness variation of displacements and stresses. It shows that the results obtained using Host7 and Host8 are in good agreement and are very close to exact values. The

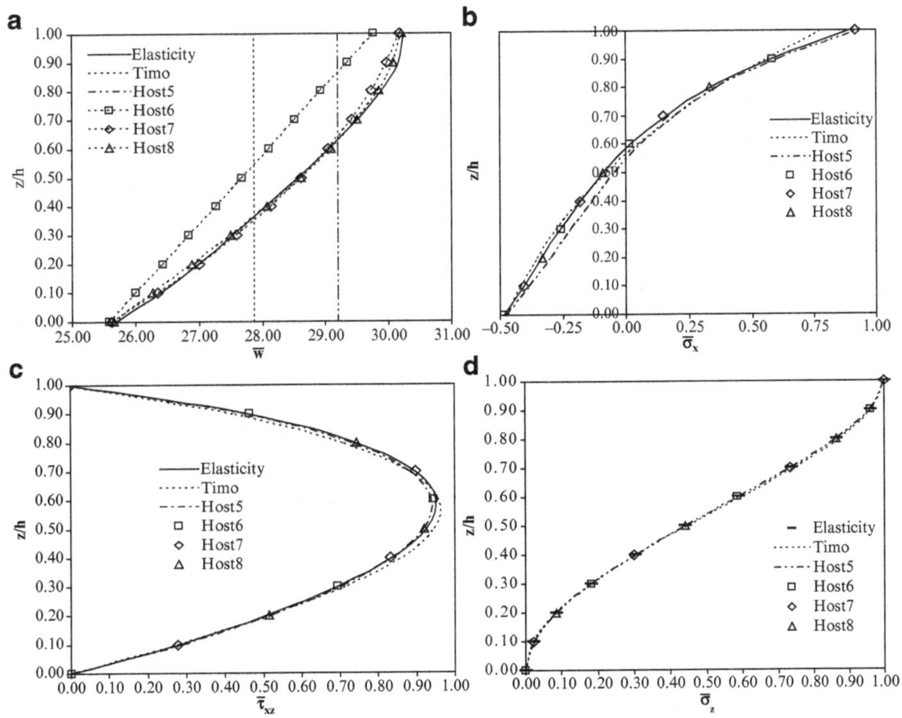

Fig. 4 Through thickness variations of normalized displacements and stresses of a SS FG beam under sinusoidal transverse load

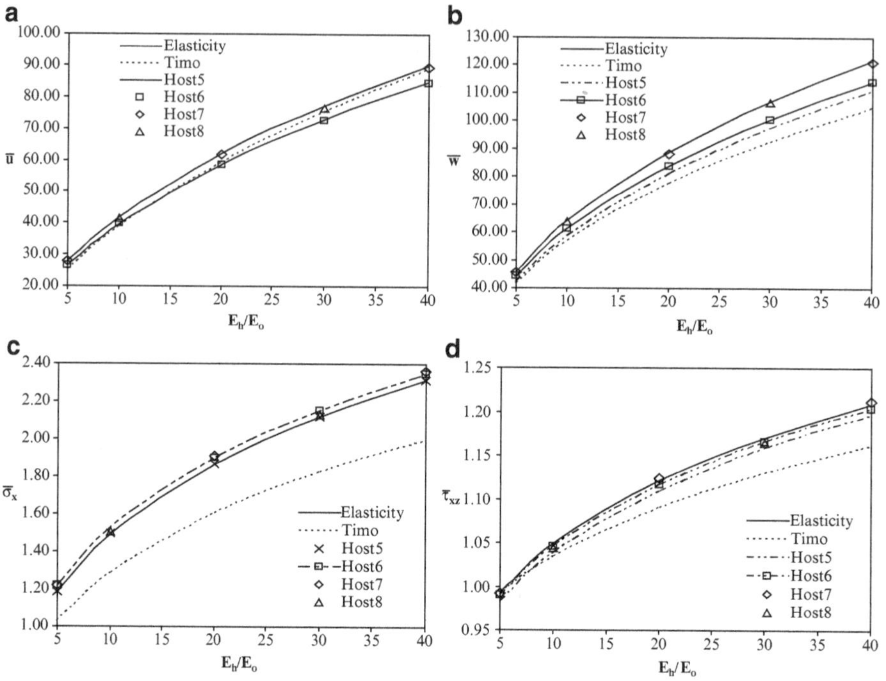

Fig. 5 Variation of normalized displacements and stresses versus Eh/Eo of a SS FG beam under sinusoidal transverse load

results show that even at slenderness ratio as low as two, Host7 and Host8 are giving results with better accuracy compared to other displacement models. The accuracy of all models increases with the increase in the L/h ratio.

Example 4. A simply supported functionally graded beam under sinusoidal and uniformly distributed transverse load is considered. Material set 2 is used. The numerical results are obtained for different Eh/Eo values. The results of maximum in-plane and transverse displacements and stresses for different material proportions and slenderness ratios under sinusoidal transverse load are plotted in Fig. 5. These results show that the Host7 and Host8 provide displacement quantities closest to the elasticity solution even at the higher material proportions compared to other displacement models.

4 Conclusions

The concept of FG eliminates the abrupt change in material properties and hence reduces the interlaminar stresses and delamination of the composite laminates. In this paper, stress analysis of FG beams and plates, simply supported and subjected to mechanical surface traction at top of the plate is carried out by HOSNTs. Various blending schemes are discussed in this paper. Two examples are presented with power law blending schemes in FG plates and the present theory demonstrates an accurate and economical approach for the static analysis of FG plates. In relatively thick FG beams, the effects of transverse components of stresses and strains cannot be neglected and thus, higher order models (Host7 and Host8) are seen to be effective for thick FG beams.

References

1. Aboudi J, Pindera MJ, Arnold SM (1995) Thermo-inelastic response of functionally graded composites. Int. J. Solids and Structures 32:1675–1710.
2. Aboudi J, Pindera MJ, Arnold SM (1999) Higher-order theory for functionally graded materials. Composites Part B-Engineering 28:93–108.
3. Croce LD and Venini P (2004) Finite elements for functionally graded Reissner-Mindlin plates. Comput. Methods Appl. Mech. Engrg 193:705–725.
4. Kant T (1982) Numerical analysis of thick plates. Computer Methods in Applied Mechanics and Engineering 31:1–18.
5. Kant T and Manjunatha BS (1988) An unsymmetric FRC laminate C^o finite element model with 12 degrees of freedom per node. Engineering Computations 5:300–308.
6. Kant T and Swaminathan K (2002) Analytical solutions for the static analysis of laminated composite and sandwich plates based on a higher order refined theory. Composite Structures 56:329–344.
7. Koizumi M. (1993) Concept of FGM. Ceramic Transactions. 34:3–10.
8. Koizumi M. (1997) FGM activities in Japan. Composites Part B. 28B:1–4.
9. Matsunaga H (2009) Stress analysis of functionally graded plates subjected to thermal and mechanical loadings 87(4):344–357.
10. Mindlin RD (1951) Influence of rotatory inertia and shear on flexural motions of isotropic, elastic plates. ASME J. Appl. Mech 18:31–38.
11. Mori T and Tanaka K (1973) Average stress in matrix and average elastic energy of materials with misfitting inclusions. Acta Metal 21:571–574.
12. Nguyen TK, Sab K and Bonnet G (2008) First-order shear deformation plate models for functionally graded materials. Composite Structures 83:25–36.
13. Pan E (2003) Exact solution for functionally graded anisotropic elastic composite laminates. J. Composite Materials 37:1903–1920.
14. Pandya BN and Kant T (1988) A refined higher-order generally orthotropic C^o plate bending element. Computers and Structures 28: 119–133.
15. Praveen GN and Reddy JN (1998) Nonlinear transient thermoelastic analysis of functionally graded ceramic-metal plates. Int J. Solids Structures 35:4457–4476.
16. Ramirez F, Heyliger PR and Pan E (2006) Static analysis of functionally graded elastic anisotropic plates using discrete layer approach. Composites Part B-Engineering 37: 10–20.
17. Ray MC and Sachade HM (2006) Exact solutions for the functionally graded plates integrated with a layer of piezoelectric fiber-reinforced composite. ASME J. Appl. Mech. 73:622–632.
18. Reddy JN (2000) Analysis of functionally graded plates. International Journal for Numerical Methods in Engineering 47: 663–684.
19. Reddy J N and Cheng ZQ (2001) Three-dimensional thermomechanical deformations of functionally graded rectangular plates. Eur. J. Mech.A/Solids 20:841–855.
20. Reissner E (1945) The effect of transverse shear deformation on the bending of elastic plates. ASME J. Appl. Mech 12:69–77.

21. Sankar BV (2001) An elasticity solution for functionally graded beams. Composites Science and Technology 61:689–696.
22. Vel SS and Batra RC (2004) Three-dimensional exact solution for the vibration of functionally graded rectangular plates. Journal of Sound and Vibration 272:703–730.
23. Wetherhold RC, Seelman S, Wang J (1996) The use of functionally graded materials to eliminate or control thermal deformation. Composites Science and Technology 56:1099–1104.

Section II

New Materials B: Nano Materials and Composites

A Strain Sensor from a Polymer/Carbon Nanotube Nanocomposite

Ning Hu, Yoshifumi Karube, and Hisao Fukunaga

Abstract In this work, the piezoresistive behavior of a strain sensor made from a polymer/carbon nanotube nanocomposite has been investigated. To underpin the working principle of the sensor, we propose an improved 3D statistical resistor network model by incorporating the tunneling effect among randomly distributed carbon nanotubes (CNTs) in a polymer matrix. Moreover, the strain sensor has been experimentally fabricated from a polymer nanocomposite with multi-walled carbon nanotube (MWNT) fillers. The piezoresistivity of this nanocomposite strain sensor has been experimentally investigated. The numerical results obtained from the 3D statistical resistor network model combined with the fiber reorientation model agree very well with the experimental measurements. From numerical and experimental results, as compared to the traditional strain gauges, much higher sensitivity can be obtained in this nanocomposite sensor. The tunneling effect is highlighted to be the major working mechanism of the sensor under small strains. By using the proposed successful numerical model, the influences of various parameters on the sensitivity of the new sensor have been numerically investigated in detail. The influence of various experimental parameters on the sensitivity of the sensor has also been studied. Both results have shown that higher resistance of the sensor leads to higher sensitivity of the sensor. Furthermore, the different behaviors of the sensor under tension and compression have been investigated.

Keywords Carbon nanotubes · Nanocomposites · Strain sensor · Tunneling effect

---

N. Hu (✉), Y. Karube, and H. Fukunaga
Department of Aerospace Engineering, Tohoku University, Aramaki-Aza-Aoba 6-6-01, Aoba-ku, Sendai 980-8579, Japan
e-mail: huning@faculty.chiba-u.jp; karube@ssl.mech.tohoku.ac.jp; Fukunaga@ssl.mech.tohoku.ac.jp

B. Dattaguru et al. (eds.), *IUTAM Symposium on Multi-Functional Material Structures and Systems*, IUTAM Bookseries 19, DOI 10.1007/978-90-481-3771-8_8,

1 Introduction

CNTs based nanocomposites are increasingly being reviewed as a realistic alternative to conventional smart materials, offering higher sensitivity and superior electrical properties. In particular, great interest has recently been aroused in building strain sensors with these nanocomposites. Generally, CNTs are able to be blended with a polymer to make a strain sensor providing piezoresistivity, i.e., a relationship between mechanical strain and resistance change. As compared to conventional sensors, higher sensitivity has been observed in this novel sensor made from polymer/CNT nanocomposites [1–5], at least at a macro-scale. In spite of these promising results, the fundamental understanding of conductivity in a polymer/CNT nanocomposite is still lacking, largely due to the less effort being put into theoretical and numerical investigations on the piezoresistive behavior in these materials. In this work, to investigate piezoresistive behavior which underpins the working principle of these sensors, we propose an improved 3D statistical resistor network model by which the tunneling effect among randomly distributed CNTs in a polymer matrix can be evaluated. The change of CNT networks in the polymer under a given strain is predicted using a fiber reorientation model. Then, the resistance change of the nanocomposites caused by the applied strain is estimated by using the 3D resistor network and the fiber reorientation models in an iterative way. To verify the numerical simulation, experiment has also been conducted on the sensors made from polymer/MWNT nanocomposites. The parameters, which influence the sensor sensitivity, were also studied in detail.

2 Theory

To predict the electrical conductivity of the nanocomposite, a 3D resistor network containing randomly distributed CNTs in the polymer has been constructed. Aggregation of CNTs is neglected in this study. The CNT cylinders are of length L and diameter D, and are allowed to penetrate each other [6]. This assumption can lead to the tremendous reduction in the computational cost. The ideal state of uniformly dispersed straight CNTs in matrix is shown in Fig. 1. To construct the 3D resistor

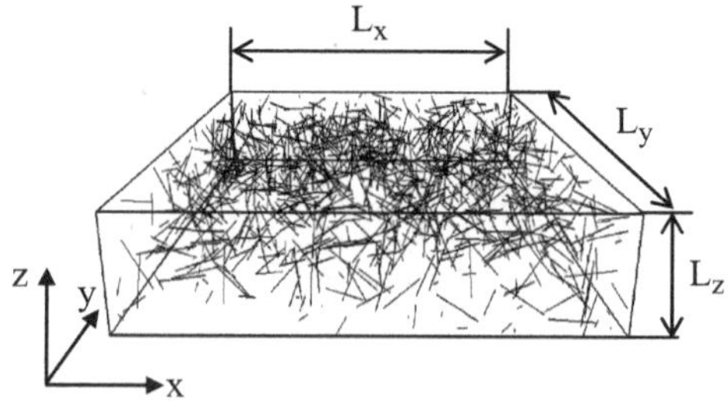

Fig. 1 A 3D representative unit element

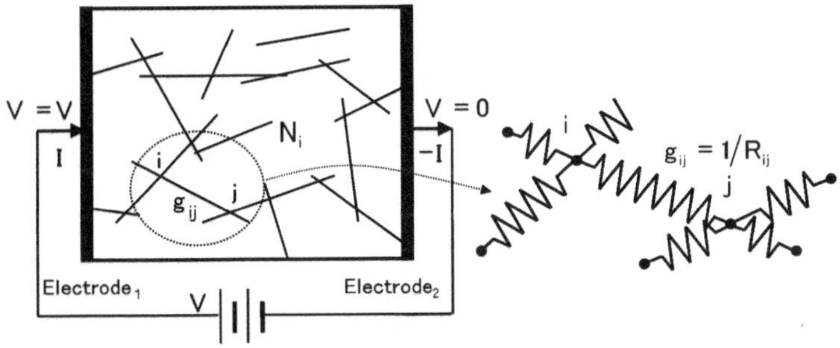

Fig. 2 A resistor model with randomly distributed CNTs

network model as shown in Fig. 2 (only a 2D model is shown), for a CNT with two contacting points i and j with neighbouring CNTs, the conductance g_{ij} between i and j (the inverse of resistance R_{ij}) can be evaluated as:

$$g_{ij} = \sigma_{CNT} \frac{S_{CNT}}{l_{ij}} \tag{1}$$

where l_{ij} is the length between the points i and j, and σ_{CNT} and S_{CNT} are electrical conductivity and cross section area of the CNT, respectively.

The tube-tube contacts among CNTs are assumed to be perfect here with zero resistance. Based on the well-known matrix representation for a resistor network [7] and Kirchhoff's current law, the total current I under an applied voltage can be estimated. An iterative equation solver, i.e., the incomplete Cholesky conjugate gradient method (ICCG) has been employed to solve this large linear system for obtaining the total current I. Then the macroscopic electrical conductivity of nanocomposites can be evaluated by the Ohm's law.

Due to the short inter-distance between adjacent CNTs, it is necessary to investigate the possible tunneling effect among the CNTs and its effects on the electrical conductivity. The random distribution and possible tunneling paths between adjacent CNTs have been examined using scanning electron microscopy (SEM), as shown in Fig. 3. This physical picture is simplified as a model for evaluating the resistance, as schematically shown in Fig. 4. The tunneling resistance between two neighbouring CNTs can be approximately estimated as [8],

$$R_{tunnel} = \frac{V}{AJ} = \frac{h^2 d}{Ae^2\sqrt{2m\lambda}} \exp\left(\frac{4\pi d}{h}\sqrt{2m\lambda}\right) \tag{2}$$

where J is tunneling current density, V the electrical potential difference, e the quantum of electricity, m the mass of electron, h Planck's constant, d the distance between CNTs, λ the height of barrier (for epoxy, 0.5–2.5 eV), and A the cross sectional area of tunnel (the cross sectional area of CNT is used here).

Fig. 3 Possible tunneling effect among CNTs

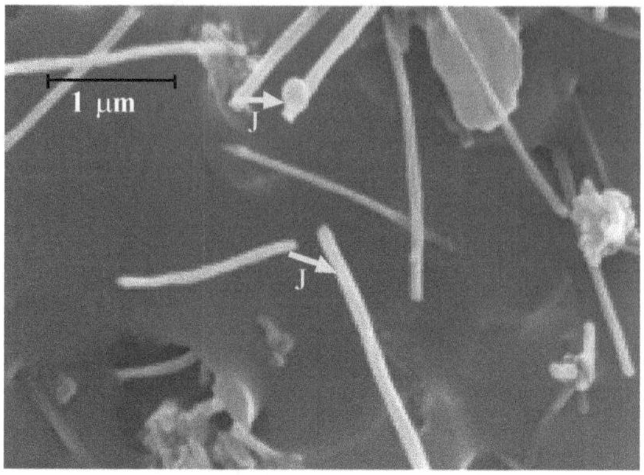

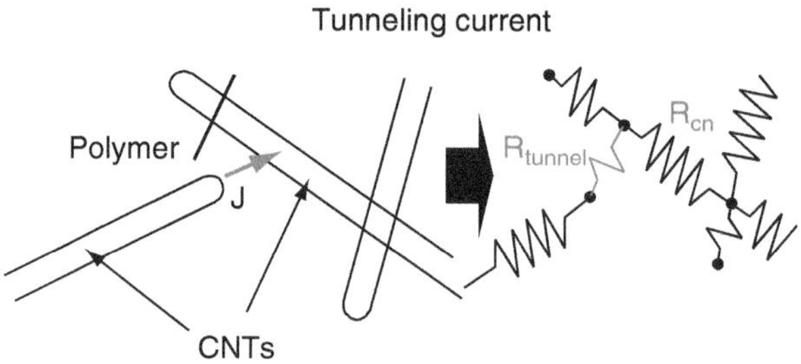

Fig. 4 Modelling of tunneling resistance

To consider the tunneling effect among CNTs, if the shortest distance d between two CNTs is: $D < d \leq D + d_t$, where D is the diameter of CNT and d_t is the cutoff distance of tunneling effect, two new nodes are added as shown in Fig. 4. By using Eq. (2), we can evaluate the tunneling resistance. The above cutoff distance d_t is set to be 1.0 nm here, which relates to very low tunneling conductivity (lower than 10^2 S/m) compared with that of CNTs, i.e. 10^4 S/m used here by referring to [9].

Here, the piezoresistivity is evaluated via numerical simulation, incorporating the change of inter-filler distances and possible breakup of conductive networks when subjected to strains. The resistance change of CNTs themselves under elastic strain is ignored. Considering the rigid-body movement of the CNTs, the change of position and orientation of the CNTs under the effects of strain and Poisson's ratio are evaluated using the 3D fiber reorientation model [10]. Corresponding to an updated distribution of the CNTs under a prescribed strain, a new network of CNTs can be formed by re-calculating the possible intersections between CNTs and tunneling resistances between CNTs within the cutoff distance. Then, the resistance of the nanocomposite can be re-evaluated using the 3D resistor network.

3 Verifications

To verify the numerical simulation, the resistance variation with strain has been investigated using a piece of the nanocomposite with a thickness of about 170 μ m in Fig. 5. In experimentation, the polymer/CNT nanocomposite was fabricated by *in situ* polymerization. MWNTs (060125-01K) were obtained from the Nano Carbon Technologies Co. (NCTC) in Japan. They had a purity of higher than 99.5%. The average diameter and length of the MWNTs were 50 nm and 5 μm, respectively. An insulating bisphenol-F epoxy resin (jER806, Japan Epoxy Resins Co., Ltd.) and an amine hardener (Tomaido 245-LP, Fuji Kasei Kogyo Co., Ltd.) were used. The nanocomposite was prepared by mixing the epoxy and the hardener using a planetary mixer at 2,000 rpm for 20 s. Then, the MWNTs were added into the mixture and mixed again at 2,000 rpm for 1 min. The final mixture was poured into a silicone mold, and cured in a vacuum oven at 80°C for 3 h.

Note that, in experiments, this nanocomposite sheet was attached to the top surface of an insulating cantilevered beam of the thickness of 2.0 mm, and a traditional strain gauge was glued to the bottom surface of the beam, in a symmetrical position to the nanocomposite sheet. The position of two sensors was close to the clamped end of the beam. The numerical results of piezoresistivity of the sensor are shown in Fig. 6. In addition, the electrical resistance of the nanocomposite sensor was measured using a LCR meter and the results are shown in Fig. 7. Compared to the results in Fig. 6, it is clear the numerical simulation which incorporates the tunneling effect among the CNTs can qualitatively catch the variation of resistance with the strain. Furthermore, in comparison to a traditional strain gauges whose gauge factor (sensitivity) is two, higher gauge factors can be observed in these polymer/CNT sensors with different CNT loadings. For instance, the gauge factor of the sensor with 1.0 wt% CNT loading is about eight times higher than that of the traditional strain gauges. When the CNT loading becomes lower, the gauge factor of the new sensors increases remarkably. However, for high weight fractions of CNTs, such as 5.0 wt%, the stability of sensor is higher. Weak nonlinear piezoresistivity is observed in the nanocomposite sensors for the cases of lower CNT loading (Fig. 6). In addition, it was found that tunneling effect was a key factor for the new sensors for small strain.

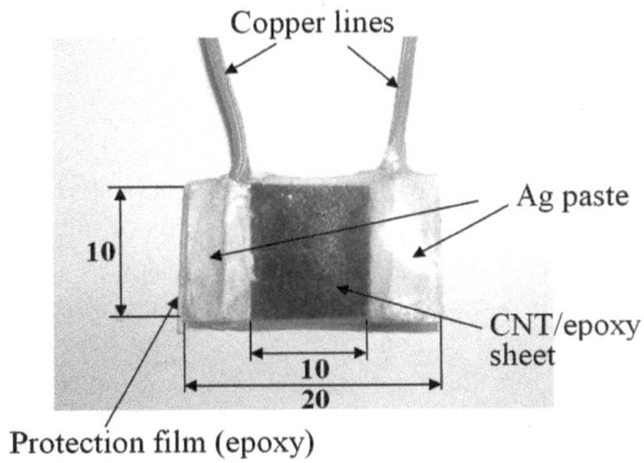

Fig. 5 A polymer/CNT strain sensor

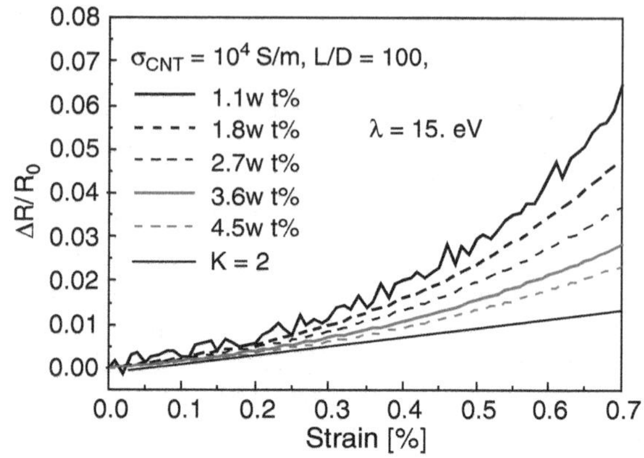

Fig. 6 Numerical piezoresistivity

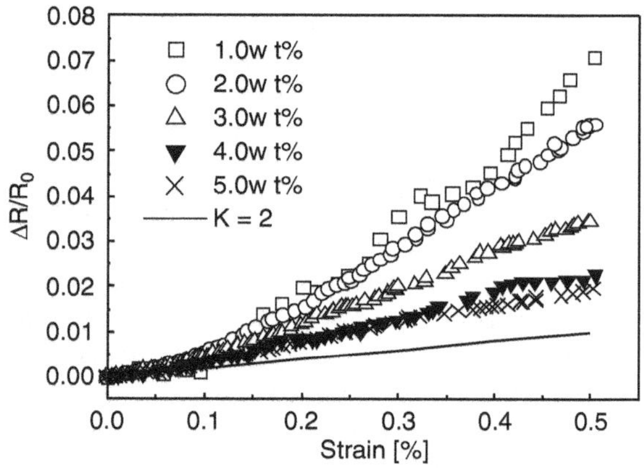

Fig. 7 Experimental piezoresistivity

4 Influence of Various Parameters on Piezoresistivity Using Numerical Model

After verifying the effectiveness of the present numerical model, we investigated the influences of various parameters on piezoresistivity using it. First, by considering Eq. (2), two important parameters are chosen, i.e., λ: the height of barrier and A: the cross sectional areas of tunnel. Here, the diameter of MWNTs used in experiments is used as a standard value. The influence of various cross sectional area of CNTs on piezoresitivity is shown in Fig. 8 when keeping λ to be 1.5 eV and CNT loading to be 2.0%. Observation of Fig. 8 reveals that with the decrease of the cross sectional area A of CNTs, the piezoresistivity of the sensor increases remarkably. Next, we keep the cross sectional area of tunnel as $1.96 \times 10^{-15}\,\mathrm{m}^2$ and CNT loading as 2.0% to investigate the influence of the height of barrier, i.e., λ. As shown in Fig. 9, we can find that with the increase of λ, the piezoresistivity of the sensor increases, which corresponds to the increase of tunneling resistance by viewing Eq. (2). Another factor is the orientation of CNTs in matrix. To generate numerical models with the various CNT orientations, the orientation is defined within an angle θ between a CNT rod and x-axis, as shown in Fig. 1. When $\theta = 90°$, the numerical

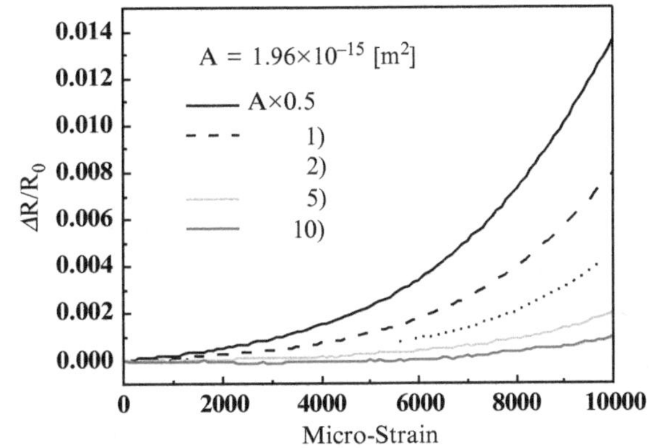

Fig. 8 Influence of A

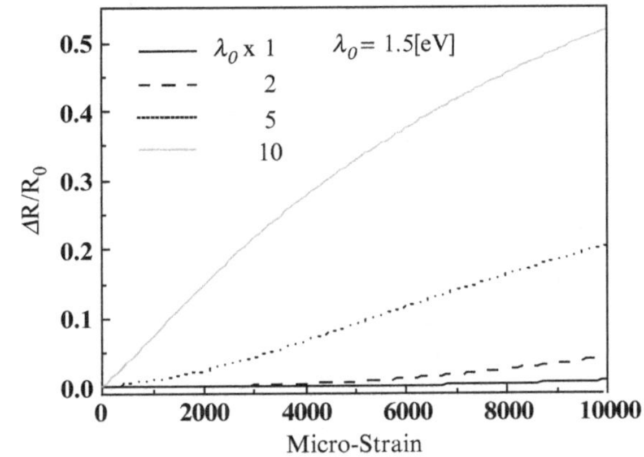

Fig. 9 Influence of λ

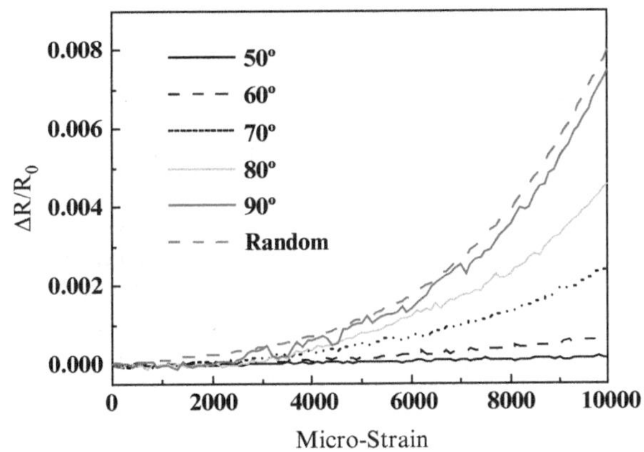

Fig. 10 Influence of θ

model corresponds to complete random distribution of CNTs. Figure 10 shows the influence of θ on the piezoresistivity (CNT loading: 2.0%).

We can find that the increase of θ leads to the higher piezoresistivity. It means that a higher mixing speed, when dispersing CNTs into matrix, is desirable.

When $\theta = 90°$, the present result agrees with that of a model with randomly distributed CNTs very well. The present numerical results demonstrate that the higher resistance of nanocomposite can lead to higher sensitivity of the sensor.

5 Influence of Various Parameters on Piezoresistivity Using Experiments

In this section, the sensor properties will be studied by experiments. The nanocomposite was prepared by mixing the epoxy and the hardener using a planetary mixer at 2,000 rpm for 30 s. Then, the MWNTs were added into the mixture and mixed again at 2,000 rpm for 1 min. The final mixture was poured into a silicon mold, and cured in a vacuum oven at 80°C for 3 h. Three sensors have been made for one CNT loading value. The piezoresistivity is shown in Fig. 11. An important feature in Fig. 11 is that the behaviors of the sensor under tension and compression states are different. The ability of the sensor for measuring compressive strains is limited to 4,000 μs. The sensitivity in a compression state is much lower than that in a tension state. However, when we focus on a region in Fig. 11 from −1,000 to 1,000 μs, we can find that the behavior of the sensor is almost anti-symmetric around the origin for tension and compression states.

For investigation of influence of various processing parameters, e.g., mixing speed, the following experiments have been performed. First, we mix epoxy and hardener at 2,000 rpm for 30 s, and then MWNTs are dispersed in the mixture using three kinds of speed, i.e., 800, 1,400 and 2,000 rpm, for 1 min. Finally, the mixture is cured at three kinds of curing temperatures, i.e., 80°C, 100°C and 120°C for 3 h. The average piezoresistivities of three sensors under a same processing condition are shown in Fig. 12 for 1.0% CNT loading. From it, we can find that the

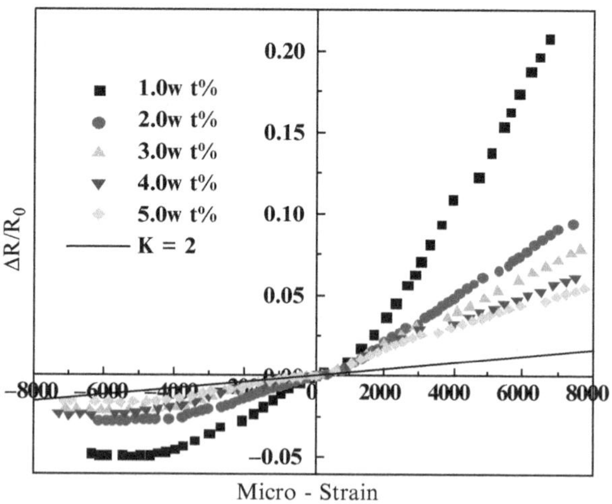

Fig. 11 Experimental piezoresistivity under tension and compression states

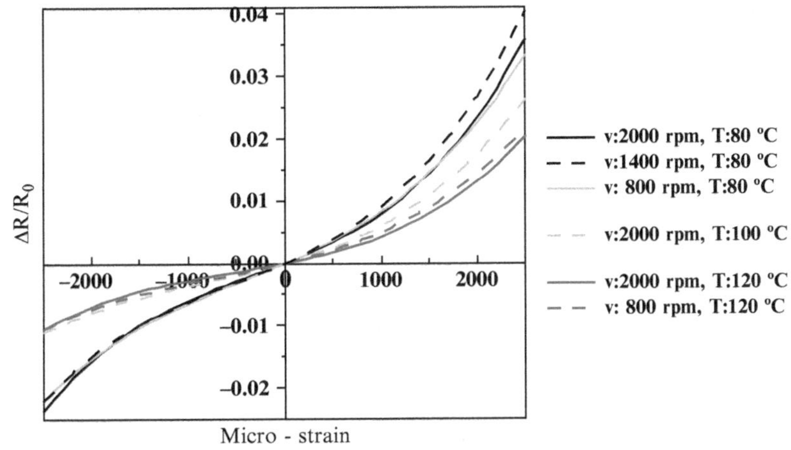

Fig. 12 Influence of experimental parameters on piezoresistivity

curing temperature has a very important impact on the piezoresistivity of the sensor. As the curing temperature decreases, the piezoresistivity increases. As shown in our previous study [11], a higher curing temperature can accelerate the formation of conductive network of CNTs and result in a higher electrical conductivity. Therefore, a low curing temperature can lead to higher resistance of the sensor, which consequently increases the sensitivity of the sensor. For the influence of mixing speed, although there are some exceptions in our results, basically, higher mixing speed is beneficial to the sensitivity of the sensor. As shown in our previous study [11], a higher mixing speed can break up the conductive network of CNTs, which leads to the higher electrical resistance and consequently higher sensitivity of the sensor. This finding is consistent with those obtained from the previous numerical investigations.

6 Conclusions

In this work, the piezoresistive behavior of a strain sensor made from a polymer/CNT nanocomposite was investigated numerically and experimentally. It was found that a 3D resistor network model incorporating the tunneling effect of CNTs can yield very good results of piezoresistivity of the sensor. We have identified that the tunneling effect plays a key role in the performance of the sensor under small strains. Much higher sensitivity of the sensor can be obtained for lower CNT loading. The behavior of the sensor is weakly non-linear due to the tunneling effect. The influence of various factors in the numerical model on the piezoresistivity of the sensor has been investigated numerically. Furthermore, the influence of various processing conditions on the piezoresistivity of the sensor has been studied. Both from the numerical and experimental results, we can find that higher resistance of the sensor can lead to the higher sensitivity of the sensor. Therefore, the factors which can result in the higher resistance of the nanocomposite are desirable in practical fabrication of this sensor although the resistance level of the sensor should not exceed the measurement limit of LCR meter.

References

1. I. Kang, M.J. Schulz, J.H. Kim, V. Shanov and D. Shi, Smart Mater. Struct., 15, 737 (2006).
2. J. Halary, J.L. Stanford, P.A. Lovell and R.J. Young, Mater. Res. Soc. Sym. Proc., 791, 379 (2004).
3. A. Ramaratnam and N.J. Jalili, J. Intell. Mater. Syst. Struct., 17, 199 (2006).
4. Y. Shimamura, T. Yasuoka and A. Todoroki, Proc. of the 16th Int. Conf. on Comp. Mater., 224503 (CD-ROM) (2007).
5. N. Hu, Z. Masuda and H. Fukunaga, Proc. of the 16th Int. Conf. on Comp. Mater., 224333 (CD-ROM) (2007).
6. I. Balberg and N. Binenbaum, Phys. Rev. A, 31, 1222 (1985).
7. S. Kirkpatrick, Phys. Rev. Lett., 20, 1722 (1971).
8. J.G. Simmons, J. Appl. Phy., 34, 1793 (1963).
9. W.A. de Heer, W.S. Bacsa, A. Châtelai, T. Gerfin, R. Humphrey-Baker, L. Forro and D. Ugarte, Science, 268, 845 (1995).
10. M. Taya, W.J. Kim and K. Ono K, Mech. Mater., 28, 53 (1998).
11. N. Hu, Z. Masuda, G. Yamamoto, H. Fukunaga, T. Hashida and J. Qiu, Composites Part A, 39, 893 (2008).

A Hysteresis Compensator Based on a Modified Dynamic Preisach Model for Conductive Polymer Nanocomposites

Mohammed Rabius Sunny and Rakesh K. Kapania

Abstract This paper proposes a compensator for strain sensors made of conductive polymer nanocomposites. These materials show hysteresis with dependence on input-rate in their variation of electrical resistance with strain. A modified Preisach model has been proposed previously by the authors to model this type of hysteretic behavior. A compensation algorithm based on this model has been proposed in this paper. Result of hysteresis compensation have been presented. Efficiency of the compensator has been discussed by comparison of the strain predicted by the compensator with the actual strain input.

Keywords Compensator · Hysteresis · Preisach model · Relaxation

1 Introduction

Hysteresis is very common in materials used as sensors and actuators, such as piezoelectric materials, ferroelectric materials, magnetoelectric materials. It makes the calibration of the materials for their use very difficult. For this reason, different mathematical models for hysteresis have been developed and depending on the model hysteresis compensators have been implemented. Compensators are generally placed after the sensors to compensate the effect of hysteresis and other nonlinearities and get back the actual data to be sensed. In case of actuators they are placed before it so that the compensator can generate the appropriate input for the actuator and the desired response from the actuator is obtained.

M.R. Sunny (✉)
215 Randolph Hall, Blacksburg, VA 24061, USA
e-mail: sunny@vt.edu

R.K. Kapania
213E Randolph Hall, Blacksburg, VA 24061, USA
e-mail: rkapania@vt.edu

B. Dattaguru et al. (eds.), *IUTAM Symposium on Multi-Functional Material Structures and Systems*, IUTAM Bookseries 19, DOI 10.1007/978-90-481-3771-8_9,

Our research is a part of our investigation into the suitability of using conductive polymers as large strain sensor. We performed conductivity test on conductive polymer nanocomposite sample and found hysteresis and relaxation in the variation of electrical resistance with strain of the sample. We first reviewed two of the several existing hysteresis models, namely the fractional calculus and Preisach models and found the necessity of modification in their traditionally used forms. We developed a modified Preisach model to represent the path-dependent behavior of conductive polymers [1]. Then by using this model we have developed a hysteresis compensator.

Preisach model, developed by Preisach in 1935 [2] to model hysteresis in magnetic materials, considers hysteresis to be a weighted combination of outputs from numbers of elementary hysteresis operators. Everett and Whitton [3] independently invented and developed the model for adsorption hysteresis. Everett surface based algorithm is very common among the different compensation algorithms for Preisach hysteresis [4, 5]. Tan et al. [6] proposed a closest match algorithm for hysteresis compensation. It is based upon the principle of determining the input whose output matches the desired output most closely among all possible outputs. Viswamurthy et al. [7] used it for hysteresis compensation in piezoelectric actuator for helicopter vibration control. Most of the presently available compensators are based on classical Preisach model and they are effective in compensating for time independent static hysteresis. Research on compensator for piezoelectric actuators with consideration to both time independent hysteresis and time dependent creep can be found in the works by Janocha et al. [8], Changhai et al. [9]. But literature on compensation of hysteresis combined with creep or relaxation is not sufficient to the best of the author's knowledge.

The main features observed in the hysteresis loop under study are that the loop keeps shifting up with change in the width of the loop with number of cycles and relaxation occurs at any constant value of input. Consequently, input-rate also has significant effect on the behavior of the material. A mathematical model developed by some modifications of the Preisach hystereis operator to model the static hysteresis and addition of a dynamic hysteresis operator to model the relaxation and input-rate dependence have been proposed in our previous work [1]. A compensator based on the above mentioned model has been proposed in this paper to compensate for both the static effect of the modified Preisach hysteresis operators and the dynamic effect of the dynamics relaxation operators.

2 Modified Preisach Model

Preisach model considers the hysteresis loop to be a weighted combination of output from different independent hysteresis operators. Each independent operator $\gamma_{\alpha\beta}$ is a mechanical unit that upswitches to a value $+1$ when the input increases and reaches a value α and downswitches to a value -1 when the input decreases and reaches a value β. Parameters α and β are called upswitching and downswitching values respectively. Output $f(t)$ due to an input $u(t)$ is given by Eq. (1)

$$f(u(t)) = \int_T \mu(\alpha,\beta)\gamma_{\alpha\beta}(u(t))d\alpha d\beta = \int_{T_1} \mu(\alpha,\beta)d\alpha d\beta - \int_{T_2} \mu(\alpha,\beta)d\alpha d\beta \tag{1}$$

Here, $\mu(\alpha,\beta)$ is the weight associated with operator with upswitching value α and downswitching value β. T is the domain of the Preisach hysteresis operators. T_1 and T_2 refer to the domains of upswitched and downswitched operators respectively.

The hysteresis operators in classical Preisach model give rise to a closed loop hysteresis in a complete cycle. Hence, it cannot model the upward shifting of the hysteresis loop and relaxation of output at a constant input. These have been taken care of by a modification of the hysteresis operators and addition of a dynamic operator. For our problem, both the input strain and the output electrical resistance always remain positive. So α and β always remain positive. Hence, in the downswitched stage, output from each operator is zero. The hysteresis operator considered in our model is shown in Fig. 1(a). Here at α, the output upswitches by a value $+1$. At β it does not downswitch fully. It downswitches by a value η which is another parameter associated with each operator. Due to this partial downswitching at β the loop keeps shifting upward with number of cycles as shown in Fig. 1(a). If the value of η is more than one the loop will shift downward. As the operators does not downswitch fully, at any time instant an operator can be either fully downswitched, upswitched or partially upswitched depending on the input history. Partially upswitched state refers to the state of the operator after it downswitches by η followed by any upswitching. We can also call it partially downswitched state. Output of an operator at partially upswitched state after n numbers of up and partial downswitching will be $n(1-\eta)$ and of an operator at upswitched state after n numbers of up and partial downswitching will be $n(1-\eta)+1$. The space spanned by the hysteresis operators is called the PM space [10] (after Preisach and Mayergoyz).

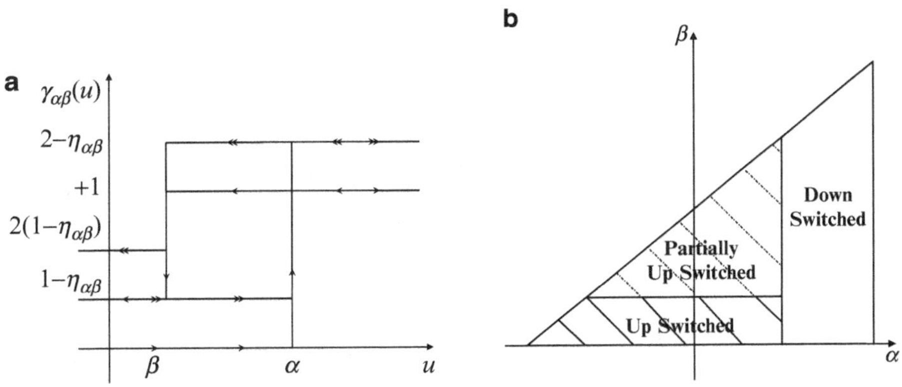

Fig. 1 Hysteresis operator and PM space in modified Preisach model

Figure 1(b) shows the operators in different states in the PM space. Output of the system due to these operators at any time t can be written as

$$f_s(u(t)) = \int_{T_1} (n(\alpha,\beta)\,(1-\eta(\alpha,\beta)) + 1)\,\mu(\alpha,\beta)d\alpha d\beta + \int_{T_3} n(\alpha,\beta)\,(1-\eta(\alpha,\beta))\,\mu(\alpha,\beta)d\alpha d\beta \tag{2}$$

In this way, partial downswitching causes a difference between the output at the end and beginning of the cycle. Consequently the hysteresis loop keeps shifting upward. The dynamic part of the output has been assumed to be the result of dynamic relaxation operators which upswitch to a value $\sum_{j=1}^{P} a_j \zeta^{d_j}$ when the input monotonically increases and reaches a value ζ and then decays exponentially as given in Eq. (4). Here, a, b, c, d are unknown parameter associated with each dynamic relaxation operator. These operators have no downswitching value. Hence, for a given input history at time t, the dynamic part of the output will be given by Eq. (4).

$$\phi_\zeta(t) = \begin{cases} 0 & \text{if } t < t_0; \\ \sum_{j=1}^{P} a_j \zeta^{d_j} e^{-b_j (t-t_0)^{c_j}} & \text{if } t \geq t_0. \end{cases} \tag{3}$$

$$f_d(u(t)) = \int_{T_d} \sum_{i=1}^{n_\zeta} \sum_{j=1}^{P} a_j \zeta^{d_j} e^{-b_j (t-t_{\zeta i})^{c_j}}\, d\zeta \tag{4}$$

T_d refers to the domain where the operators are in upswitched state, n_ζ refers to the total number of times ϕ_ζ have been upswitched and $t_{\zeta i}$ refers to the time instant at which ϕ_ζ was upswitched for the ith time. Total output of the system is the sum of f_s and f_d defined in Eq. (2) and Eq. (4). A detailed discussion on this model and the identification of unknown parameters involved can be found in [1].

3 Everett Integral and Everett Surface

Everett Integral is defined as:

$$E(h_1,h_2) = f_{h_2} - f_{h_1} = \begin{cases} 2\int_{D_1} \mu(\alpha,\beta)d\alpha d\beta & \text{if } h_1 < h_2; \\ 2\int_{D_1} -\mu(\alpha,\beta)d\alpha d\beta & \text{if } h_1 > h_2. \end{cases} \tag{5}$$

D_1 refers to the domain in PM space in which the hysteresis units are upswitched or downswtiched when input changes from h_1 to h_2. $E(h_1,h_2)$ represents the difference in output when the input reaches to a value h_2 from an extreme value h_1.

$E(h_1, h_2)$ for different (h_1, h_2) are plotted as a surface. This surface is called Everett surface. For classical Preisach model, Everett surface is antisymmetric with respect to the line $h_1 = h_2$. In case of the modified Preisach operator, Everett integral can be defined as:

$$E(h_1, h_2) = f_{h_2} - f_{h_1} = \begin{cases} \int_{D_1} \mu(\alpha, \beta) d\alpha d\beta & \text{if } h_1 < h_2; \\ \int_{D_1} -\eta(\alpha, \beta)\mu(\alpha, \beta) d\alpha d\beta & \text{if } h_1 > h_2. \end{cases} \quad (6)$$

It has the same significance in the modified Preisach model as in the Classical Preisach model. But due to the factor $\eta(\alpha, \beta)$, Everett surface is not antisymmetric with respect to the line $h_1 = h_2$ in modified Preisach model.

4 Everett surface Using Both Hysteresis and Dynamic Relaxation Operators

To solve the problem of the compensation of hysteresis and relaxation, we added one more term using the dynamic relaxation operators to the expression for the previously defined Everett integral. Suppose that the current extreme value of input is h_1. Then at time instant t_{k-1}, the value of input is h_{in}. After that at time instant t_k the value of the input rises upto h_2. Using Eq. (4) contribution of the dynamic operators upswitched in the interval $(t_{k-1}, t_k]$ can be written as $\int_{D_2} \sum_{i=1}^{n_\zeta} \sum_{j=1}^{P} a_j \zeta^{d_j} e^{-b_j (t - t_{\zeta i})^{c_j}} d\zeta$, where D_2 refers to the region of the dynamic operators which are upswitched in the time interval $(t_{k-1}, t_k]$. If we assume that $\Delta t = t_k - t_{k-1}$ is very small then it can also be written as $\int_{D_2} \sum_{j=1}^{P} a_j \zeta^{d_j} d\zeta$. By adding this quantity to the previously defined Everett integral we defined a new quantity $E_1(h_1, h_2, h_{in})$ for $h_1 > h_2$.

$$E_1(h_1, h_2, h_{in}) = f_{h_2} - f_{h_1} = \int_{D_1} \mu(\alpha, \beta) d\alpha d\beta + \int_{D_2} \sum_{j=1}^{P} a_j \zeta^{d_j} d\zeta \quad (7)$$

This quantity gives rise to a 3D surface. Input decreases monotonically when $h_1 < h_2$. In that case only the dynamic operators upswitched previously, relaxes. Apart from that, dynamic operators do not play any other role in this case. Hence, E_2 contains only the contribution from the hysteresis operators.

$$E_2(h_1, h_2, h_{in}) = f_{h_2} - f_{h_1} = -\int_{D_1} \eta(\alpha, \beta)\mu(\alpha, \beta) d\alpha d\beta \quad (8)$$

5 Forward Calculation Using Everett Surface

Consider the cyclic input (strain) shown in Fig. 2(a). At time t_k between t_0 and t_{n1} value of the output can be written as

$$f u_i = f_{u_0} + E_1(u_0, u_k, u_{k-1}) + \int_{T_d} \sum_{i=1}^{n_\zeta} \sum_{j=1}^{P} a_j \zeta^{d_j} e^{-b_j (t_k - t_{\zeta i})^{c_j}} d\zeta \quad (9)$$

T_d refers to the region of the dynamic operators upswitched in $(t_0, t_k]$. The current extreme value of input and output are zero. In between t_{n1} and t_{n2}, when the strain keeps decreasing the current extreme value of input and outputs are u_{n1} and f_{n1} respectively. Output at any time t_l between t_{n1} and t_{n2} will be given by

$$f_{u_l} = f_{u_{n1}} + E_2(u_{n1}, u_l) + \int_{T_d} \sum_{i=1}^{n_\zeta} \sum_{j=1}^{P} a_j \zeta^{d_j} e^{-b_j (t_l - t_{n1})^{c_j}} d\zeta \quad (10)$$

T_d refers to the domain of the dynamic relaxation operators upswitched in $(t_0, t_l]$ and this is same as of the operators upswitched in $(t_0, t_{n1}]$, as none of the dynamic relaxation operators are upswitched when the input decreases.

In this way, value of output can be calculated at any step by using the surfaces formed by the quantities $E_1(h_1, h_2, h_{in})$ and $E_2(h_1, h_2)$.

6 Compensation

The compensator works in two stages. At first, it removes the part of the output that results from the relaxation of the dynamic relaxation operators.

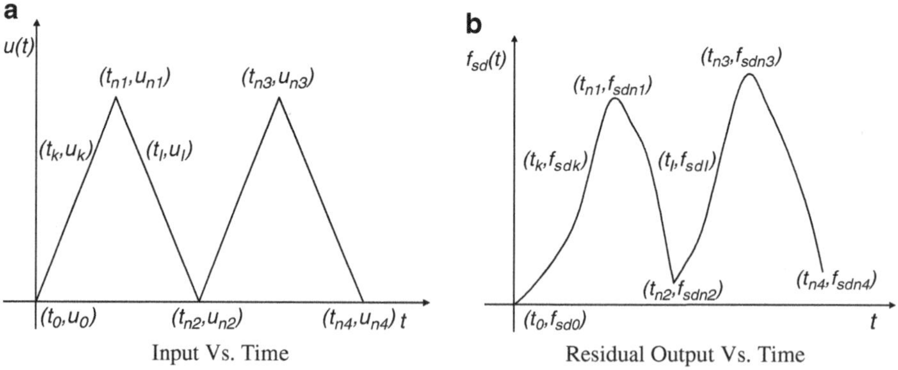

Fig. 2 Cyclic input and residual part of output after compensating the relaxation

At time instant t_k the contribution from the relaxation of the dynamic operators (upswitched in the interval $(t_0, t_{k-1}]$) to output can be written as

$$f_{d1} = \int_{T_{d1}} \sum_{i=1}^{n_\alpha} \sum_{j=1}^{P} a_j u^{d_j} e^{-b_j (t_k - t_{\xi i})^{c_j}} d\alpha \tag{11}$$

Here, T_{d1} refers to the domain containing the operators upswitched in the interval $(t_0, t_{k-1}]$. This quantity f_{d1} is subtracted from output at each time instant. The residual part of the output is the result of the static hysteresis and the contribution from the dynamic operators upswitched in $(t_{k-1}, t_k]$. Compensation for this residual part f_{sd} is done by inverse interpolation of the surfaces of $E_1(h_1, h_2, h_{in})$ and $E_2(h_1, h_2)$. At time t_0 both input and output are zero. Input at time $t = t_1$ can be found by inverse interpolation on the surface E_1 by considering $h_1 = 0$, $h_2 = u_1$, $h_{in} = 0$ and $E_1 = f_{sdl} - f_{sdO} = f_{sdl}$. Similarly at any time step t_k ($t_0 < t_k < t_{n1}$), input can be calculated by considering $h_1 = 0$, $h_2 = u_k$, $h_{in} = u_{k-1}$ and $E_1 = f_{sdk}$. After $t = t_{n1}$, the value of f_{sd} starts decreasing. That means the value of input starts decreasing. So, the current extreme value of input changes to u_{n1} and the surface E_2 needs to be considered for inverse interpolation. At any time step t_l ($t_{n1} < t_l < t_{n2}$), input can be calculated by considering $h_1 = u_{n1}$, $h_2 = u_l$ and $E_2 = f_{sdl} - f_{sdn1}$.

7 Results and Discussion

Figure 3(a) shows the comparison of the hysteresis curve obtained from the modified Preisach model with that obtained from experiment. The proposed compensation algorithm was applied to the output (change in electrical resistance) variation obtained from experiment to predict the input (applied strain). Comparison of the strain

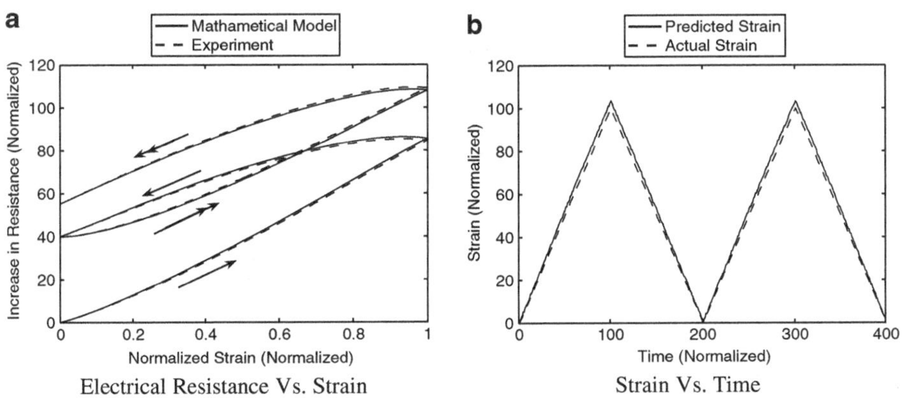

Fig. 3 Comparison of the results obtained from mathematical model with the experimental data

predicted by the compensator with the actual strain applied during experiment is in Fig. 3(b). The error in compensation was 4%. The error was calculated by using the expression $\frac{\sum_{i=1}^{i=n}(\varepsilon_p(t_i)-\varepsilon_a(t_i))^2}{\sum_{i=1}^{i=n}\varepsilon_a(t_i)^2}$, where ε_p means predicted strain, ε_a means actual strain and n means number of data points. The comparison shows a good accuracy of the compensator model.

8 Conclusion

In this paper a compensation algorithm for rate dependent dynamic hysteresis has been proposed. A modified version of the classical Preisach hysteresis operator along with a dynamic operator has been used to model the hysteresis. Relaxation of the dynamic operators has been compensated at each time step. Compensation of residual of the output has been done by using a surface generated by using the modified Preisach hysteresis operator and the dynamics relaxation operator. Comparison of the input predicted by the compensator with the actual input proves the efficiency of the compensator. Conductive polymer sensor with the proposed compensator put in series after it can be used for large strain sensing.

Acknowledgements The research was performed under project grant NNL05AA29G from NASA Langley Research Center (NASA LaRC) and by Institute for Critical Technologies and Sciences (ICTAS) at Virginia Polytechnic Institute and State University, Blacksburg, VA.

References

1. Sunny MR, Kapania RK (2009) A Modified Dynamic Preisach Model for Hysteresis. 50th AIAA/ASME/ASCE/AHS/ASC Structures, Structural Dynamics, and Materials Conference.Palm Springs, California
2. Preisach FZ (1935) Uber die magnetische nachwirkung. Z. Phys. 94:277–302
3. Everett DH, Whitton WI (1952) A General Approach to Hysteresis. Transactions of the Faraday Society 48:749–757
4. Venkataraman R, Krishnaprasad PS, Visonec C (2000) A novel algorithm for the inversion of the Preisach operator. Proceedings of SPIE. 3984:404–414
5. Natalea C, Velardib F, Visonec C (2001) Identification and compensation of Preisach hysteresis models for magnetostrictive actuators. Physica B. 306:161–165
6. Tan X, Venkataraman R, Krishnaprasad PS (2001) Control of hysteresis: theory and experimental results. Proceedings of SPIE – Smart Structures and Materials: Modeling, Signal Processing and Control in Smart Structures. 4326:101–112
7. Viswamurthy SR, Ganguli R (2007) Modeling and compensation of piezoceramic actuator hysteresis for helicopter vibration control. Sensors and Actuators A: Physical. 135:801–810
8. Janocha H, Kuhnen K (2000) Real-time compensation of hysteresis and creep in piezoelectric actuators. Sensors and actuators A: Physical. 79:83–89
9. Changhai R, Lining S (2005) Hysteresis and creep compensation for piezoelectric actuator in open-loop operation. Sensors and actuators A:, Physical. 122:124–130
10. Guyer RA, McCall KR, Boitnott GN (1997) Quantitative Implementation of Preisach-Mayergoyz Space to Find Static and Dynamic elastic moduli in rock. Journal of Geophysical Research. 102:5281–5293

Multi-Axial Behavior of Ferroelectrics with Two Types of Micro–Macro Mechanical Models

K. Jayabal, A. Arockiarajan, A. Menzel, and S.M. Sivakumar

Abstract Ferroelectric ceramics exhibit a significantly different nonlinear behavior with external electric and mechanical fields applied at angles to the initial poled direction. This angle dependent response of the ferroelectric polycrystals are predicted by two types of models based on irreversible thermodynamics and physics of domain switching. The first type is a uniaxial model dealing with simultaneous evolution of three variants at a given instant. The back stress and electric fields, assumed as linear functions of remnant strain and polarization developed by the domain switching process, are introduced in the model to assist or resist further switching process. The second type is a three dimensional model that considers all six variants of a tetragonal crystal in each grain and the dissipation associated with grain boundary constraints are brought into the model through switching criterion. The pressure dependent constraints imposed by the surrounding grains on the grain of interest at its boundary during domain switching process is correlated with the resistance experienced by a ferroelectric single crystal on its boundary during domain switching. Taking all the domain switching possibilities, the volume fractions of each of the variants are tracked and homogenized for macroscopic behavior. Numerical simulations were carried out for the behavior of ferroelectrics using both the models and the outcome was found to be qualitatively comparable with experimental observations given in literature.

Keywords Ferroelectrics · Domain switching · Boundary dissipation · Back fields · Multi-axial loading

---

K. Jayabal
Indian Institute of Technology, Chennai, India Institute of Mechanics, TU Dortmund, Germany

A. Arockiarajan and S.M. Sivakumar (✉)
Indian Institute of Technology, Chennai, India
e-mail: mssiva@iitm.ac.in

A. Menzel
TU Dortmund, Germany

B. Dattaguru et al. (eds.), *IUTAM Symposium on Multi-Functional Material Structures and Systems*, IUTAM Bookseries 19, DOI 10.1007/978-90-481-3771-8_10,

1 Introduction

Ferroelectrics employed in sensors and actuators in their piezoelectric region may exhibit appreciable nonlinear response with high external fields. Due to complex geometries upon demand from the device designers, ferroelectrics can experience external fields to act at angles to their poled direction. These kind of loadings can make the nonlinear response more complicated causing ferroelectric devices providing erroneous results. Hence, the prediction of multi-axial response of ferroelectric ceramics is essential in order to improve the performance of the devices.

The constitutive models for ferroelectric ceramics relating the electromechanical forces with the response are, in general, classified as micromechanical and phenomenological models [2]. The micromechanical models include the microstructural changes that accompany the processes of polarization and switching from a physical angle. This varies from simple approach [3] to advanced self-consistent homogenization method [4]. However, these methods are computationally intensive and slow owing to a large number of internal variables which can be overcome by phenomenological models.

Phenomenological models are commonly derived from a thermodynamic framework to describe the kinetics of the internal variables derived from experimental observations. These models are attractive as the state variables are few and hence, easily traceable; however, these models lack in bringing the underlying physics in microstructural changes. In this paper, two types of models combining the benefits of both phenomenological and micromechanical approaches are discussed. The first type is a one dimensional simplistic model that considers three variants as internal variables adequately representing the state of the ferroelectric polycrystal. Simultaneous evolution of more than two variants at a given instant is considered in this model. The second type of model deals with the nonlinear response in a three dimensional perspective with a suitable switching criterion embedded in the model for boundary dissipation of a grain in the ferroelectric polycrystal.

2 Model I: Simultaneous Evolution Model

This is an uniaxial model that deals with the response of ferroelectric polycrystal along a direction of interest. In a thermally depoled state, on cooling ferroelectrics below Curie temperature in the absence of external loads, a tetragonal unit cell can orient its spontaneous polarization in any one of six possible ways with respect to crystallographic axis as shown in Fig. 1(a). Each orientation is referred to a crystal variant. Based on the orientation, the variants 1, $1'$, 3 and $3'$ are referred to as in-plane variants, and 2 and $2'$ as out of plane variants respectively as shown in Fig. 1(b). Two out of plane variants 2 and $2'$ and any one of the four in-plane variants, for instance 1, would suffice for reasonably representing the state of the ferroelectric single crystal for uniaxial modeling with the direction of interest lying along the variant 2. In the case of ferroelectric polycrystal, a large number of single crystals, referred to as grains, exist with their crystallographic axes oriented randomly in the

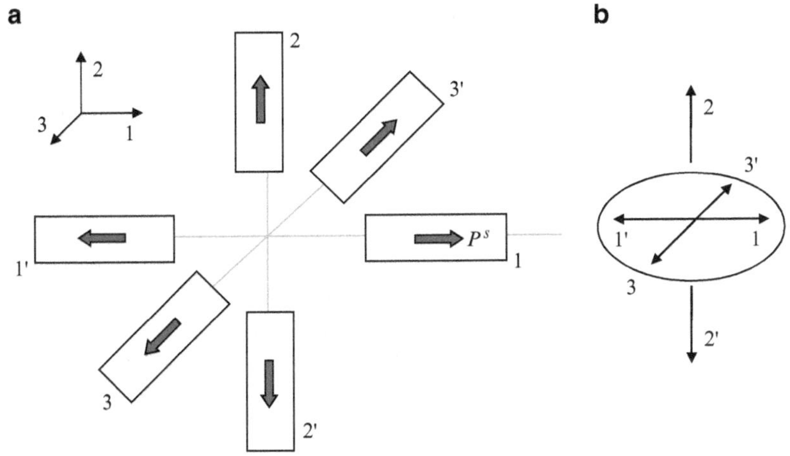

Fig. 1 (**a**) Six variants in a tetragonal crystal structure (**b**) in-plane (1,1′,3 and 3′) and out of plane (2 and 2′) variants in a ferroelectric single crystal

material. Hence, the crystal variants are also distributed in the polycrystal in all the directions. For uniaxial representation of the ferroelectric polycrystal behavior along the vertical direction, the projections of all the crystal variants are grouped into any one or more of the three uniaxial variant directions, represented by, $\uparrow$, $\downarrow$ and $\rightarrow$. While, for strain evolution, $\uparrow$ and $\downarrow$ have to be grouped together, all the non-contributing polarizations along the uniaxial direction chosen are grouped into $\rightarrow$. Each of these projections contribute to the strain and polarizations along the uniaxial direction. The volume fractions of the two out of plane variants, $\uparrow$, $\downarrow$ and the in-plane variant $\rightarrow$ respectively are indicated by $\xi^{\uparrow}$, $\xi^{\downarrow}$ and $\xi^{\rightarrow}$ respectively. These three volume fractions play the role of internal variables in adequately representing the averaged microscopic state of the ferroelectric polycrystal from uniaxial modeling perspective.

2.1 Driving Force and Evolution Equations

The total polarization and strain of ferroelectric material at a given instant of loading can be additively decomposed as the reversible (R) and irreversible or remnant (I) polarization and strain respectively. The macroscopic response of the crystal in terms of the total polarization and strain can be approximated from the microscopic evolution of the individual variants as,

$$\dot{P}_i = \sum_n \left[\dot{P}_i^{R(n)} + \dot{P}_i^{I(n)} \right]; \; \dot{\varepsilon}_{ij} = \sum_n \left[\dot{\varepsilon}_{ij}^{R(n)} + \dot{\varepsilon}_{ij}^{I(n)} \right]; \; n = 1, 2, ..., 6 \qquad (1)$$

where, n represents the six variants of the tetragonal crystal structure for a generalized case. For uniaxial cases, the number of state variables are reduced to three as discussed in Section 2. From the first and second laws of thermodynamics dealing with the balance of energy and entropy, the dissipation potential in terms of the Gibbs energy density, g, can be derived as,

$$\sum_n \left[\sigma \Delta H^{(m,n)} + E \Delta N^{(m,n)} - \rho g_{,\xi^{(m,n)}}\right] \dot{\xi}^{(m,n)} \geq 0; \quad m,n = \uparrow, \downarrow, \rightarrow \ \& m \neq n. \tag{2}$$

where σ, E, N, H and ξ indicate the applied external stress and electric field, spontaneous polarization and strain, and the volume fraction of variants respectively. The superscript (m,n) indicate that the transformation takes place from variant m to variant n and hence $\Delta(\cdot)^{(m,n)} = (\cdot)^{(n)} - (\cdot)^{(m)}$. The driving forces for the three forward transformation systems, $(\rightarrow, \uparrow)$, $(\rightarrow, \downarrow)$ and $(\downarrow, \uparrow)$ can be derived after normalizing with ΔH in order to get stress like terms,

$$D^{(m,n)} = \sigma + \frac{N^{(n)}}{\Delta H} E - \beta^{\varepsilon} - \beta^{P}; \quad m = \rightarrow, n = \uparrow, \downarrow \tag{3}$$

$$D^{(m,n)} = 2E \frac{N^{(n)}}{\Delta H} - 2\beta^{P}; \quad m = \downarrow, n = \uparrow \tag{4}$$

where

$$\beta^{\varepsilon} = \rho \frac{\partial g}{\partial \varepsilon^{I}}; \quad \beta^{P} = \frac{N^{(\uparrow)}}{\Delta H} \rho \frac{\partial g}{\partial P^{I}} \tag{5}$$

represent the back fields that depend on the remnant strain and polarization of the ferroelectric ceramic respectively. The change in the volume fraction of distinct variants, on being tracked for each load step increment, can be related to the back fields,

$$\dot{\beta}^{\varepsilon} = h_1 \dot{\varepsilon}^{I} = h_1 \sum_n H^{(n)} \dot{\xi}^{(n)}; \quad \dot{\beta}^{P} = h_2 \dot{P}^{I} = h_2 \sum_n N^{(n)} \dot{\xi}^{(n)}; \quad n = \uparrow, \downarrow, \rightarrow. \tag{6}$$

The parameters $h_1 := h_1(\xi^{(n)})$ and $h_2 := h_2(\xi^{(n)})$ are referred to as the hardening parameters that depend in general on the current state of the material but, here considered as constants. These parameters can be determined experimentally. The evolution equation of a variant for a specific transformation system can be derived using Eqs. (3–6) as,

$$\dot{\xi}^{(n)} = \frac{D^{(m,n)} - k_1}{h_1 \Delta H + h_2 N^{(\uparrow)}}; \quad m, n = \rightarrow, \uparrow \ \& \ m \neq n \tag{7}$$

$$\dot{\xi}^{(n)} = \frac{D^{(m,n)} - k_1}{h_1 \Delta H + h_2 N^{(\uparrow)}}; \quad m, n = \rightarrow, \downarrow \ \& \ m \neq n \tag{8}$$

$$\dot{\xi}^{(n)} = \frac{D^{(m,n)} - k_2}{4 h_2 N^{(\uparrow)}}; \quad m, n = \downarrow, \uparrow \ \& \ m \neq n \tag{9}$$

The above evolution equations were obtained based on the assumption that only one transformation system would be active for the given load increment.

However, in ferroelectric materials, there are cases where more than one transformation system would be set off by a combination of the external loading and the arrangement of domains. For such simultaneous evolution cases considering both the transformation systems that are active by using Eqs. (3, 5 and 6),

$$\dot{\xi}^{(n)} = \frac{1}{2}\left[\frac{\dot{\beta}^{\varepsilon}}{h_1 \Delta H} + \frac{\dot{\beta}^{P}}{h_2 N^{(n)}}\right]; \; n = \uparrow, \downarrow \tag{10}$$

$$\dot{\xi}^{(\rightarrow)} = 1 - \sum_n \dot{\xi}^{(n)}; \; n = \uparrow, \downarrow. \tag{11}$$

Based on the number of transformation systems, the evolution equations are identified and the change in volume fractions are determined. Once the present state of the crystal is specified in terms of the volume fraction of the variants, the macroscopic response of the crystal is expressed by the sum of the reversible and irreversible parts of the polarization and strain for the given load as,

$$P = \sum \left[P^{R(n)} + P^{I(n)}\right] \xi^{(n)} = \sum_n \left[d\sigma + \chi E + N^{(n)}\right] \xi^{(n)}; \; n = \uparrow, \downarrow, \rightarrow \tag{12}$$

$$\varepsilon = \sum \left[\varepsilon^{R(n)} + \varepsilon^{I(n)}\right] \xi^{(n)} = \sum_n \left[C\sigma + dE + H^{(n)}\right] \xi^{(n)}; \; n = \uparrow, \downarrow, \rightarrow \tag{13}$$

where C, d and χ denote the compliance modulus, the piezoelectric coefficient and the dielectric permittivity respectively. The developed model was set to predict the response of a ferroelectric polycrystal under a multi-axial electric loading conditions (refer Fig. 2) as demonstrated by [1,5] by suitably choosing the initial state of internal variables. Despite the minimalistic nature of the model, it is able to reproduce the characteristics observed in the ferroelectric responses under polarization rotation tests.

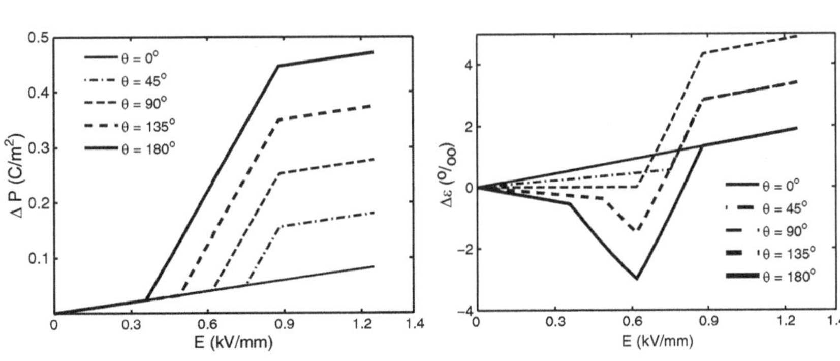

Fig. 2 Simulated response of the considered ferroelectric polycrystal under polarization rotation tests using the simultaneous evolution uniaxial model. When the domain switching always improves the polarization for all angle of rotations, it provides a mixed response for strain depending upon the angle of rotation

3 Model II: Pressure Dependent Model

This model deals with the response of the ferroelectric ceramics in a three dimensional perspective. Since all the six variants of tetragonal crystal structure are considered in each grain, there would be as many as 30 transformation systems possible in the switching process. The dissipation inequality can be derived from taking all switching possibilities as,

$$0 \leq \sum_{n=1}^{6} \sum_{m=1}^{6} \left[\sigma_{ij} \Delta\varepsilon_{ij}^{I(m,n)} + E_i \Delta P_i^{I(m,n)} - \rho \Delta g^{(m,n)} \right] \dot{\xi}^{(m,n)} \quad for \quad m \neq n \tag{14}$$

where the change in the Gibbs energy can be derived as,

$$\rho \Delta g^{(m,n)} = -\frac{1}{2} \left[\sigma_{ij} \Delta C_{ijkl}^{(m,n)} \sigma_{kl} + E_i \Delta \chi_{ij}^{(m,n)} E_j + 2\sigma_{ij} \Delta d_{kij}^{(m,n)} E_k \right]. \tag{15}$$

The change in the Gibbs energy associated with the domain switching is due to the variation in the material properties of the domains involved in the switching process. When the variation in the compliance tensor C_{ijkl} and the dielectric permittivity tensor χ_{ij} may be neglected between the domains involved in the switching process, the difference in the piezoelectric coupling tensor d_{ijk} can not be ignored [6]. Hence, the dissipation inequality is reduced to

$$0 \leq \sum_{n=1}^{6} \sum_{m=1}^{6} \left(\sigma_{ij} \Delta\varepsilon_{ij}^{I(m,n)} + E_i \Delta P_i^{I(m,n)} + \sigma_{ij} \Delta d_{kij}^{(m,n)} E_k \right) \dot{\xi}^{(m,n)} \quad for \quad m \neq n \tag{16}$$

The terms within brackets in the above equation refer to the driving force developed for the given load which on reaching certain critical values initiate the corresponding domain switching.

3.1 Modified Switching Criterion with Boundary Constraints

A switching criterion is important in the micromechanical modeling since it determines the onset of domain switching for the given external fields. Several quantities have been considered in the switching criterion such as the work done, the Gibbs free energy, the total potential and the internal energy density. The switching criterion used in this work is a refinement of the one earlier proposed [7]. The critical value that should be attained by the driving force for the occurrence of 90° and 180° switching can be derived from Eq. (16) as,

$$f_c^{90} = \sqrt{2} E_c^{90} \Delta P^I; \quad f_c^{180} = 2 E_c^{180} \Delta P^I \tag{17}$$

Above critical values were found reasonably reproducing the ferroelectric behavior in the absence of external stress. However, they would not suffice under combined electromechanical loading conditions since the pressure dependent boundary dissipations are not included [7]. In this model, the resistance arising out of boundary constraints due to external compressive stress is brought into the switching criterion in a straight forward way as discussed below. In the experiments conducted by [8], cyclic electric field was applied in the presence of compressive stress on single crystal barium titanate. It was argued from the observations that 90° domain switching exhibited a dissipation associated with friction between the crystal surface and the loading platen due to the superimposed compressive stress.

It may be reasonable, here, to draw an analogy between a constrained ferroelectric single crystal and a grain in a ferroelectric polycrystal constrained by the surrounding grains. As well, the dissipation on the grain boundary in a ferroelectric polycrystal under the influence of stress during 90° domain wall motion may be comparable to that of pressure dependent boundary dissipation in a constrained ferroelectric single crystal during domain switching. Here, the dissipation concerning to the polarization mismatch on the grain boundary arising from 180° domain switching is ignored to make the model simpler, yet found to predict the behavior of ferroelectric polycrystal reasonably. Thus the switching criterion for 90° domain switching is modified to take care of pressure dependent boundary dissipation as,

$$f_c^{90} = \sqrt{2}E_c^{90}\Delta P^I + K(< -\sigma_{ii} >)\Delta\xi \qquad (18)$$

where $\Delta\xi$ represents the increment in the volume fraction of the favored domains for the load step increment. The constant K is to be determined by comparing the model simulation with experimental results. The Macaulay bracket indicates that the influence of tensile stress on the boundary dissipation is neglected here. The changes in the volume fractions are updated for each variant based on the switching criterion and a simple Reuss averaging was used to obtain the polycrystal behavior. The response of a ferroelectric polycrystal under various polarization rotations are simulated and presented in Fig. 3 in terms of strain and polarization which is in qualitative agreement with the experimental results reported by [5].

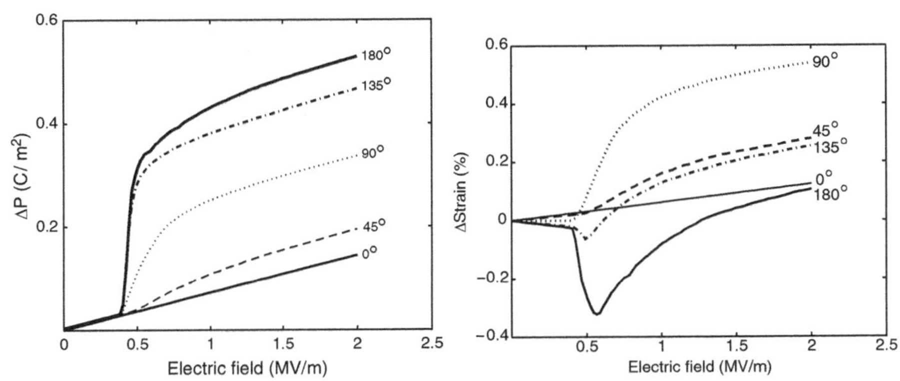

Fig. 3 Multi-axial (polarization rotation) response of ferroelectric polycrystal PZT in terms of the hysteresis and butterfly curves under electric loading case with pressure dependent model

4 Conclusions

The dependence of the nonlinear response of the ferroelectric single and polycrystals on the orientation of the electric field with regard to the initial poled direction has been studied using two different types of models. In the uniaxial model, simultaneous evolution of more than two variants for the given loading was considered and despite its simplicity, the model was able to capture the key features of electro-mechanical behavior of ferroelectric ceramics. In the three dimensional model, the dissipative mechanisms in the subgrain of a ferroelectric polycrystal, due to the resistance offered at its boundary by the adjacent grains during domain switching are considered and the dissipation thresholds have been appropriately refined. From the results, it can be seen that while the three dimensional model captures the salient features of the electromechanical behavior of the ferroelectric polycrystal, the uniaxial model offers an effective way of capturing well the essential features of the electromechanical behavior given the initial condition of the crystal.

References

1. D. Zhou, M. Kamlah, and B.Laskewitz. Multi-axial non-proportional polarization rotation tests of soft pzt piezoceramics under electric field loading. *Proc. SPIE.*, 6170:617009, 2006.
2. C. M. Landis. Non-linear constitutive modeling of ferroelectrics. *Current opinion solid state Mat. Sc.*, 8:59–69, 2004.
3. S. C. Hwang, C. S. Lynch, and R. M. McMeeking. Ferroelectric/ferroelastic interactions and a polarization switching model. *Acta Metall. Mater.*, 43:2073–2084, 1995.
4. J. E. Huber, N. A. Fleck, C. M. Landis, and R. M. McMeeking. A constitutive model for ferroelectric polycrystals. *J. Mech. Phys. Solids*, 47:1663–1697, 1999.
5. J. E. Huber and N.A. Fleck. Multi-axial electrical switching of a ferroelectric: theory versus experiment. *J. Mech. Phys. Solids*, 49:785–811, 2001.
6. M. Kamlah and Q. Jiang. A constitutive model for ferroelectric pzt ceramics under uniaxial loading. *Smart Mat. Struct.*, 8:441–459, 1999.
7. K. Jayabal, A. Arockiarajan, and S.M. Sivakumar. A micromechanical model for polycrystal ferroelectrics with grain boundary effects. *Comput. Model. Eng. Sci.*, 27:111–123, 2008.
8. E. Burcsu, G. Ravichandran, and K. Bhattacharya. Large electrostrictive actuation of barium titanate single crystals. *J. Mech. Phys. Solids*, 52:823–846, 2004.

Active Single Walled Carbon Nanotube–Polymer Composites

Sujay Deshmukh and Zoubeida Ounaies

Abstract Electric field-activated electroactive polymers (EAPs) are an attractive class of smart materials that exhibit electromechanical coupling conversion. However, they require high actuation voltages, have low blocked stresses and low operating temperatures. These current limitations are linked to inherent polymer properties such as low dielectric constant and low modulus. Our recent efforts in EAP-based nanocomposites focus on incorporating single walled carbon nanotubes (SWNTs) to significantly improve their electromechanical response. SWNTs have shown great promise as fillers in polymer-based composites, especially in terms of improving mechanical, electrical and thermal properties. In this research we evaluate the effect of single walled carbon nanotubes (SWNTs) on the electromechanical response of polymers with varying polar nature, from non-polar to highly polar. The electric, dielectric and electromechanical strain response is studied as a function of SWNT content and polymer matrix. The electromechanical response is found to be electrostrictive and it increases with SWNT content and polarity of the matrix. Dielectric measurements are also used to measure the enhanced polarization in the nanocomposites due to the presence of SWNTs.

Keywords Actuation · Electrostriction · Multifunctional carbon nanotube composites · Polarization · Polymer nanocomposites

1 Introduction

Incorporating single walled carbon nanotubes (SWNTs) into polymers has resulted in enhancements in their mechanical [1] and electrical [2] properties. Few studies however have focused on their coupled properties, such as electro-mechanical, and those that have are limited to studying the effect of SWNTs on already-active

S. Deshmukh and Z. Ounaies (✉)
Department of Aerospace Engineering, Texas A&M University, TX, USA
e-mail: ZOUNAIES@tamu.edu

B. Dattaguru et al. (eds.), *IUTAM Symposium on Multi-Functional Material Structures and Systems*, IUTAM Bookseries 19, DOI 10.1007/978-90-481-3771-8_11,

polymers. Kang et al. have demonstrated an enhancement in the d_{31} piezoelectric coefficient of (β-CN) APB-ODPA polyimide, in the presence of SWNTs [3]; Levi et al. have demonstrated an increase in the piezoelectric d_{31} coefficient of poly(vinylidene fluoridetrifluoroethylene) (PVDF-TrFE) due to SWNTs [4]. In this paper, we evaluate the effect of single walled carbon nanotubes (SWNTs) on the electromechanical response of polymers with varying polar nature, from non-polar to highly polar. Our data suggests that the use of SWNTs can create an electrostrictive response in a non-electroactive polymer and enhance it in a polar polymer. This finding offers many advantages such as the possibility of rendering any polymer electro-active. The ultimate goal is to design an electroactive response in polymers while tuning their other physical properties, resulting in multifunctional materials. Three polymers are selected as matrix materials; a non polar polyimide CP2, a weakly polar polyimide (β-CN)APB-ODPA and a strongly polar poly(vinylidene fluoride)(PVDF). Selecting polymers with varying polarity helps us evaluate the effect of matrix dipole moment along with SWNT content on the electromechanical properties.

2 Experimental Studies

Thermoplastic polymer poly(vinylidene fluoride)(PVDF) and thermosetting polyimides like CP2(APB-6FDA) and (β-CN)APB-ODPA are selected as matrix materials. PVDF is a strongly polar material which exhibits a piezoelectric response when simultaneously poled and stretched under a DC electric field. CP2 polyimide has a low dipole moment and is generally non-polar, while (β-CN)APB-ODPA has a stronger dipole moment and is a weakly piezoelectric material. Nanoinclusions used in this study are HIPCO single walled carbon nanotubes (SWNTs). The polymer nanocomposites were processed using a solvent-based film casting technique [5, 6]. Details on the experimental procedures for the polyimides and PVDF-based nanocomposites can be found in [5, 6] and [7] respectively.

Electrical measurements are carried out to quantify the dispersion of nanotubes and determine the percolation transition. Dielectric measurements are also used to investigate the polarization inherent to the structure of the polymer and the enhanced polarization due to presence of nanotubes. AC electrical conductivity and dielectric constant are measured as a function of frequency (20 Hz–1 MHz) using a QuadTech precision LCR meter. A Sun Systems EC1x environmental chamber is used to add temperature capability. The nanocomposite films are coated with a thin silver layer by a vapor deposition process. The thickness of the samples ranges from 30–60 μm, and that of the silver layer is kept at 100 nm. The film samples are cut into strips of 3×0.5 cm, which are then used for the electromechanical tests. Figure 1 shows the experimental set-ups. For thickness actuation, a small area on the bottom face is constrained and the electric field is applied through the thickness (Fig. 1(a)). The change in thickness is measured by a MTI fiber optic sensor. For the bending experiment, the top of the sample is sandwiched between glass plates

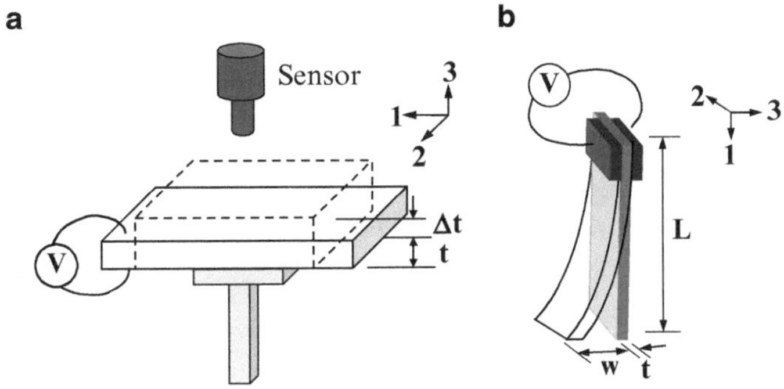

Fig. 1 (**a**) Thickness actuation, (**b**) bending actuation

with copper leads. This setup is then suspended vertically in an acrylic box chamber (Fig. 1(b)). The leads allow the application of electric field (DC or AC) to the strip. The bending of the sample is captured by a Photron Fastcam PCI R2 high speed camera setup. The captured videos are analyzed using Photron image analysis software. This software allows measurement of the sample displacement by analyzing the sample position in successive video frames.

3 Results

Figure 2 shows the DC conductivity of CP2 and (β-CN)APB-ODPA as a function of SWNT content. The conductivity is given by the scaling law below

$$\sigma = A(v - v_c)^t \qquad (1)$$

where σ is the conductivity of the nanocomposite, v is the volume fraction of the SWNT in the nanocomposite, v_c is the percolation volume fraction, A and t are fitted constants. A best fit to the data for CP2 –SWNT nanocomposites results when $A = 5 \times 10^{-2}\,S/cm$ and $t = 1.7$, resulting in a $v_c = 0.04$ vol% while a similar analysis shows $A = 10^{-4} S/cm$ and $t = 2$ resulting in a $v_c = 0.06\%$ for (β-CN)APB-ODPA-SWNT.

The dielectric constant is measured as function of temperature to study the dielectric relaxation in the nanocomposites. The dielectric relaxation results from movement of permanent and induced dipoles called dipolar relaxation and electric charges called ionic relaxation. The dielectric relaxation strength of dipoles is measured by Δε, which is the difference between the static (low frequency) and high frequency limit on dielectric constant. This value also quantifies the remnant polarization when the Clausius Mossotti equation is used to relate the dielectric constant to the dipole moment in the material given by [8]:

$$P_R = \Delta\varepsilon\, \varepsilon_0 E \qquad (2)$$

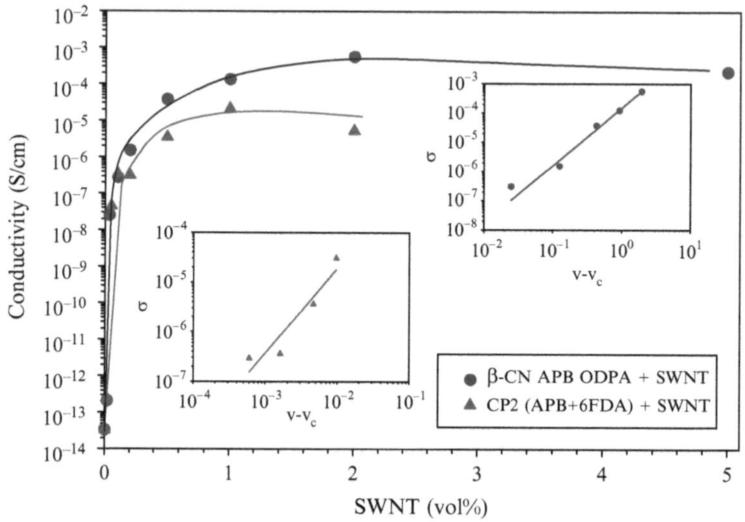

Fig. 2 Variation of conductivity with SWNT loading for both polyimide nanocomposites (The solid lines in the main figure is added to ease the reading of the data; the solid lines in the insets are the linear trend exhibited by the data)

Table 1 Comparison of $\Delta\varepsilon$ as a function of SWNT loading for CP2 and (β-CN) APB-ODPA based nanocomposites

SWNT loading (Vol%)	$\Delta\varepsilon_{@21.5,Hz}$ (CP2)	$\Delta\varepsilon_{@20\,Hz}$ ((β-CN)APB-ODPA)
0	0.69	4.41
0.02	0.72	9.69
0.035	–	10.59
0.05	21	–
0.075	–	48
0.1	24	–

where ε_0 is the permittivity of free space and E is the applied electric field. This polarization is due to contributions by dipoles present in the system. The $\Delta\varepsilon$ value can also be measured by using temperature in place of frequency, where the behavior of dipoles under high temperature is analogous to that at low frequency, while the value of dielectric constant at low temperature is analogous to that at high frequency. Table 1 shows the $\Delta\varepsilon$ for different SWNT loading for CP2 and (β-CN)APB-ODPA. The value displays an increase with SWNT loading. It also shows higher values for (β-CN)APB-ODPA than CP2 for the same SWNT concentration. (β-CN)APB-ODPA has a higher dipole moment (8.8 D) [9, 10] than CP2 which is estimated at 2.93 D [11–13]. This indicates that the polarization in the nanocomposites increases with SWNT content as well as with dipole strength of the polymer matrix.

Figure 3 shows constant longitudinal strain S_{11} along the length due to bending as a function of applied electric field. The strains (calculated assuming a uniform radius of curvature) increase with SWNT content. One significant finding is that the

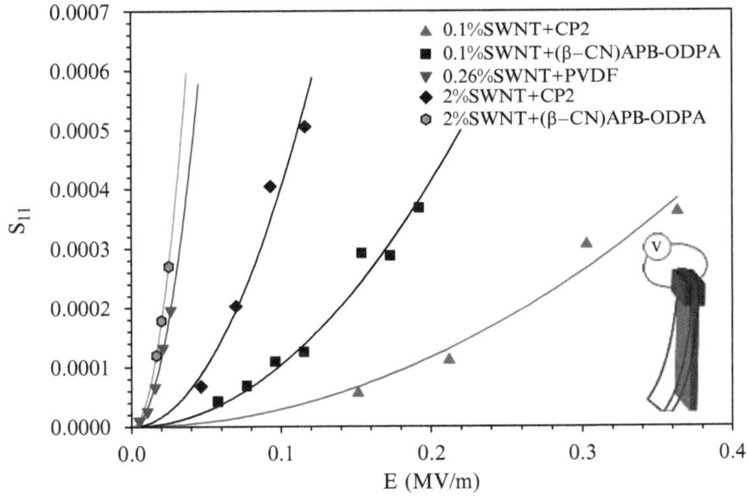

Fig. 3 Longitudinal strains from bending as a function of electric field, SWNT content and polymer matrix

(β-CN)APB-ODPA-SWNT composites show a higher electromechanical response than CP2-SWNT. Since (β-CN)APB-ODPA has a higher dipole moment compared to CP2 polyimide, it can be inferred that as the dipole moment and SWNT content in the nanocomposite increases, a higher electromechanical response is measured. Thus both the nature of the polymer and the content of SWNTs affect the electromechanical response of the nanocomposites. These observations are also supported by the fact that 0.26% SWNT-PVDF composite shows higher response than both 2% SWNT-CP2 and 2% SWNT-(β-CN)APB-ODPA composites; PVDF has the highest dipole moment of all the polymers under study. Similar to the bending case the thickness strain response increases as the SWNT content and dipole moment in the polymer matrix increases. The magnitude of electric field required for the actuation was a low 0.01–0.4 MV/m which is significantly lower than that required to drive current electronic Electroactive Polymers. This result demonstrates that we can manipulate the strain response and electric field required to achieve it by choosing the nanoinclusion content and the type of polymer matrix used.

Next we investigate the mechanism driving the actuation response. The strain has a quadratic dependence on electric field indicating an electrostrictive behavior. The coefficient of electrostriction **M** is the material property that relates the strain and applied electric field through $S_{ij} = M_{ijkl}E_kE_l$. This coefficient which determines the strength of an electrostrictive material is measured for both bending and thickness actuation as a function of SWNT content and type of polymer matrix. Figure 4 is a plot of M_{3333}, measured from the thickness experiment, as a function of SWNT content for CP2 and (β-CN)APB-ODPA nanocomposites. The coefficient increases with SWNT content until 1% SWNT. Above this concentration the dielectric loss values of the samples increase due to increase in conductivity, and no further increase in the electrostrictive coefficient is observed. Nanocomposites made from

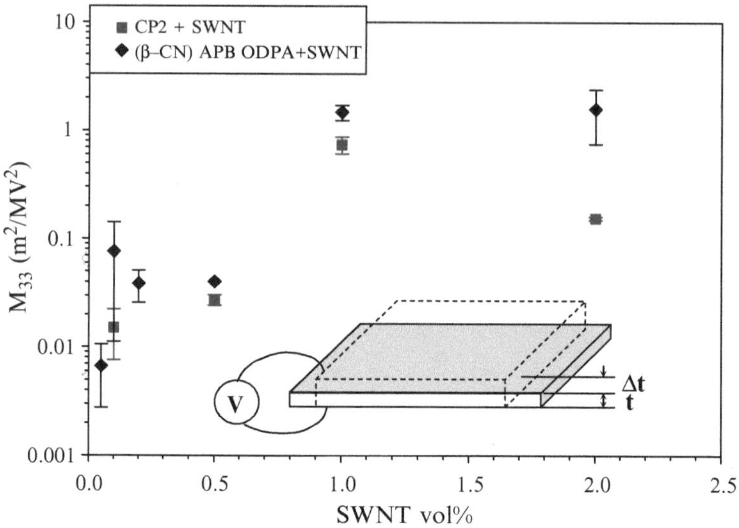

Fig. 4 Thickness coefficient of electrostriction for CP2 and (β-CN)APB-ODPA as a function of SWNT content

(β-CN)APB-ODPA, which is more polar, shows higher values for the coefficient than CP2. Same inference is also drawn from M_{1133} (not shown here) data.

4 Discussion

The dielectric relaxation strength $\Delta\varepsilon$ quantifies the remnant polarization in the material under an applied field as illustrated in Eq. (2). Dielectric relaxation spectroscopy results show an increase in the $\Delta\varepsilon$ value with SWNT content (see Table 1). Further inspection of Table 1 also highlights higher values for (β-CN)APB-ODPA than for CP2 for the same SWNT content. Thus the results indicate an increase in the remnant polarization with the SWNT content and the dipole moment in the polymer matrix. The enhanced electrostrictive effect is ascribed to the enhancement in the polarization of the nanocomposites. This enhancement in the polarization could be (a) inherent to the SWNTs, (b) driven by the polymer dipoles, or (c) a result of an interaction between the SWNTs and polymer dipoles. It has been shown by Dumitrica et al. [14] that a polarization exists in a SWNT due to a shift in the sp^2 hybridization. It can be inferred from their work that bending deformation of the tube, would yield a non-zero global polarization in the SWNT structure. The SEM images of our composites (not shown here) reveal the presence of bent or deformed SWNT bundles throughout the polymer matrix.

In our previous publication, we stated that the low percolation is thought to result from an electron hopping phenomenon called quantum tunneling, rather than from physical contact between the SWNTs [2]. Thus, presence of nano and micro-capacitors in between the SWNTs is possible. This would result in a local field enhancement which would make the dipole on the polymer chain more mobile. CP2 has a dipole strength (μ) of around 2.93 D [11–13] while that of (β -CN)APB-ODPA is 8.8 D [9, 10]. PVDF on the other hand has a dipole moment of 2.1 D but an ultimate polarization between 50–100 $mC/m^2(\mu^*N)$ [15], much higher than the 40 mC/m^2 possible for (β-CN)APB-ODPA. Due to the local field enhancement in the presence of SWNTs, PVDF would demonstrate high polarization followed by (β-CN)APB-ODPA and then CP2. This possibility supports dielectric relaxation results and the observation of high electrostrictive nature for PVDF nanocomposites followed by (β-CN)APB-ODPA and CP2 nanocomposites.

Enhanced polarization could also exist due to SWNT – polymer interaction resulting from presence of a secondary non-covalent interaction between the polymer and SWNTs. This can occur due to a charge donor-acceptor relationship between the electronegative group on the monomer and the SWNTs. Researchers have suggested such a relationship between the polymer and SWNTs [5, 16]. This interaction could also contribute to the enhanced polarization observed in the nanocomposites.

5 Conclusion

This study investigates the effect of presence of SWNTs and dipolar polarity of polymers on the electromechanical properties of nanocomposites. The polyimide based nanocomposites showed a bending electrostrictive response above percolation. Larger displacements are measured with higher SWNT content and for the more polar (β-CN)APB-ODPA. PVDF+0.26%SWNT showed a much stronger actuation response than both polyimides. Dielectric relaxation spectroscopy indicated that, in the polyimides increasing the SWNT content increased the polarization in the samples. Also the more polar (β-CN)APB-ODPA-based composites had a higher polarization compared to the less polar CP2-based composites; therefore, the electrostrictive response was also polymer dependent and polymers with high dipole moments showed higher response. Through judicious selection of nanoinclusions and polymers, we can tailor the electromechanical response of polymer-based nanocomposites. This is a step towards engineering a given electromechanical response in polymers.

References

1. Dufresne A, Paillet M, Putaux JL et al. (2002) Processing and characterization of carbon nanotube/poly(styrene-co-butyl acrylate) nanocomposites. J Mater Sci 37:3915–3923.
2. Ounaies Z, Park C, Wise KE et al. (2003) Electrical properties of single wall carbon nanotube reinforced polyimide composites. Compos Sci Technol 63:1637–1646.

3. Kang JH, Park C, Gail SJ et al. (2006) The effect of single wall carbon nanotubes on the dipole orientation and piezoelectric properties of polymeric nanocomposites. NANO: Brief Reports and Reviews 1:77–85.
4. Levi N, Czerw R, Xing S et al. (2004) Properties of polyvinylidene difluoride-carbon nanotube blends. Nano Lett 4:1267–1271.
5. Wise KE, Park C, Siochi EJ et al. (2004) Stable dispersion of single walled carbon nanotubes in polyimide: the role of noncovalent interactions. Chem Phys Lett 391:207–211.
6. Park C, Ounaies Z, Watson KA et al. (2002) Dispersion of single wall carbon nanotubes by in situ polymerization unders sonication. Chem Phys Lett 364:303–308.
7. Almasri A, Ounaies Z, Kim YS et al. (2008) Characterization of solution processed double walled carbon nanotube/poly(vinylidene) nanocomposites. Macromol Mater Eng 293:123–131.
8. Hilczer B, Malecki J (1989) In: *Electrets*; studies in electrical and electronic engineering, Elsevier, New York.
9. Harrison JS, Ounaies Z (2001) Piezoelectric polymers. ICASE NASA langley research center: 1–26.
10. Young JA, Farmer BL, Hinkley J (1999) Molecular modeling of the poling of piezoelectric polyimides. Polymer 40: 2787–2795.
11. Dean JA (1999) In: Lange's handbook of chemistry, 15th edn, McGraw-Hill INC.
12. Hinkley JA, Dezern JF (2002) Crystallization of stretched polyimides: a structure-property study, NASA langley research center CASI 301.
13. Makowski MP (1993) Conformational analysis of a series of bridged phenyl compounds using AM1, Comput Polym Sci 3:1–8.
14. Dumitric T, Landis CM, Yakobson BI (2002) Curvature-induced polarization in carbon nanoshells. Chem Phys Lett 360:182–188.
15. Nalwa HS (1995) In: Ferroelectric Polymers: Chemistry, Physics, and Applications, CRC Press.
16. Owens FJ, Jayakody JRP, Greenbaum SG (2006) Characterization of single walled carbon nanotube: Polyvinylidene difluoride composites. Compos Sci Technol 66:1280–1284.

Modeling of Fibre Formation and Oxygen Permeability in Micro-fibrillar Polymer-Polymer Composites

D. Bhattacharyya and R.J. Shields

Abstract Micro- or nano-fibrillar composite (MFC or NFC) materials can be created by blending two homopolymers with distinctly different melting temperatures, such as polyethylene (PE) and poly(ethylene terephthalate) (PET), and processing the blend under certain thermo-mechanical conditions. This processing creates in-situ fibrils of the polymer that has the higher melting temperature and the other constituent polymer surrounds the fibrils. The resulting composites possess excellent mechanical properties and can have wide ranging applications that require superior mechanical and/or functional properties. However, the properties and applications depend very much on the morphology of created polymer composites and the fibril orientations. The present paper studies the development of microfibril formation in PE/PET blends and reports their mechanical and oxygen barrier characteristics. It is revealed that the initial extrusion and subsequent mechanical processing create fibrillar morphology with high aspect ratios, thus resulting in not only superior mechanical performance but also in significant reduction of oxygen permeability, both of which can have profound impacts on the food packaging technology. The importance of selecting a proper processing method and temperature is highlighted and a model is proposed to theoretically predict the oxygen permeability through different MFC films with varying morphology. The predicted values are compared with the theoretical predictions and suitable amendments of the formulation are suggested.

Keywords Draw ratio · Mechanical properties · Microfibrillar composites · Oxygen permeability · Polymer blending

1 Introduction

Micro- and Nanofibril Reinforced Composites (MFCs) form a novel group of polymer-polymer composites with the potential for exploitation in a wide variety of

D. Bhattacharyya (✉) and R.J. Shields
Centre for Advanced Composite Materials and Department of Mechanical Engineering, University of Auckland, Private Bag 92019, Auckland, New Zealand
e-mail: d.bhattacharyya@auckland.ac.nz

B. Dattaguru et al. (eds.), *IUTAM Symposium on Multi-Functional Material Structures and Systems*, IUTAM Bookseries 19, DOI 10.1007/978-90-481-3771-8_12,

both high and low tech applications. MFCs utilise several key principles of polymer blend processing to gain several of the mechanical and physical advantages. This paper investigates several of these characteristics, as found in MFCs derived from blends of polyethylene (PE) and poly(ethylene terephthalate) (PET). These polymers are used throughout the global packaging industry, primarily for bottling milk and carbonated soft drinks, respectively. This research is based around exploring the potential for MFCs to utilise the advantages inherent in their design to rival homopolymers and layered or coated polymers in low cost, high volume applications.

An MFC contains relatively high strength and stiffness microfibrils from a *drawn polymer*, that by virtue of the unique manufacturing process retain properties significantly superior to their bulk polymer counterparts. They are created via a three stage process using standard industrial polymer processing equipment. Initially, the constituent polymers are mixed, melted, compounded and extruded into a continuous blend. This disperses the reinforcing polymer homogeneously throughout the matrix and creates a morphology of spheres and/or elongated reinforcing particles (*microfibres*). The extruded filament is allowed to cool in a water bath at room temperature. Next, the blend is molecularly aligned by a drawing process during which the spheres and microfibres are transformed into highly oriented *microfibrils* that will be present in the final composite. Drawing is performed around the glass transition temperature (T_g) of the reinforcing polymer to achieve maximum alignment. The drawn blend can be used in its filament form or can be pelletised to manufacture composite products using standard processes, such as injection or compression moulding. This processing subjects the blend to a thermal treatment at a temperature between the melting points of the constituent polymers, which reverts the matrix back into an isotropic state, leaving the reinforcing microfibrils both molecularly aligned and well dispersed within the manufactured product.

2 Experimental Details

The blends and composites were produced using a matrix polymer of linear low density PE (LLDPE by Qamar FC21HS) with a melting temperature of 122°C, a melt flow rate of 1.0 g/10 min and a density of 0.918 g/cm^3. This was blended with PET (OZPET 3180) which has a melting temperature of 248°C and a density of 1.41 g/cm^3 – blends were mixed with 0–50 wt% PET. Extrusion was performed in a SEW Eurodrive single screw extruder using a heating profile of 250°C, 250°C, 270°C, 280°C, 290°C and 285°C (from the feed zone to the die). Screw speed was 50 RPM and blends were extruded into a water bath using a conical die of 1 mm exit diameter (with an extrusion ratio of 25:1), solidifying into a continuous filament. Subsequent to extrusion, the blend filament was elongated under tension, while in an air tunnel heated to 80–90°C. Draw ratios (final drawn length/initial undrawn length) were generally in the vicinity of five to seven. All mouldings were performed at 180°C. Injection moulding was performed on a BOY-50A injection moulding machine. Thin films (150–200 μm) for oxygen permeability tests were compression moulded. Tensile modulus and yield strength were determined via tensile tests in accordance with ASTM D-638. A gauge length of 50 mm was used and tests were performed at a crosshead speed of 5 mm/min for modulus and 50 mm/min for strength.

The oxygen permeability of each specimen was evaluated using a permeation cell that operated via the concentration increase principle developed by Davis and Huntington [1]. Chambers on either side of the specimen were purged with nitrogen gas, before one side of the cell was filled with oxygen. The oxygen permeability coefficient was then calculated via the rate of permeation of oxygen through the specimen. A minimum of four tests were performed for each sample. More details of the experimental and analytical work may be found in [2] and [3].

3 Results and Discussion

These polymer blends after extrusion can contain high aspect ratio microfibres, Fig. 1(a), and although they, without any significant degree of polymer chain alignment, do not improve the mechanical properties of the blend, their sizes and shapes influence the formation of aligned reinforcing microfibrils during drawing, Fig. 1(b), improving both mechanical and functional properties.

Figure 2 shows that the stiffness of the undrawn filaments varies between 0.1–0.9 GPa, increasing with the fraction of PET. Neat PE and the 10% blend appear to have two distinct stiffness values, dominated by the properties of either the undrawn (ratio below threshold of 3–3.5) or drawn materials (ratio above 3–3.5). Thirty percent and 50% blends have gradual increases in stiffness for draw ratios up to 3.5, indicating greater PET content is improving the material stiffness even at these lower ratios. Above draw ratios of 3.5 (where complete drawing has occurred along the entire filament) these blends are consistently between two and four times stiffer than their undrawn counterparts, with the 50/50 blend achieving a stiffness of between 4–4.5 GPa. The observance of a threshold ratio indicates the natural draw ratio of the blend is being attained at this point. Drawing above this ratio can provide further

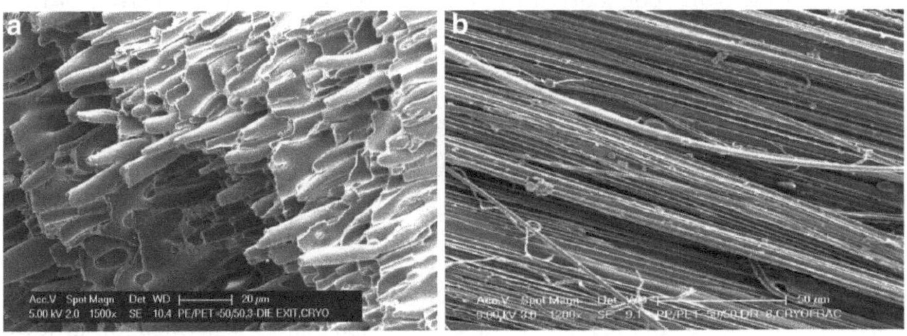

Fig. 1 Morphology of PE/PET blend (50/50): (**a**) after extrusion only; (**b**) after drawing

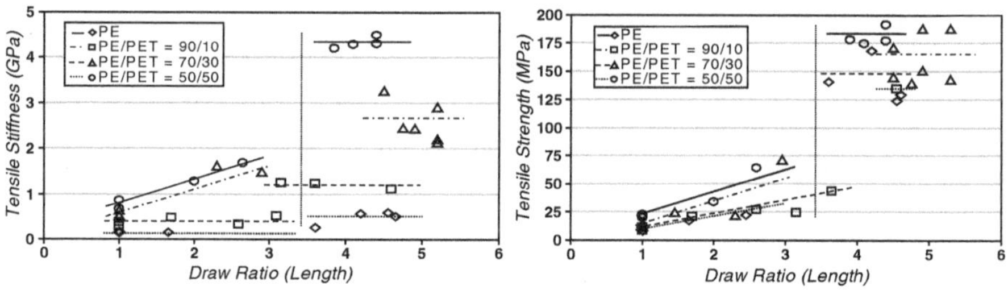

Fig. 2 Relationship between tensile properties and mechanical draw ratio for reinforced PE filaments

molecular chain unfolding but little more in the way of chain alignment. This is why mechanical properties do not improve significantly with further drawing above the threshold ratio [4].

Blend tensile strength trend, Fig. 2, is similar in nature to that of the stiffness, with increases of approximately 25–50 MPa occurring below a ratio of 3–3.5 and much higher strengths are obtained above this ratio. All reinforcement loadings from 0–50% produce filaments that yield between 125–200 MPa, with more PET generally giving slightly higher results.

The orientation of the microfibrils within the matrix after moulding is another factor that exerts significant influence on the mechanical properties and isotropy of an MFC. This can be altered according to variations in the post-production techniques (injection and compression moulding) and the different material states prior to moulding (pellets or continuous filament). Like in a typical composite laminate, different fibril configurations are possible using MFC filaments. As expected, specimens taken from the principal direction of uniaxial plates give the most impressive results with the tensile strength, Fig. 3, nearing 70 MPa in some cases (compares favourably to those of neat PE and PET). The modulus results are no less impressive, with 2.1 GPa for the uniaxial (primary) MFC compared to 0.4 and 2.3 GPa for PE and PET, respectively.

Films from pellets with 0–60% microfibril contents were produced and tested. Blends with 40% PET or greater were unable to consolidate into films, with voids occurring at the pellet boundaries. In Fig. 4, the oxygen barrier properties of films with 0–30% PET content are compared with the expected values for the films based upon a ROM approach, using values of 112 ml.mm/m^2.24 h.atm for LLDPE (from testing) and 1 mlmm/m^2.24 h.atm for oriented PET. Significant improvement in oxygen barrier properties occurs with increasing microfibrillar content – they become two and almost three times higher with 20% and 30% PET contents, respectively. These improvements are due mainly to the favourable reinforcement morphology caused by careful selection of extrusion and moulding conditions.

Although investigations in the present paper were performed using thin films in an oxygen atmosphere, results are envisaged to be applicable to thicker walled items, such as bottles and storage containers, and for permeation in lower concentration environments, such as air. Decreasing thickness and increasing oxygen concentration serve to expedite the testing process. Films were also made from

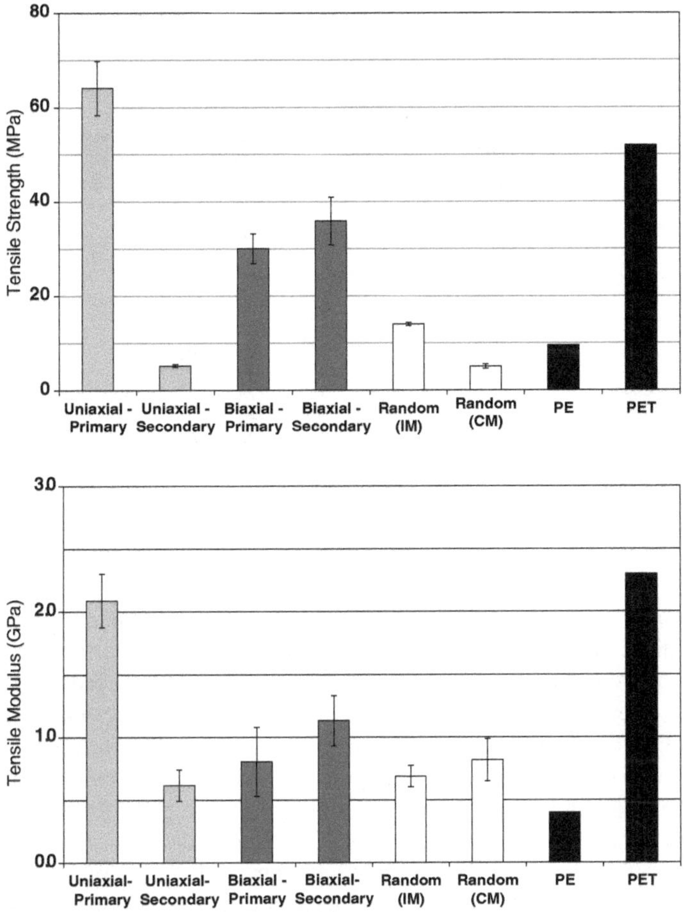

Fig. 3 Effect of fibril orientation on tensile strengths and moduli of MFCs from PE/PET (70/30)

MFCs and neat polymer under different draw ratios. Figure 4 also shows that the barrier properties are not greatly influenced by draw ratio for either type of film. This constancy of properties may be attributed to the formation of microfibres during extrusion, which are as effective as microfibrils formed after drawing in enhancing the tortuosity of the composite. This has a positive connotation from the product manufacturing point of view.

4 Modeling

The permeation of gases and liquids through filled polymer systems can be an extremely complex event, with many factors needing consideration during the explanation and modeling of experimental data. Ideally, factors would include the

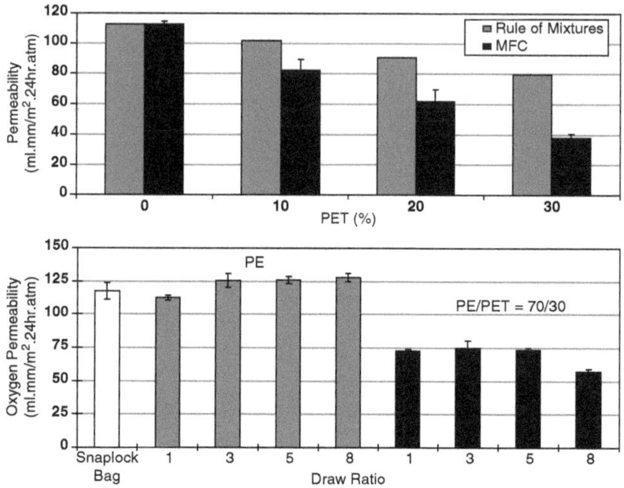

Fig. 4 Effects of PET content and draw ratio on permeability properties

dispersed phase's aspect ratio, orientation, dispersion, shape and volume fraction, as well as the density and crystallinity of the matrix and the affinity between the constituent polymers and diffusing species [5]. Despite all these factors, barrier predictions are often made based upon relatively simple tortuosity arguments that are able to give a good indication of the improvements possible for a particular reinforcement/matrix system without taking many of these complexities into account. Much of the research into permeation modeling of filled polymer systems is based on the work of Nielsen [6]. More recently, Lu and Mai [7] have identified three factors which they described as the main geometric influences on permeability: (i) filler exfoliation/intercalation (and thus aspect ratio); (ii) filler orientation; and (iii) inter-particle dispersion distance.

Taking this into consideration, MFCs becomes a natural choice of barrier material for several reasons. Firstly, in-situ formation of microfibrils remove the need for exfoliation or intercalation of the reinforcing medium while creating a near perfect fibrillar dispersion. Furthermore, an element of control over the filler shape and aspect ratio can be achieved by changing the extrusion and drawing parameters. Finally, manipulation of the film preparation technique can produce films with either well aligned or randomly oriented fibrils.

The limiting cases for permeation represent the two scenarios in which the reinforcing phase is oriented across the direction of permeation (Series Model), Eq. (1), and parallel to the direction of permeation (Parallel Model), Eq. (2). These models provide the lower and upper bounds of permeability modeling respectively.

$$P = \frac{P_m P_r}{\phi_m P_r + \phi_r P_m}, \tag{1}$$

where P denotes permeability, ϕ is the volume fraction and the subscripts $_m$ and $_r$ represent the matrix and reinforcing polymers respectively. The Parallel Model is analogous to the Rule of Mixtures:

$$P = \phi_m P_m + \phi_r P_r \tag{2}$$

A theoretical model developed by Nielsen [8] to predict the elastic modulus of two-phase systems has also been applied to predict permeability, Eq. (3):

$$P = P_m \left[\frac{1 + AB\phi_r}{1 - B\psi\phi_r} \right], \tag{3}$$

where $A = k_e - 1$, and k_e is the Einstein coefficient and is equal to 1.5 for fibres and 2.5 for spheres. Furthermore,

$$B = \frac{\left(\frac{P_r}{P_m}\right) - 1}{\left(\frac{P_r}{P_m}\right) + A} \quad \text{and} \quad \psi = 1 + \left[\frac{1 - \phi_p}{{\phi_p}^2} \right] \phi_r, \tag{4}$$

where ϕ_p represents the maximum fibre packing fraction. This value is 0.785 for square packing, 0.82 for random packing and 0.907 for hexagonal packing.

If one assumes a random distribution of phases, the film permeability can be estimated using the weighted geometric mean of the polymer permeabilities via the Geometric Mean Model:

$$P = {P_m}^{\phi_m} + P_r^{\phi_r} \tag{5}$$

In addition to these, Maxwell [9] developed a theory for the permeability of a polymer filled with a random dispersion of homogeneous spheres. There also exists the Rayleigh relation for cylinders. This differs from Maxwell's equation only by the value of the shape factor defined in the following generalised equation as f, which is equal to 1 for Rayleigh's relation and 2 for Maxwell's relation [10].

$$P = P_m \left[\frac{1 + (f + 1)\left(\frac{P_r}{P_m} - 1\right)\phi_r}{\left(\frac{P_r}{P_m} + f\right) - \left(\frac{P_r}{P_m} - 1\right)\phi_r} \right] \tag{6}$$

Figure 5 combines all the prediction models. It can be seen that the data points fall well within the upper and lower bounds predicted by the Series and Parallel Models. Both the Rayleigh and Maxwell relations appear to over-predict the effective film permeability of the MFCs, as does the Nielsen model for all three fibre packing fractions. The Geometric Mean Model is the closest of the existing models, although it too slightly underestimates the barrier effects of these composite films. Fitting of the general model Eq. (6) to the experimental data points is the best when the shape

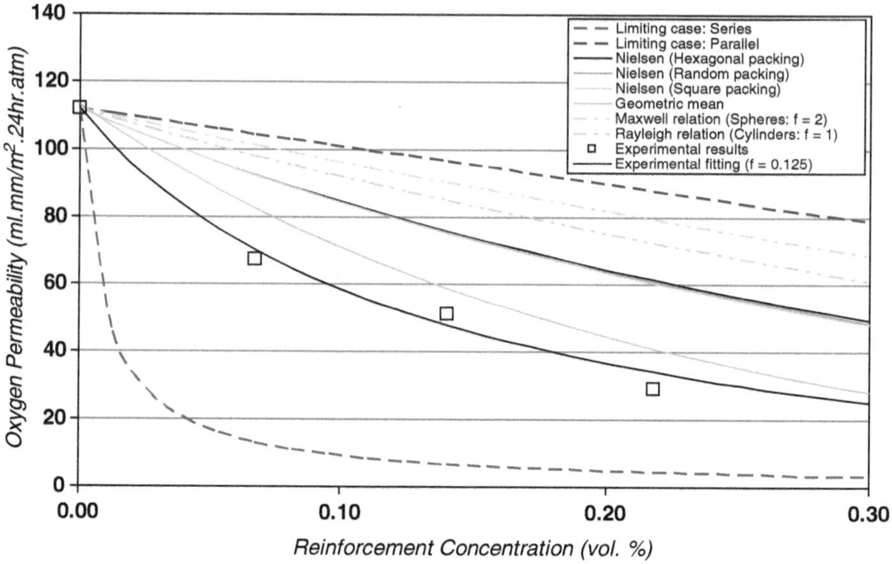

Fig. 5 Applicable permeability models for filled and two phase polymer systems

factor $f = 0.125$. This means that MFC permeability for PE/PET = 70/30 blends is most accurately predicted by:

$$P = P_m \left[\frac{1 + (1.125)\left(\frac{P_r}{P_m} - 1\right)\phi_r}{\left(\frac{P_r}{P_m} + 0.125\right) - \left(\frac{P_r}{P_m} - 1\right)\phi_r} \right] \quad (7)$$

5 Conclusions

It may be concluded from the results presented in this paper that microfibrillar composites from blends of PE/PET have the potential to be well suited to packaging applications that require improved strength, stiffness and oxygen barrier ability when compared to plain PE but are still derived from low-cost, easily processed polymers.

MFCs are manufactured using an extrusion, drawing and matrix consolidation process that creates highly drawn microfibrils homogeneously dispersed throughout a secondary polymer matrix. The formation of elongated microfibres in PE/PET blends right after extrusion only has been demonstrated and discussed. Tensile testing of blend filaments reveals that a ratio of about 3–3.5 is the threshold between undrawn and drawn characteristics. Drawn blends can be as much as ten times stronger and four times stiffer than their undrawn counterparts. Upon conversion to MFCs, directional effects become clear, with uniaxial MFCs displaying much better mechanical properties compared to those of PE matrix and they even rival or exceed the properties of the reinforcing polymer PET.

Finally, the oxygen permeability of MFCs has been measured as significantly superior to that of neat PE, with 30% reinforcement providing almost three times higher barrier. Apparently the draw ratio is of little significance to barrier properties because of the formation of microfibres directly during extrusion. The Geometric Mean Model appears to best describe the permeability properties.

Acknowledgments The authors wish to acknowledge the support of the Foundation for Research, Science and Technology New Zealand for their sponsorship of this work through Grant #UOAX0406. They would also like to acknowledge the experimental help received from Mr. Rex Halliwell.

References

1. Davis EG, Huntington JN. New cell for measuring the permeability of film materials. CSIRO Fd Res. Q. 1977;37:55–59.
2. Shields, RJ, Bhattacharyya D, Fakirov S, Oxygen permeability analysis of microfibril reinforced composites from PE/PET blends, Composites Part A: Applied Science and Manufacturing, 2008;39:940–949.
3. Shields RJ, Bhattacharyya D, Fakirov S, Fibrillar polymer-polymer composites: morphology, properties and applications, Journal of Materials Science, 2008; Vol. 43, 2008; 6758–6770.
4. Seguela, R. On the Natural Draw Ratio of Semi-Crystalline Polymers: Review of the Mechanical, Physical and Molecular Aspects, Macromolecular Materials and Engineering, 2007; 292, 235–244.
5. Massey LK. Permeability properties of plastics and elastomers - a guide to packaging and barrier materials. 2nd ed. William Andrew Publishing, Norwich, 2003.
6. Nielsen LE. Models for the permeability of filled polymer systems. Journal of Macromolecular Science, Part A 1967;1(5): 929–942.
7. Lu C, Mai Y-W. Influence of aspect ratio on barrier properties of polymer-clay nanocomposites. Physical Review Letters 2005;95(8):0883031–0883034.
8. Robeson LM. Gas permeability in polymer blends. In: Simon GP, editor. Polymer characterisation techniques and their application to blends. Oxford University Press: Oxford 2003; 288–312.
9. Maxwell C. Treatise on electricity and magnetism. vol. 1. Oxford University Press, London, 1973.
10. Kamal MR, Jinnah IA, Utracki LA. Permeability of oxygen and water vapor through polyethylene/polyamide films. Polymer Engineering and Science 1984; 24(17):1337–1347.

Section III
Multifunctional Material Systems

Multiscale Computational Analysis of Biomechanical Systems

V.U. Unnikrishnan, G.U. Unnikrishnan, and J.N. Reddy

Abstract The material properties of biological materials, often derived from experiments, are found to vary by orders of magnitude. This disparity in experimentally-derived mechanical properties can be understood only by mathematical models that correlate the structural constituents to its mechanical response. New mechano-biological computational models that consider the effect of microstructural constituents on the response of biological materials are considered in this paper. Various mathematical models are presented to study the macroscopic effects, such as deformation and diffusion in tissues, using multi-scale computational models. The implementation of the computational models for the determination of mechanical behaviour in pathological conditions like cancer progression, cardiovascular diseases, and gynaecological conditions are discussed. The significance of this work lies in the use of a multi-physical modelling of the complex material geometry as well as physical processes representing physiological systems, thereby establishing a suitable and efficient multi-scale computational framework.

Keywords Bio-materials · Biphasic finite elements · Micromechanics · Soft tissue

1 Introduction

Biological materials are complex hierarchical systems subjected to a wide range of external and internal stimuli in the form of mechanical forces, chemical potentials, and electrical signals. Soft tissues like skin, cartilage, and artery constitute an important part of human body performing various vital physiological functions. A deeper understanding of the behavior of these biological materials is required for the response characterization of healthy and diseased tissues as well as for the design of

V.U. Unnikrishnan, G.U. Unnikrishnan, and J.N. Reddy (✉)
Texas A&M University, College Station, TX 77843-3123, USA
e-mail: unnithan@tamu.edu; ginu@tamu.edu; jnreddy@tamu.edu

B. Dattaguru et al. (eds.), *IUTAM Symposium on Multi-Functional Material Structures and Systems*, IUTAM Bookseries 19, DOI 10.1007/978-90-481-3771-8_13,

artificial organs. In recent times, the variation in the mechanical response properties of cells and tissues are used to determine the pathological condition, as in the detection of malarial infection through mechanical extension test and compression of tissues for tumor detection. In this paper, we present mathematical models to predict the behavior of normal and diseased tissues to help in developing diagnostic tools with close integration with experimental methods. To this effect, the major emphasis is placed in providing constitutive models for the soft tissue and also on its application to capture the healthy and diseased conditions.

2 Modeling of Soft Tissue

Soft tissues are anisotropic non-homogenous materials having their properties dependent on external stimuli as well as physiological activities like growth, diseases, etc. The structural makeup of soft tissues like artery and cartilage essentially consists of a solid phase and a fluid phase dispersed with varying volume fractions of solutes. Previous works on soft tissues have used nonlinear elastic solid, viscoelastic solid and multiphase material models for mechanical analysis. The composition of soft tissues varies according to the type of tissue, its location and differs from person to person [1]. In general, a soft tissue consists of fibrous materials like collagen and elastin and amorphous materials like hydrated matrix of proteoglycans, and it can be treated as a fiber reinforced composite material. Diseases like tumor induce drastic changes in the physiological characteristics of the tissue and alter its overall mechanical properties. Constitutive modeling of soft tissues by microstructure based computational models offers more flexibility in modeling normal as well as pathological conditions of various tissues. In this study, microstructure based modeling of the tissues is carried out considering the distinct fluid and solid phase and 'theory of mixtures' approach.

2.1 *Biphasic Soft Tissues*

Even though various finite element studies have been carried out using biphasic models, their applicability across different boundary conditions are not investigated due to varying boundary conditions for different tissues. In this study, a finite element model based on biphasic continuum formulation is developed to study the mechanical response of tissues in the presence of fluid flow. Assuming that the biphasic material is composed of intrinsically incompressible constituents that are chemically inert, a set of governing equations are derived, as described next.

Consider a domain Ω with closed boundary Γ and total volume V, which is the sum of the fluid volume V^f and solid volume V^s. The fluid is assumed to be viscous and incompressible, while the solid is assumed to be linearly elastic and isotropic. The volume fractions of the fluid and solid phases are represented as ϕ^f and ϕ^s, respectively, where $\phi^\alpha = V^\alpha/V$, $\alpha = s, f$ refers to the solid and fluid phases, respectively. It is to be noted that $\phi^s + \phi^f = 1$. The continuity equation for the biphasic material is

$$\nabla.\left(\phi^f \mathbf{v}^f + \phi^s \mathbf{v}^s\right) = 0 \tag{1}$$

Neglecting inertia and body forces, the principle of conservation of linear momentum yields the following equations:

$$\begin{aligned}
&\nabla.\boldsymbol{\sigma}^s + \Pi^s = 0\\
&\nabla.\boldsymbol{\sigma}^f + \Pi^f = 0\\
&\Pi^s = -\Pi^f = -p\nabla\phi^f - K\left(\mathbf{v}^s - \mathbf{v}^f\right)
\end{aligned} \tag{2}$$

where $\boldsymbol{\sigma}^\alpha$ is the Cauchy stress tensor of α phase, Π^α is the diffusive momentum exchange between the two phases, p the apparent pressure, $\mathbf{v}^\alpha$ is the velocity field vector, and K is the diffusive drag coefficient [2].

2.2 *Fluid Tissue Interface Modeling*

Fluid flow and soft tissue interactions are very common in the field of biomechanics, with common examples being the blood flow through an artery, and synovial fluid interaction with cartilage. Theoretical study of tissue-fluid interaction is complex due to the presence of nonlinearity of the tissue and the fluid and their interactions. A method to treat fluid–solid coupling is to use the theory of mixtures approach, which treats tissues as fluid–solid mixtures. The solid phase of the tissue consists of solid organic matrix having elastic fibers in a gel of proteoglycans and cells, and the liquid phase is predominantly water. The fluid–solid mixture model of a tissue assumes the coexistence of solid and fluid phases at a point. The physical phenomenon of the mixture is governed by the balance equations of the individual constituents with an additional term influencing the interactions between the phases.

In the numerical solutions of fluid-structure interaction, the solid and fluid sub-domains are decoupled and the interactions are treated as external boundary conditions on the individual sub-domains. In the finite element model considered in this study, the entire domain is assumed to be *regular*, i.e. all the field variables are smooth and continuous. In a tissue-fluid computational domain a *singular* surface relative to the fluid phase volume fraction occurs at the interface between the solid and tissue sub-domain. A *singular* surface relative to ϕ^f occurs when ϕ^f is smooth in the tissue and fluid domains and suffers a *jump* at the interface. A *jump*, relative to ϕ^f is defined as $[[\phi^f]] = \phi_T^f - \phi_F^f$, where ϕ_T^f and ϕ_F^f are the one-side limits for fluid volume fraction from the tissue and fluid sub-domain separated by the interface surface. For an incompressible solid and incompressible fluid phase, the discontinuity ϕ^f introduces the jump condition on conservation of mass at interfacial surface as $[[\phi^f\left(\mathbf{v}^f - \mathbf{v}^s\right)]] \cdot \hat{\mathbf{n}} = 0$ (see [3]), where, $\hat{\mathbf{n}}$ is the unit outward normal of the interface surface. A "pseudo-no slip" boundary condition, which allows different velocities for the solid and fluid phase at the interface can be defined [3]. This leads to the following jump condition for the mass conservation on the surface:

$$\left[\left[\phi^f \mathbf{v}^f + \phi^s \mathbf{v}^s\right]\right] = 0 \tag{3}$$

For a tissue-fluid interface, the continuity equation would give the following kinematic boundary condition at the interface region:

$$\mathbf{v}_F^f = \left(\phi_T^f \mathbf{v}_T^f + \phi_T^s \mathbf{v}_T^s\right) \text{ at } \Gamma_I \tag{4}$$

where subscripts F and T denote the fluid and tissue sub-domains, respectively. The above kinematic boundary condition shows that the momentum of the fluid in the fluid sub-domain is transferred to the solid and fluid phase in the tissue subdomain. In this study, a new approach to satisfy Eq. (4) based on the physiologically observed transition between the fluid and surrounding tissue is developed. We assume that the solid phase volume fraction in the fluid and tissue sub-domain is smooth, and

$$\lim_{(x,y)^T \to \Gamma_I^T} \phi_T^s = 0 \tag{5}$$

where, $(x, y)^T$ denotes the coordinates in the tissue sub-domain. Thus, in the limit as we approach the interface surface Γ_I^T from the tissue sub-domain, the above assumption leads to the continuity equation over the boundary surface as

$$\lim_{\phi_T^s \to 0|_{\Gamma_I^T}} \left(\phi_T^f \mathbf{v}_T^f + \phi_T^s \mathbf{v}_T^s\right) = \mathbf{v}_T^f \tag{6}$$

Substitution of Eq. (6) into Eq. (4) yields the boundary condition for fluid velocity at the interface, $\mathbf{v}_T^f = \mathbf{v}_F^f$ at Γ_I.

2.3 *Blood Flow Through Atherosclerotic Artery*

Atherosclerosis is a common form of disease affecting large and medium sized arteries. The formation of atherosclerosis is primarily caused by the blood flow conditions and arterial wall properties, which induce uptake of atherogenic substances into the artery wall. When flowing through a narrowed artery, the blood velocity increases leading to a negative pressure in the constricted region. Thus the study of blood flow through stenotic artery requires coupling of fluid and arterial wall mechanics [4, 5].

The complex mechanical formation of the artery wall is simplified and modeled in this work as an isotropic bilayer having an inner media and outer adventitia layers.

The axial length of the domain is taken as 0.085 m, with the stenotic region modeled at a distance of 0.035 m from the inlet. The thickness of the wall is assumed to be 0.4 cm with equal thickness for media and adventitia. The porosity of inner layer is taken as 0.25 and outer layer as 0.05. An elastic modulus of 67.5 kPa and 6.75 GPa is assumed for the inner and outer layer, respectively. Viscosity of fluid in the arterial wall is taken as 0.72×10^{-3} Pa.s and the blood is assumed to be viscous and incompressible. To analyze the influence of stenosis in the blood flow, simulation is carried out for various stenotic blocks. The variation of fluid velocity and solid displacement and pressure drop due to the blocks are noted.

A linear quadrilateral element with 4 degrees of freedom (two velocities and two volume fractions) per node is used in an isoparametric formulation. The tissue-fluid interface lies along an interelement boundary region and a finer mesh is provided at the entrance and at the stenotic region of the artery. A tolerance limit of 0.001 was used for convergence in the analysis. At the lumen inlet, a fully developed velocity profile is prescribed with a central velocity of 0.24 m/s, and at the outlet of the lumen, free boundary conditions are given. The arterial wall was constrained in both directions at the two ends. The axial velocity profile of the blood for the various cases of blocks is shown in Fig. 1. The white dotted line shows the interface region between the fluid and arterial wall. With a change in the stenosis the velocity profile is significantly altered in the lumen. The maximum velocity increases with an increase in the degree of block. Also, it is observed that the fluid velocity for 60% block has negative velocities in the post-stenosis region showing fluid re-circulation. The recirculation effects of blood flow alter the flow and increase the intensity of atherosclerosis through other physiological events.

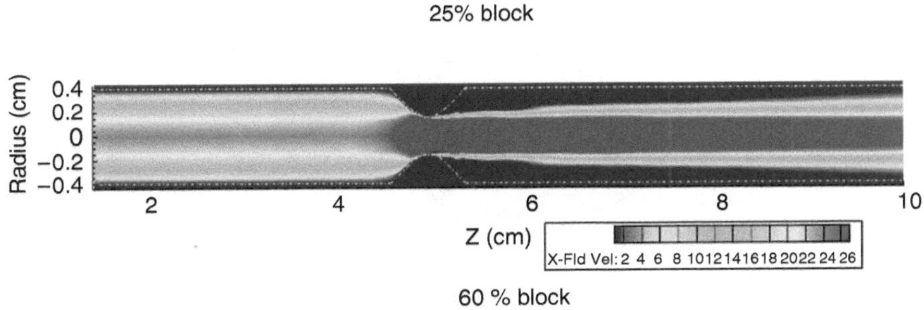

Fig. 1 Axial fluid velocity profile (cm/s) for arterial tube with various degrees of constrictions

3 Constitutive Modeling of Solid Tumor

Solid tumors are abnormal mass of tissues that do not contain cysts or liquid areas, having distinct mechanical properties compared to normal healthy tissues. Diagnostic procedures like image guided tools, ultrasound procedures, magnetic resonance imaging, are clinically used to monitor differential tissue characteristics, like mechanical, electrical properties, to detect the presence of solid tumors in healthy tissues [6, 7]. In image guided detection of tumor, the changes in the appearance of the tumors in a medical image data is used for diagnosis, while in ultrasound imaging (US) tumors are detected from the difference in the acoustic properties of normal and cancerous tissues. In the mechanical diagnostic procedures, like clinical palpation of breast and liver tissues, external mechanical force is used to identify tumor by analyzing the qualitative variation in the tissue stiffness. These methods are restricted to masses which are comparatively very stiff, large and superficial in nature [8]. Combination of imaging tools and mechanical diagnostic procedures has motivated the development of diagnostic tools by deriving tissue elastic modulus in a quantitative manner. To accurately predict the mechanical parameters like strains or displacements for complex geometry and hierarchical material like tissues, computational methods like the finite element method (FEM) are being extensively used [7]. Recent advances in the imaging of biological tissues have considerably improved the accuracy in reproducing complex geometric features of tumor for mathematical analysis. With the geometrical model developed, the next step is to develop appropriate constitutive model of tumor tissues.

To develop the constitutive relationship it is required to identify the tumor tissue microstructure. Studies have shown that the tumor tissues are heterogeneous

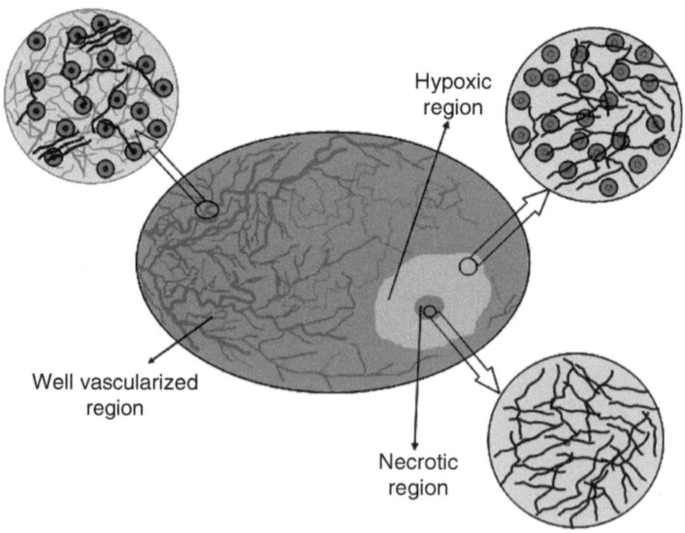

Fig. 2 Schematic representation of vascularised tumor

structures, consisting of a distinct inner core of dead cells, an outer rim of highly proliferating cells, and an intermediate region of non-proliferating cells with a non-uniform distribution of vascular tissues spanning the entire tumor. The outer proliferating region is well vascularized having abnormal vessels with increased branching and tortousity, the inner core and central portions of tumors are often hypoxic in nature [9] (see Fig. 2).

4 Biomechanical Analysis of Uterus During Parturition

A Cesarean delivery is a frequently used surgical procedure of parturition where an incision is made in the abdominal wall through the uterus. This procedure is carried out at the lower uterine segment and this procedure has gained wide acceptance because of the lower blood loss and fewer uterine rupture during subsequent pregnancies. However, rupture of the uterine segment during future pregnancies has often resulted in severe maternal and fetal morbidity and therefore more precise procedures needs to be developed. It is for the development of these medical procedures that mechanics-based theoretical studies can be used to improve the efficacy and optimization that could lead to effective and safe medical practices. In this study, mechanical stiffness and integrity of the scar tissue is developed for the uterine myometrium to counter the forces generated by the growing fetus. The ultimate aim of such a study is to develop diagnostic tools for the determination of the probability of uterine rupture in critical situations like during a vaginal birth after cesarean delivery (VBAC).

4.1 Mathematical Homogenization of Myometrial Tissue

As size of uterus increases considerably during pregnancy to accommodate the growing fetus, there is a corresponding change in the elastic properties. Mechanical modeling of tissues would be extremely useful in deriving mechanical properties of tissues from experimental setup. In vivo response of these tissue samples would require mathematical modeling of the embedding environment like fibers and fluid making the ECM. Various approximations techniques, like variational bounds or mean field method, are invoked to obtain the concentration tensors. The variationally consistent Hashin–Shtrikman bounds are one of the best possible bounds for macroscopically homogeneous, isotropic, two-phase materials. The Hashin-Shtrikman (H-S) upper bounds for the shear and bulk modulus of a two-phase cylindrical composite materials are given by

$$K_{HS}^{-}=K_0+\frac{v_1}{\frac{1}{K_1-K_0}+\frac{1-v_1}{K_0+G_0}}; \quad G_{HS}^{-}=G_0+\frac{v_1}{\frac{1}{G_1-G_0}+\frac{(K_0+2G_0)(1-v_1)}{2G_0(K_0+G_0)}} \tag{7}$$

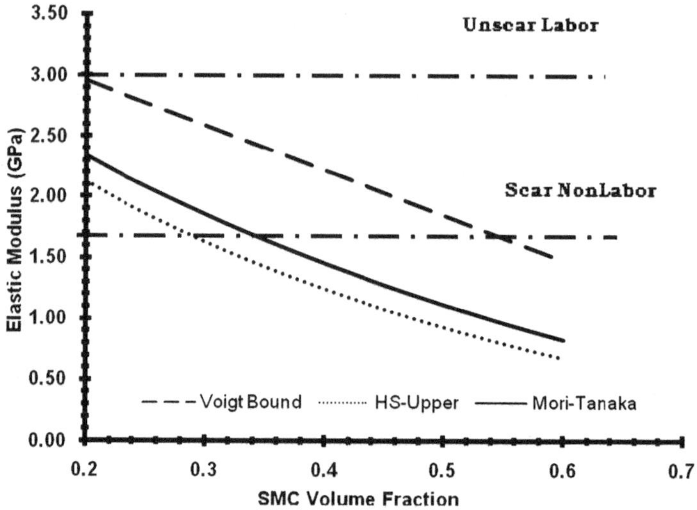

Fig. 3 Variation of elastic modulus of myometrium with collagen volume fractions (with respect to SMCs) and compared with the modulus from clinical study

$$K^+_{HS} = K_1 + \frac{1 - \nu_1}{\frac{1}{K_0 - K_1} + \frac{\nu_1}{K_1 + G_1}}; \quad G^+_{HS} = G_1 + \frac{1 - \nu_1}{\frac{1}{G_0 - G_1} + \frac{(K_1 + 2G_1)\nu_1}{2G_1(K_1 + G_1)}} \tag{8}$$

where K^+_{HS} & G^+_{HS} are the H-S upper and K^-_{HS} & G^-_{HS} are the H-S lower bounds of the bulk and shear modulus respectively.

Let us consider a cylindrical RVE of myometrium with SMC as the matrix phase and collagen as the fiber phase. The averaged modulus of collagen is taken as 3.7 GPa, and smooth muscle cells is taken as 1.8 kPa with an average Poisson's ratio as 0.3. The variation of the elastic modulus of the myometrium by the Voigt bound, the Hashin-Shtrikman upper bounds and the Mori-Tanaka homogenization with respect to various volume fractions of the collagen are compared in Fig. 3 with the clinical studies by Buhimschi et al. [10], where the elastic stiffness of scarred and unscarred myometrium in patients with and without labor was studied. It can be seen that the variation of the effective properties of the myometrium is a function of the amount of collagen in the tissue. This variation of the volume of the collagen was found to the primary reason for the difference in the stiffness of the uterine tissue for various physiological conditions [11].

5 Conclusions

Constitutive models for normal and diseased soft tissues are developed and a biphasic finite element model to analyze the behavior of blood through a constricted artery is presented. Considerable negative pressure, negative displacement and flow separation is found to occur at downstream regions of the stenosis. These observed characteristics have a direct relation on the behavior of the plaque, and ultimately responsible for the plaque cap to rupture, leading to arterial blockage. A multiscale modeling strategy to obtain the effective elastic modulus of the scarred and unscarred myometrium is also presented. This study also highlights the biomechanical analysis on the effect of scarring tissue formed as a result of wound healing of a uterine tissue. These studies have been shown to be good diagnostic tools in the determination of conditions like tumor tissue, artery disease and uterine rupture, and also for the determination of sites for possible incisions during assisted labor.

Acknowledgments The research reported herein was supported by the Oscar S. Wyatt Endowed Chair account at Texas A&M University, College Station.

References

1. Mow VC, Holmes MH, Lai WM (1984) Fluid transport and mechanical properties of articular cartilage: A review. Journal of Biomechanics 17: 377–394
2. Spilker RL, A MT (1990) A mixed-penalty finite element formulation of the linear biphasic theory for soft tissues. International Journal for Numerical Methods In Engineering 30: 1063–1082
3. Hou JS, Holmes MH, Lai WM, Mow VC (1989) Boundary conditions at the cartilage-synovial fluid interface for joint lubrication and theoretical verifications. Journal of Biomechanical Engineering 111: 78–87
4. Tang D, Yang J, Yang C, Ku DN (1999) A nonlinear axisymmetric model with fluid-wall interactions for steady viscous flow in stenotic elastic tubes. Journal of Biomechanical Engineering 121: 494–501
5. Theodorou G, Bellet D (1986) Laminar flows of a non-Newtonian fluid in mild stenosis. Computer Methods in Applied Mechanics and Engineering 54: 111–123
6. Sciarretta J, Samani A, Bishop J, Plewes DB (2002) MR validation of soft tissue mimicing phanton deformation as modeled by nonlinear finite element analysis. Med Phys 29: 65–72
7. Plewes DB, Bishop J, Samani A, J S (2000) Visualization and quantification of breast cancer biomechanical properties with magnetic resonance elastography. Physics in Medicine and Biology 45: 1591–1610
8. Newcomb P, Weiss S, Storer B, Scholes D, Young B (1991) Breast self examination in relation to the occurrence of advanced breast cancer. J Natl Cancer Inst 83: 260–265
9. Kuszyk BS, Corl FM, Franano NF, Bluemke DA, Hofmann LV, Fortman BJ, Fishman EK (2001) Tumor transport physiology: Implications for imaging and imaging-guided therapy. American Journal of Roentgeology 177: 747–753
10. Buhimschi CS, Buhimschi IA, Yu C, Wang H, Sharer DJ, Diamond MP, Petkova AP, Garfield RE, Saade GR, Weiner CP (2006) The effect of dystocia and previous cesarean uterine scar on the tensile properties of the lower uterine segment. American Journal of Obstetrics and Gynecology 194: 873–883
11. Buhimschi CS, Buhimschi IA, Patel S, Malinow AM, Weiner CP (2005) Rupture of the uterine scar during term labour: contractility or biochemistry? BJOG: An International Journal of Obstetrics and Gynaecology 112: 38–42 DOI:10.1111/j.1471-0528.2004.00300.x

Effect of Magnetic-Field on Stress–Strain Behavior of Magneto-Sensitive Elastomers

Prashant Kumar Srivastava and Kartik Venkatraman

Abstract Magneto-sensitive materials are a class of active materials whose mechanical properties are sensitive to an externally applied magnetic field. These materials exhibit non-linear and elastic/viscoelastic behavior and can undergo large deformation. In the present work, we study the effect of the external magnetic field on dynamic stress–strain behavior of a system consisting of linear polymer chains with uniformly dispersed ferrite particles in it. Molecular dynamics simulation is used as a numerical tool for this study. At moderate magnetic field strengths, change in the constitutive behavior is not significant, but as we increase the field strength, material becomes stiffer. We also study the effect of the magnetic particles on constitutive behavior by varying the volume fraction of the same.

Keywords Hysterisis · Magneto-sensitive elastomers · Molecular dynamics · Viscoelasticity

1 Introduction

Magneto-sensitive elastomers are class of active materials which exhibit instantaneous change in their mechanical behavior when subjected to external magnetic field. These materials typically consist of a polymer embedded with magnetic particles. Volume fraction of magnetic particles varies between 10% and 50% and size of the particles range from 0.1 to 10 μ. The sensitivity of these materials to the external magnetic field depends on the interaction of the magnetic particles among themselves and their interaction with the united atoms in the polymer chain. Magnetic particles have the tendency to align themselves in the direction of the field. As a result an attractive force acts between all pairs of magnetic particles and hence material becomes stiffer. This phenomenon is called magneto striction. In addition to this, these materials also show viscoelastic behavior due to relaxation in the configuration of the polymer chains. The molecular structure of the polymer chain and

P.K. Srivastava (✉) and K. Venkatraman
Department of Aerospace Engineering, Indian Institute of Science, Benguluru, India
e-mail: prashant@aero.iisc.ernet.in; kartik@aero.iisc.ernet.in

B. Dattaguru et al. (eds.), *IUTAM Symposium on Multi-Functional Material Structures and Systems*, IUTAM Bookseries 19, DOI 10.1007/978-90-481-3771-8_14,

their flexibility in changing the configuration allows them to undergo non-linear and large deformation. Due to their wide variety of properties and increasing number of applications in active controls, it is of interest to study the behavior of the same.

Maxwell and Kelvin-Voigt [11, 12] models are most commonly used in understanding the viscoelastic behavior of the materials. But they do not provide the relation between the internal structure and the constitutive behavior. A little detail description is given in continuum mechanics models [3, 8] where the constitutive relation is expressed in terms of free energy. But these models suffer from the arbitrary forms of the free energy function. Few statistical mechanics models [4] such as freely jointed chain model, reptation model, etc., do provide the correlation of the internal structure to their constitutive behavior, but are applicable to specific cases alone. Also, these models often suffer from assumptions that are physically non-admissible.

In spite of the limitations of molecular dynamic (MD) simulation in terms of spatial and temporal scales, it is known to capture the evolution of the molecular structure and changes in the behavior of the material due to change in the configuration. The extended ensemble dynamics introduced by Anderson [1] enabled molecular dynamic simulation to simulate the behavior of the system in constant pressure ensemble. Further Nosé [9] prescribed an extended ensemble method to fix the temperature. Yang et al. [13] studied the stress–strain behavior of the amorphous glass under uniaxial extension. Further study on the stress–strain behavior of the polymeric network was done by Bergström and Boyce [2] where they compared the result of molecular simulation with classical statistical mechanics model of rubber elasticity. Zhu et al. [14] developed a multi-chain model to analyze the effect of external field on shear modulus. A dissipative particle dynamics (DPD) based computer experiment was performed by Raos et al. [10] to study the viscoelastic properties of rubber with filler particles. They found that modulus increases with the addition of filler particles and also depends on the strain.

The present work is a study of the effect of external magnetic field on the stress–strain behavior of the magneto-sensitive elastomers. In addition, effect of the volume fraction of the magnetic particles on the constitutive behavior of the material is also studied. The stress–strain simulations are all uniaxial. Different types of dynamic strain histories are imposed on the magneto-sensitive elastomers and the resultant stress is computed. The effect of magnetic particles and applied magnetic field on the dynamic stress–strain behavior is investigated. The results show significant effect of the magnetic field on the stress–strain behavior.

2 Mathematical Modeling

Use of MD simulation in Canonical (NVT) ensemble requires a modification in the equations of motion. This is achieved by introducing additional degree of freedom [7, 9] in the equations of motion.

$$m_i \frac{d^2 \mathbf{r}_i}{dt^2} = -\frac{\partial U}{\partial \mathbf{r}_i} - \eta \mathbf{v}_i, \quad i \in [1, N] \tag{1}$$

where m_i is the mass of the united atom i and $\mathbf{r}_i$ is the position, $\mathbf{v}_i$ is the velocity and U is the total interaction potential. η is a parameter used to control the kinetic energy of translational degrees of freedom. It evolves as

$$\dot{\eta} = \frac{1}{Q}\left[\sum_{i=1}^{N} m_i \mathbf{v}_i^2 - f k_B T\right], \tag{2}$$

where f is the number of translational degrees of freedom of the system and Q is the kinetic mass. Taking the analogy of the equations of motion for translational degrees of freedom (Eq. 1), we write the equations of motion for the rotational degrees of freedom of the magnetic particles as

$$\mathscr{I}_j \dot{\boldsymbol{\omega}}_j = \Gamma_j - \zeta \omega_j, \; j = 1, 2, \ldots, N_m, \tag{3}$$

where Γ_j is the torque on the particle j. $\mathscr{I}_j$ is the moment of inertia tensor of the magnetic particle j and $\boldsymbol{\omega}_j$ is the angular velocity of the same. In this study the magnetic particles are assumed to be spherical and hence moment of inertia tensor is diagonal with equal diagonal elements. ζ is a dynamic parameter similar to η to control the kinetic energy of the rotational degrees of freedom in order to satisfy the equipartition theorem. The evolution of ζ is given by

$$\dot{\zeta} = \frac{1}{Q}\left[\sum_{j=1}^{N_m} \boldsymbol{\omega}_j \cdot \mathscr{I}_j \boldsymbol{\omega}_j - 3N_m k_B T\right] \tag{4}$$

The Velocity Verlet algorithm is the most commonly used algorithm to integrate the equations of motion in MD and is given by

$$\mathbf{r}_i(t + \delta t) = \mathbf{r}_i(t) + \delta t \mathbf{v}_i + \frac{\delta t^2}{2}\mathbf{a}_i(t), \tag{5}$$

$$\mathbf{v}_i(t + \delta t) = \mathbf{v}_i(t) + \frac{\delta t}{2}[\mathbf{a}_i(t) + \mathbf{a}_i(t + \delta t)]. \tag{6}$$

Constraining the temperature in the simulation imposes a constraint on the direct use of the Velocity Verlet algorithm. Hence, the Velocity Verlet algorithm is modified in the following way in which Eq. (1) is written as

$$m_i \mathbf{a}_i(t + \delta t) = -\frac{\partial U(t + \delta t)}{\partial \mathbf{r}_i} - \eta(t + \delta t)\mathbf{v}_i(t + \delta t). \tag{7}$$

Solving Eqs. (6) and (7) we get

$$\mathbf{v}_i(t + \delta t) = \left(1 + \frac{\eta(t + \delta t)\delta t}{2m_i}\right)^{-1}\left\{\mathbf{v}_i(t) - \frac{\delta t}{2m_i}\frac{\partial U(t + \delta t)}{\partial \mathbf{r}_i} + \frac{\delta t}{2}\mathbf{a}_i(t)\right\} \tag{8}$$

Following the similar steps for the rotational degrees of freedom of a linear molecule

$$\mathbf{s}_i(t+\delta t) = \mathbf{s}_i(t) + \delta t\, \dot{\mathbf{s}}_i(t) \tag{9}$$

with

$$\dot{\mathbf{s}}_i(t) = \mathbf{s}_i(t) \times \boldsymbol{\omega}_i(t) \tag{10a}$$

$$\boldsymbol{\omega}_i(t+\delta t) = \boldsymbol{\omega}_i(t) + \frac{\delta t}{2}(\dot{\boldsymbol{\omega}}_i(t) + \dot{\boldsymbol{\omega}}_i(t+\delta t)) \tag{10b}$$

where **s** is the orientation of the dipole. Modified Velocity Verlet applied to the rotational degrees of freedom results in

$$\boldsymbol{\omega}_i(t+\delta t) = \left(\mathbf{I} + \frac{\delta t}{2}\zeta(t+\delta t)\mathscr{I}^{-1}\right)^{-1} \left\{\boldsymbol{\omega}_i(t) + \frac{\delta t}{2}\dot{\boldsymbol{\omega}}_i(t) + \frac{\delta t}{2}\mathscr{I}^{-1}\boldsymbol{\Gamma}_i(t+\delta t)\right\} \tag{11}$$

and

$$\mathcal{I}\dot{\omega}_i(t+\delta t) = \Gamma_i(t+\delta t) - \zeta(t+\delta t)\omega_i(t+\delta t) \tag{12}$$

Then the equations of motion are integrated in the manner given by Bergström and Boyce [2] for both rotational as well as translational degrees of freedom.

3 Simulation

Initial configuration to start the simulation is obtained by taking the uniformly spaced points in the simulation box as the center of mass of each polymer chain. Chain growth is performed by generating the random bond vectors by varying the dihedral angle randomly between 0° and 360° from the position of previous atom in the chain. Bond lengths and bond angles are kept at their mean values to have a initial structure very close to the one corresponding to minimum energy configuration. Magnetic particles are randomly dispersed in the system while performing the check for the occupancy of the sites. In random growth of the chain, partial overlap of the monomers is allowed and is later removed by subjecting them to a soft potential. Velocities of the united atoms is initialized to follow Gaussian distribution. Accelerations of all the united atoms are initially set to zero. To model the non-bonded interaction we use the shifted and truncated Lennard-Jones potential which acts between all pairs of monomers

$$U_{LJ}(r) = \begin{cases} 4\varepsilon_{LJ}[(\frac{\sigma}{r})^{12} - (\frac{\sigma}{r})^{6} - (\frac{\sigma}{r_c})^{12} + (\frac{\sigma}{r_c})^{6}] & \text{if } r < r_c \\ 0 & \text{otherwise} \end{cases} \tag{13}$$

Here $\sigma = 0.401$nm is the LJ length parameter and $\varepsilon = 0.468$kJ/mol is the LJ energy parameter. To describe the different types of bonded interactions, we use bond stretching, bond bending and bond torsional potentials. To model the bond stretching in the elastomeric networks we use finitely extended non-linear elastic (FENE) potential

$$U_{FENE}(r) = \begin{cases} -\frac{1}{2}kR_0^2 \ln\left[1-\left(\frac{r}{R_0}\right)^2\right] & \text{if } r < R_0 \\ \infty & \text{otherwise} \end{cases} \tag{14}$$

Here $k = 30\varepsilon/\sigma^2$ and $R_0 = 1.5\sigma$. Bond bending potential is function of the bond angle θ

$$U_{bend}(\theta) = \frac{k_\theta}{2}(\cos\,\theta - \cos\,\theta_0)^2, \tag{15}$$

where $k_\theta = 520$ kJ/mol is the bond bending stiffness and $\theta_0 = 110°$. The torsional potential is a four body potential and is function of dihedral angle ϕ

$$U_{torsion}(\phi) = \sum_{l=0}^{3} w_l \cos^l\,\phi, \tag{16}$$

where $w_0 = 14.477$ kJ/mol, $w_1 = -37.594$ kJ/mol, $w_2 = 6.493$ kJ/mol and $w_3 = 58.499$ kJ/mol. Magnetic interaction among the magnetic particles is modeled by classical dipole-dipole interaction [6]

$$U_m(r_{ij}) = \frac{\mu_0}{4\pi}\frac{1}{r_{ij}^3}\left[(\boldsymbol{\mu}_i \cdot \boldsymbol{\mu}_j) - 3(\hat{\mathbf{r}}_{ij} \cdot \boldsymbol{\mu}_i)(\hat{\mathbf{r}}_{ij} \cdot \boldsymbol{\mu}_j)\right], \tag{17}$$

where $\boldsymbol{\mu}$ is the magnetic moment vector and $\hat{\mathbf{r}}_{ij}$ is the unit vector along the line joining the two particles i and j. Further a potential due to externally applied magnetic field is also added in the interaction

$$U_B = -\boldsymbol{\mu} \cdot \mathbf{B}, \tag{18}$$

where $\mathbf{B}$ is the externally applied magnetic field. Total torque on a magnetic particle is the sum of torque due to externally applied magnetic field and torques applied by all other magnetic particles present in the system

$$\mathbf{n_i} = \boldsymbol{\mu}_i \times \mathbf{B} + \sum_{j,j\neq i} \frac{\mu_0}{4\pi}\frac{1}{r_{ij}^3}[3(\hat{\mathbf{r}}_{ij} \cdot \boldsymbol{\mu}_j)(\boldsymbol{\mu}_i \times \hat{\mathbf{r}}_{ij}) - \boldsymbol{\mu}_i \times \boldsymbol{\mu}_j] \quad i, j = 1, 2, \ldots, N_m. \tag{19}$$

Calculation of stress is done by using the *virial theorem* [5] of statistical mechanics. For a system of atoms interacting with two body potentials the expression for the stress is given by

$$\boldsymbol{\tau} = -\frac{1}{V}\left[(Nk_BT)\mathbf{I} - \left\langle \sum_{m>n} \frac{1}{|\mathbf{r}|_{mn}}\frac{\partial U}{\partial |\mathbf{r}|_{mn}} \mathbf{r}_{mn} \otimes \mathbf{r}_{mn} \right\rangle\right] \quad m, n = 1, 2, \ldots, N \tag{20}$$

where τ is the Cauchy stress tensor, k_B is the Boltzmann constant, T is the temperature, $\mathbf{I}$ is unit tensor and V is the current volume of the simulation cell.

4 Results and Discussion

Figure 1 shows the response to the constant strain rate loading. Stress in the initial part of the loading increases non-linearly and subsequently becomes almost constant. This is because in the initial part of the deformation, in addition to the unfolding of the chains, bond deformations also take place, but, in the later part of the loading unfolding of the chains dominates over the bond deformations and hence stress does not increase much in this phase of deformation.

We take the stress response function, G, in the form of Prony series

$$G = \sum_{i=1}^{N} g_i e^{-\frac{t}{\tau_i}} \tag{21}$$

Then, for the system loaded with constant strain rate and with zero initial strain, stress response could be written as [12]

$$\sigma(t) = 3\dot{\varepsilon}_0 \int_0^t G(\tau) d\,\tau = 3\dot{\varepsilon}_0 \sum_{i=1}^{N} g_i \tau_i \left(1 - e^{-\frac{t}{\tau_i}}\right) \tag{22}$$

From the $\sigma(t)$ values obtained from the MD simulation we fit the values of g_i and τ_i. Strain–strain curve fit by using the above method is shown in Fig. 2. We find that the modulus is higher in the initial part of the loading in case of large magnetic field. This is more clearly seen in modulus vs. strain curve shown in Fig. 3. In the later part of the loading the modulus is lower in the case of large magnetic field because the deformation is dominated by the uncoiling of the chains.

Figure 4 shows the response to the sinusoidal strain loading with and without magnetic particles. We clearly see that in the presence of the magnetic particles material becomes stiffer. In addition to this, we find that there is phase lag between strain and stress which affirms the viscoelastic nature of such materials.

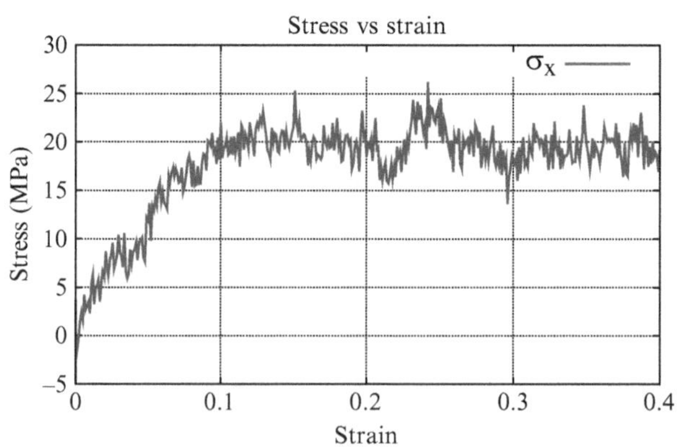

Fig. 1 Constant strain rate loading with $\dot{\varepsilon} = 5 \times 10^8$

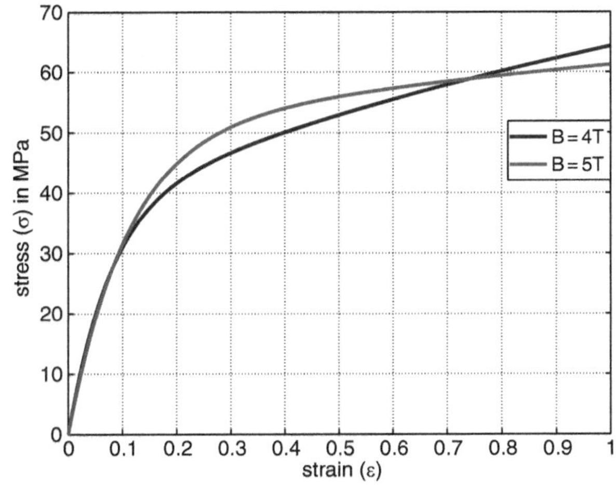

Fig. 2 Effect of magnetic particles: Stress vs. strain

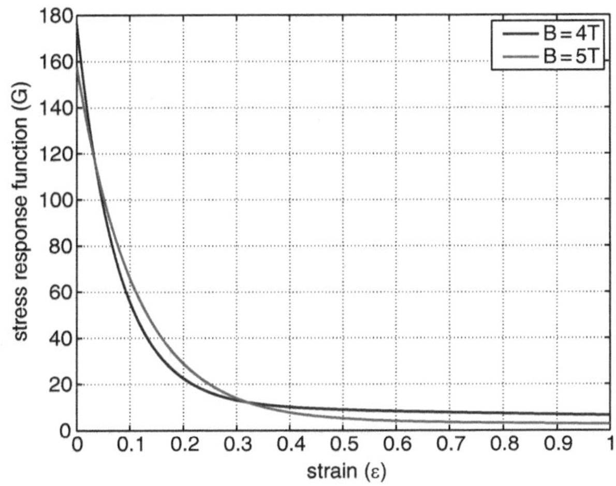

Fig. 3 Effect of magnetic particles: Modulus vs. strain

Further the system is subjected to moderate magnetic field. Response in Fig. 5 shows that there is no significant difference in the stiffness in two cases which is because at moderate field strength the thermal energy of the magnetic particles dominates over the energy associated with the magnetic interaction.

Further we increase the field strength to 4.0 T and corresponding response is shown in Fig. 6. Here we find that in the presence of the magnetic field the material becomes stiffer.

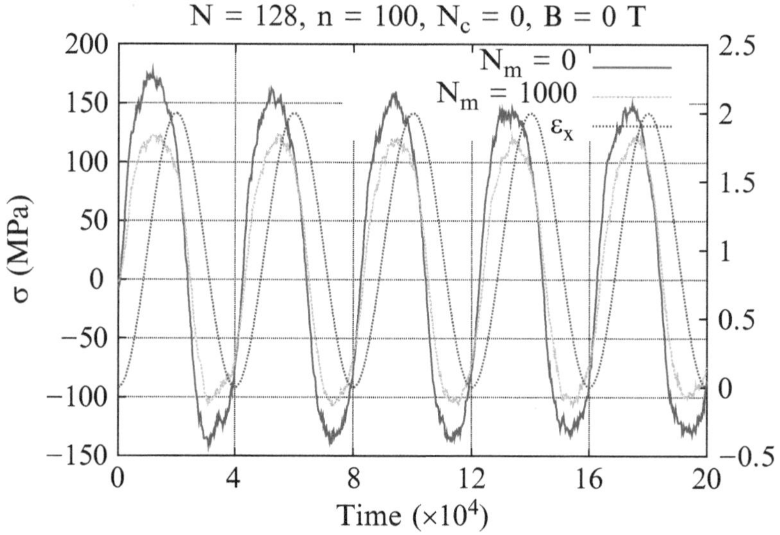

Fig. 4 Effect of magnetic particles: sinusoidal response

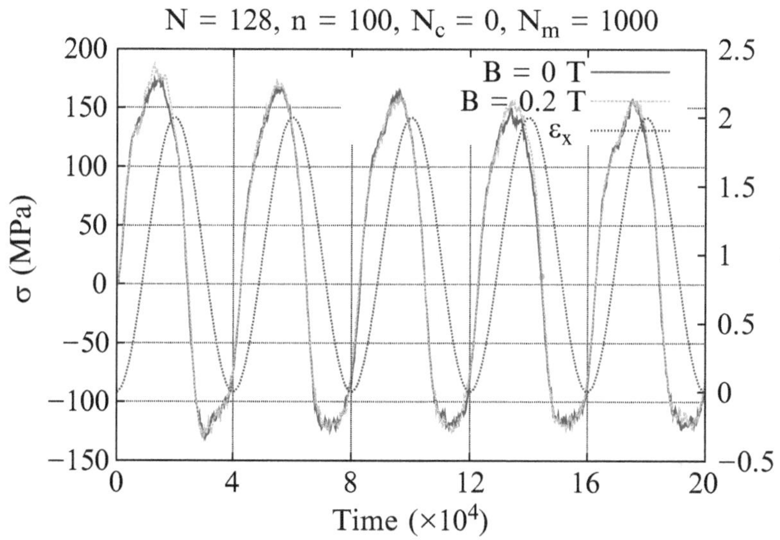

Fig. 5 Effect of magnetic field: sinusoidal response

In Fig. 7 response of the triangular strain loading on the material is shown. During the loading phase the stress increases non linearly with decreasing modulus and during the unloading phase stress comes back to zero before strain. This is because of the relaxation in the structure. When strain becomes zero there is residual compressive stress in the material.

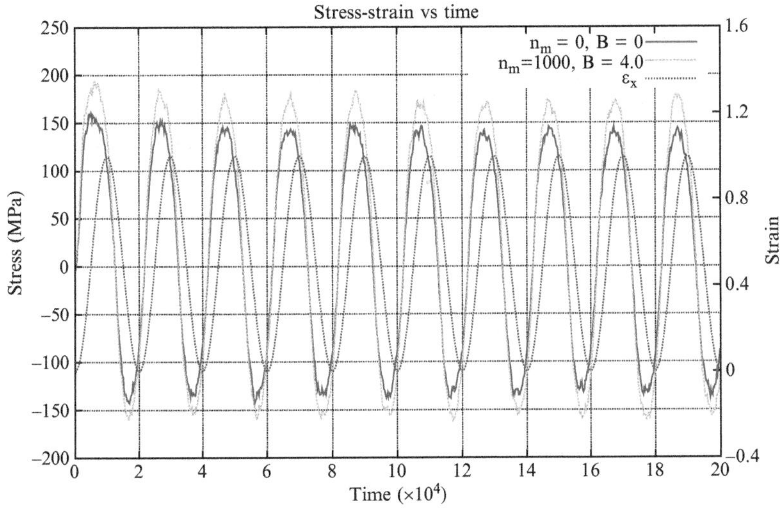

Fig. 6 Effect of magnetic field: sinusoidal response

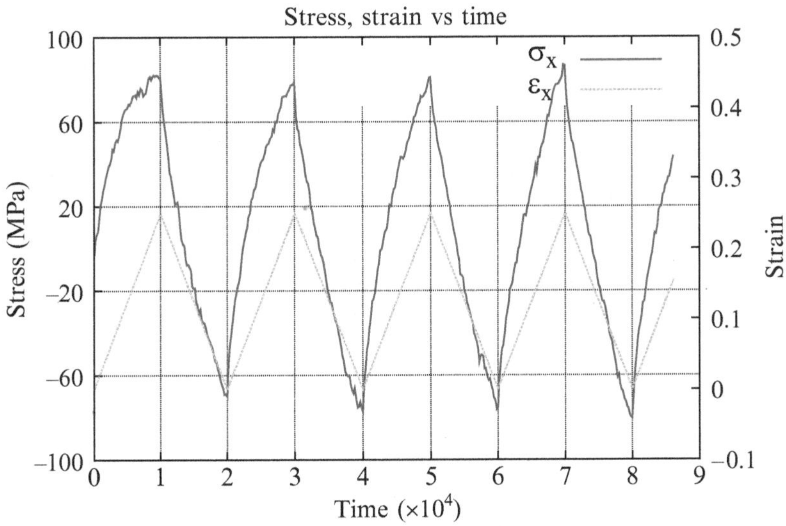

Fig. 7 Effect of magnetic particles: triangular loading

5 Conclusion

Results of the MD simulation are in good qualitative agreement with the observed behavior of such materials. Results of the simulation affirm the viscoelastic behavior of the magneto-sensitive elastomers. At moderate field strength there is no significant change in the constitutive behavior, but, as we increase the field strength the material becomes stiffer. This is due to the fact that at moderate field strengths the thermal energy dominates over the magnetic energy. Interaction between the united atom and magnetic particle is neglected in the present work which could be added for more detailed study on the same. In this work we assume that the material is incompressible, but, this assumption can easily be dropped. Further, magnetic interaction could be improved by using more sophisticated models.

References

1. Anderson HC (1980) Molecular dynamics simulation at constant pressure and/or temperature. J Chem Phys 72(4):2384–2393
2. Bergström JS, Boyce M (2001) Deformation of elastomeric networks: Relation between molecular level deformation and classical statistical mechanics models of rubber elasticity. Macromolecules 34(1):616–626
3. Brigadnov IA, Dorfmann A (2003) Mathematical modeling of magneto-sensitive elastomers. Int J Solids Struct 40(3):4659–4674
4. Doi M, Adwards SF (1986) The theory of polymer dynamics. Clarendon Press, Oxford
5. Greiner W, Neise L, Stöcker H (1997) Thermodynamics and statistical mechanics. Springer-Verlag, New York
6. Griffiths DJ (1991) Introduction to electrodynamics. Printice Hall International, Inc., U.S.A
7. Hoover WG (1985) Canonical dynamics: Equilibrium phase space distributions. Phy Rev A 31(3):1695–1697
8. Kankanala SV, Triantafyllidis N (2003) On finitely strained magnetorheological elastomers. J of mech and phy of solids 52(3):2869–2908
9. Nosé S (1983) A molecular dynamics method for simulation in the canonical ensemble. Molecular Physics 52(2):255–268
10. Raos G, Moreno M, Elli S (2006) Computational experiment on filled rubber viscoelasticity: What is the role of particle-particle interactions. Macromolecules 39(19):6744–6751
11. Thien NP (2002) Understanding Viscoelasticity: Basics of Rheology. Springer, Berlin
12. Wineman AS, Rajagopal KR (2000) Mechanical Response of Polymers: An Introduction. Cambridge University Press, Cambridge
13. Yang L, Srolovitz DJ, Yee AF (1997) Extended ensemble molecular dynamics method for constant strain rate uniaxial deformation of polymer systems. J Chem Phys 107(11): 4396–4407
14. Zhu Y, Gong X, Dang H, Zhang X, Zhang P (2006) Numercal analysis of magnetic induced shear modulus of magneto-rheological elastomers based on multichain model. Chinese J Chem Phy 19(2):126–130

Effects of Functionalization on the Morphology, Cure Kinetics and Mechanical Behavior of Thermosetting Polymers

Mahesh V. Hosur, Merlin Theodore, and Shaik Jeelani

Abstract Effects of various functionalized multi-walled carbon nanotubes (MWCNT) on cure kinetics and flexural properties of an epoxy based nanocomposite system were investigated. Chemical functionalization of MWCNT by oxidation (MWCNT-COOH), direct-fluorination (MWCNT-F) was confirmed by FTIR, and Raman spectroscopy. One weight percentage loading of MWCNT was used to prepare epoxy-based nanocomposites through in-situ polymerization. Curing kinetics showed that MWCNT-COOH and MWCNT-F would catalyze the curing process altering activation energy, enthalpy of reaction, and reaction rate. Compared to neat resin, nanocomposites fabricated with MWCNT-COOH and MWCNT-F showed a 48%, and 33% increases in flexural strengths, and 55% and 16% increases in flexural moduli, respectively. SEM analysis of the fracture surface of the composites confirmed an improvement in dispersion.

Keywords Carbon nanotubes · Cure kinetics · Functionalization · Nanocomposites

1 Introduction

CNTs are commonly used as fillers in base polymers to fabricate composite materials with superior mechanical, electrical, and thermal properties. Unfortunately, the fabrication of polymer nanocomposites (PNC) with superior strengths as high as those of the individual CNTs themselves has not been made possible so far. The achievement of such a superior PNC, reinforced with CNTs is hindered by the fact that CNTs have non-reactive surfaces, lack interfacial bonding with matrix, poor dispersion within the composite matrices [1]. They are also easily damaged with high shear mixing rate, and have very low solubility in most solvents. Overcoming these problems will promote better usage of CNTs in PNC.

M.V. Hosur (✉), M. Theodore, and S. Jeelani
Center for Advanced Materials, Tuskegee University, Tuskegee, AL 36088, USA
e-mail: hosur@tuskegee.edu; Merlin.Theodore.ctr@wpafb.af.mil; jeelanis@tuskegee.edu

B. Dattaguru et al. (eds.), *IUTAM Symposium on Multi-Functional Material Structures and Systems*, IUTAM Bookseries 19, DOI 10.1007/978-90-481-3771-8_15,

Chemical functionalization of CNTs has been reported by various groups to overcome these challenges promoting increase in overall properties of epoxy nanocomposites system [1–4]. Chemical functionalization alters the CNTs morphology and provides an abundance of reactive sites, thus increasing the possibility of crosslinking with the polymer matrix. Chemical functionalization also promotes exfoliation of agglomerates resulting in a homogeneous dispersion when incorporated into a solvent or polymer matrix, and also improves solubility of the CNTs in an organic solvent. Several researchers have reported non-covalent and covalent bonding of CNTs and its effects on their properties or PNC properties. Composite materials, reinforced with functionalized CNTs, displaying significant enhancements in dispersion, interfacial interaction with matrix, mechanical, electrical, and thermal properties have been reported [5–8]. To date, there is very limited research has been done on direct comparison of epoxy nanocomposites, reinforced with functionalized MWCNTs in a particular epoxy matrix.

The objective of this research is to improve fundamental understanding of how surface functionalized MWCNTs affect the curing process, morphology, and flexural properties of epoxy based nanocomposites. This research focuses on two different types of functionalization, namely, oxidation and direct-fluorination to attach carboxylic acid (COOH) and fluorine atoms to their sidewalls. Pristine and functionalized MWCNTs were infused into epoxy resin system and characterized for the cure behavior and flexural properties.

2 Experimental Studies

2.1 Materials

Epon 862 (Diglycidyl Ether of Bisphenol F, Part A), a low viscosity liquid epoxy resin and Epicure W, an aromatic diamine, were used in the study. Unmodified multi-wall carbon nanotube s was purchased from MER Cor., in Tucson, Arizona. These unmodified carbon nanotubes were synthesized by Chemical Vapor Deposition (CVD) with an average diameter of 140 ± 30 nm, average length of $7 \pm 2\,\mu$m, and purity greater than 90%.

2.2 Functionalization by Oxidation

Functionalization of MWCNT was done by placing 1 g of MWCNT in a 3:1 solution by volume of H_2SO_4/HNO_3 solution. The MWCNT in solution was sonicated in an ultrasonic bath for 3 h at ambient temperature at a frequency of 20 kHz. After sonication, the solution was diluted with distilled water (1:5), and filtered with large amounts of distilled water to remove all the traces of acids. The MWCNT was then dried for 24 h in a vacuum oven.

2.3 *Functionalization by Fluorination*

Modification of MWCNT by fluorination was done by placing 1 g of oxidized MWCNT, 10 ml of glyme (aka Diethylene Glycol Dimethyl Ether; 2-methoxyethyl Ether), and 1.66 g of 4-aminophenol (or 3.1 ml of 4-flourianiline) in a flame-dried three-neck round bottom flask. The mixture was then purged with nitrogen, which was used to maintain an inert atmosphere in the mixture. Then 4 ml of amyl nitrate was slowly added to the mixture. The mixture was stirred at room temperature for 1 h (small gas evolution). The temperature was then raised to 70°C and the mixture was continuously stirred for an additional 3 h. Subsequently the mixture was cooled, diluted with diethyl ether, filtered, and washed first with diethyl ether and then with copious amount of water. The MWCNTs were then dried for 24 h in a vacuum oven.

2.4 *Processing of Epoxy Nanocomposites*

One weight percentage loading of MWCNTs was mixed with acetone and sonicated for 30 min in an ultrasonic bath. After the sonication process, the solution of MWCNTs and acetone was added to the Epon 862 and mixed using a digital mechanical stirrer at a constant speed of 500 rpm. During the last 15 min of mixing the temperature was raised to 70°C. This was done to ensure complete removal of acetone. The curing agent, which was maintained at the same temperature as that of the resin, was added and mixing was continued for an additional 30 min. The mixture was then degassed under high vacuum at 60°C. The mixture was then poured into desired molds that were preheated at 121°C for 1 h. The mixture was subsequently cured in an oven at 121°C for 4 h after which it was left to cool to ambient temperature. The same process was implemented for all systems.

2.5 *Characterization*

2.5.1 Interfacial Interaction

Evidence of chemical modification by oxidation and direct fluorination of the pristine MWCNT was confirmed using Raman Spectroscopy and Fourier Transform Infrared Spectroscopy (FTIR). Raman Spectroscopy measurements were performed on a Renishaw in Via Raman Microscope, equipped with an ultra low noise Rencam CCD (charge-coupled device) detector and a 785 diode laser. FTIR measurements were performed using a Nicolet spectrometer equipped with a Specac Golden Gate Attenuated Total Reflectance Accessory (ATR) using 64 scans with a 4 cm^{-1} resolution step. Transmission mode was implemented for scans of various MWCNTs.

2.5.2 Flexural Characterization

Flexural tests under three-point bend configuration were performed according to ASTM D790-02 using Zwick-Roell materials testing system. The instrument was run under displacement control mode at a crosshead speed of 2.0 mm/min, and all the tests were performed at room temperature.

2.5.3 Morphological Characterization

Microstructures of neat and nanocomposite samples were examined under a Field Emission Scanning Electron Microscope (FE-SEM Hitachi S-900). Sputtering was done using Hummer 6.2 system to coat the samples. A 25 kV accelerating voltage was applied to accomplish desired magnification.

2.5.4 Curing Kinetics

A modulated Differential Scanning Calorimeter (TA Instruments DSC Q1000) was used to characterize thermal behavior during the curing process using ASTM E698 method. The samples were heated from 25°C to 350°C at different heating rates of 2°C, 5°C, 10°C, 15°C/min.

3 Results and Discussion

3.1 Functionalization of MWCNTs

3.1.1 Effects of Functionalization on Morphology

Chemical modification by oxidation and direct fluorination significantly alters the physical morphology of pristine MWCNTs, enhances their chemical reactivity, and reduces the concentration of agglomeration. Modification involves covalent bonding of functional groups along the surface or side walls of the MWCNTs. Oxidation of MWCNTs decrease the length of the carbon nanotubes by more than half of their original length from 7.0 to 2.0 μm, decrease the diameter from 126 to 105 nm, decreases the aspect ratio from 156 to 19, burns out amorphous carbon, and opens the end caps with carboxylic acid (-COOH) covalently attached to the sidewall. Oxidation is usually the first reaction step (purification) in any functionalization method to create a reactive site for replacement by other functional groups of atoms or molecules such as amine, fluorine, and so on. It is known that epoxy groups can react directly with the carboxylic acid functionalities to form esters. Unlike the oxidation process, direct fluorination of MWCNTs covalently bonds fluorine atoms along the surface of the CNTs and changes the hybridization from sp2 to sp3, increases the length of the MWCNT from 7.0 to 7.8 μm, and increases the aspect ratio from 56 to 62.

3.1.2 Surface Interaction and Properties (FTIR and Raman Spectroscopy Analysis)

Evidence of chemical modification on the surface of the pristine MWCNT was confirmed using Raman spectroscopy and FTIR. Figure 1 shows FTIR spectra of the functionalized MWCNTs. In a typical FTIR image of MWCNT-COOH there are three distinctive peaks (C = O, O–H, C–O) that emerge as a result of oxidation, forming COOH groups on the end caps of the CNTs. Normally the adsorption band for carboxylic acid is around 1,700 cm^{-1} region. The absorption band in Fig. 1 for MWCNT-COOH is seen at 1,735 cm^{-1}. The band at 1,644 cm^{-1} corresponds to the stretching vibration of the C–O group. The broad band in the 3,100–3,600 cm^{-1} region is assigned to the O–H stretch of the terminal carboxyl group. The absorption band around 1,572 cm^{-1} is more than likely from the C = C stretching mode of nanotubes activated by sidewall attachment. As seen in a typical fluorine substitution, interaction effects are at their highest, and C–F stretching absorptions occur anywhere within the range 1,000–1,400 cm^{-1}, depending on the nature and degree of fluorination. The absorption band of the C–F stretch appeared at 1,232 cm. Fluorine substitution results in considerable shifts in high frequency adjacent to

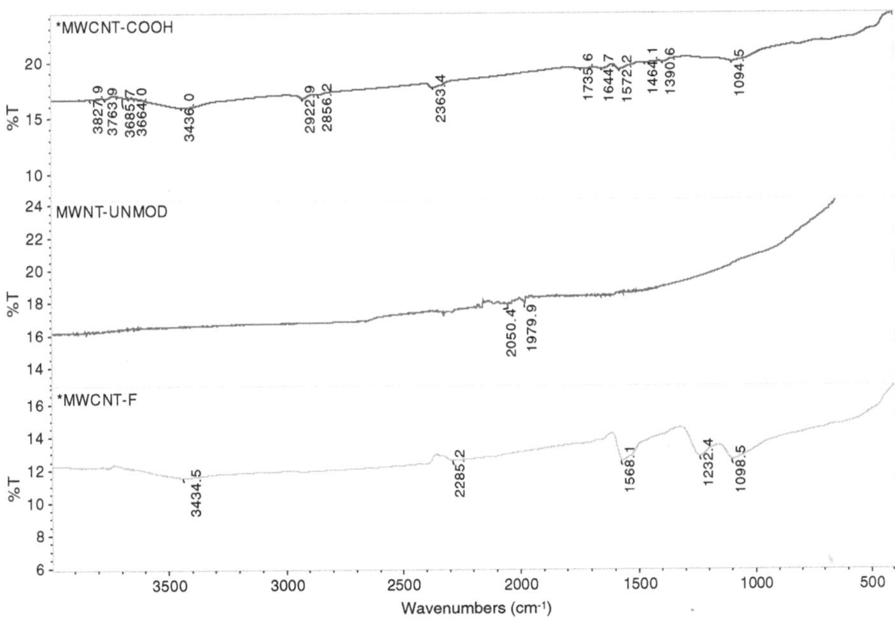

Fig. 1 FTIR spectroscopy image of MWCNTs

C–H, C = C, and C = O stretching vibrations. MWCNT-COOH and MWCNT-F show a fairly large absorption bandwidth around the 3,400 cm^{-1}. From these results, it can determine that the surface or sidewall of the MWCNTs does contain the desired functional groups or atoms.

Raman spectroscopy, shown in Fig. 2, illustrates the radial mode band (100–400 cm^{-1}), D-band (~1,350 cm^{-1}), G-band (1,500–1,600 cm^{-1}), and G*-band (~2,600 cm^{-1}) of CNTs. The D-band represents the measure of degree of disorder if the intensity of the G or G*-band is not affected by disorder or defects. The G-band represents crystalline graphitic structure whereas the G*-band represents the overtone of disorder in the functionalized MWCNTs. Chemical functionalization of pristine CNTs dramatically increases the intensity of these bands. The ratio of G/D-bands or D/G-bands is used to characterize the order/disorder in the graphite structure. The G-band is less affected by defects from functionalization as compared to the D-band. An increase in I_D/I_G(intensity of D to G-band), or decrease in I_{G*}/I_G indicates, also listed in Table 1, higher defect concentration or a higher degree of disorder as well. Furthermore, the lower the I_D/I_G ratio is, the higher will be the concentration of amorphous phase. It is very obvious that the MWCNT-F has a higher degree of disorder compared to the MWCNT-COOH. The degree of disorder depends on the reaction, and the fluorination process occurs at an elevated temperature than the oxidation process.

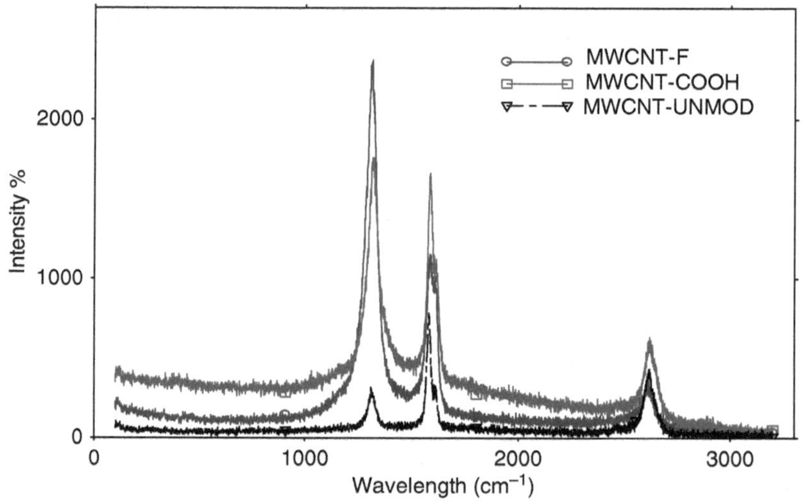

Fig. 2 Raman spectroscopy image of MWCNTs

Table 1 Disorder/order ratios of MWCNT

CNT	I_D/I_G	I_{D*}/I_G
MWCNT-UNMOD	0.39	0.60
MWCNT-COOH	1.06	0.37
MWCNT-F	2.00	0.33

3.2 Curing Kinetics by DSC and FTIR

The curing kinetics of the uncured epoxy nanocomposites were studied at different scanning rates (2°C, 5°C, 10°C, 15°C/min) from 25°C to 350° C using thermal stability kinetics on the DSC. The heat evolved during the curing process was measured and used to calculate kinetic results such as activation energies, rate constants, enthalpy, and percentage conversion using the equations:

$$d\alpha/dt = Ze^{EaRT}(1-\alpha) \quad (1)$$

$$\Delta H_{reaction} = \Sigma\Delta H_f(P) - \Sigma\Delta H_f(R) \quad (2)$$

where, $d\alpha/dt$ = reaction rate (s^{-1}); α = fractional conversion; Z = pre-exponential factor (s^{-1}); E_a = activation energy (J/mol), R is the universal gas constant; T is the temperature (ΔK); and ΔH = standard enthalpy change of reaction; $\Delta H_f(P)$ = enthalpy of products; $\Delta H_f(R)$ enthalpy of reactants. All systems were not 100% cured due to the crosslinking initiation in the thermoset, and around 1–3% remained uncured. Because these nanocomposites are multiphase system, the curing kinetics is solely dependent upon the interaction between the individual phases. The functionalized MWCNTs have disturbed the curing behavior by catalyzing the reaction creating a decrease in the activation energy and an increase in standard enthalpy change. The rate of reaction, listed in Table 2 is higher for the MWCNT-COOH simply because the reaction rate increased with higher concentration of reactive sites at the end-caps. The activation energy, heat of reaction, and rate of reaction were the lowest for the MWCNT-F. Unlike the MWCNT-COOH system which reacts with the resin system by opening the epoxy rings and forming a bond, the MWCNT-F can react with the resin systems in three possible ways; first, the fluorine atoms separated from the MWCNT during mixing may bond with hydrogen atoms on the polymer chain again causing separation to form hydrofluoric acid (HF); secondly, in-situ chemical reaction with amine curing agent; and finally, the fluorine atom opens up the epoxy rings on the end of the polymer chain forming covalent bonds with the polymer chain. The separation sites on the polymer chain and MWCNTs surface opens the opportunity for more reactive sites and other covalent bonding to occur. Based on the fact that the reaction rate (rate at which the epoxy rings are opened) of the MWCNT-F was lower than the MWCNT-COOH system, we assume that the MWCNT-F reacts initially with the curing agent and may also created HF, resulting in a reaction rate lower than the neat.

Table 2 Curing kinetics parameters determined by DSC

Material	Activation energy (kJ/mol)	Enthalpy (J/g)	Reaction rate (1/min)	Conversion %
Neat	57.7 ± 2.3	263.8 ± 15.3	0.0168	98
MWCNT-UNMOD	55 ± 0.6	258.3 ± 16.5	0.0174	99
MWCNT-COOH	55.2 ± 1.8	304 ± 4.8	0.0182	99
MWCNT-F	54.5 ± 3.8	267.5 ± 0.5	0.0147	97

Table 3 Flexural properties of epoxy nanocomposites

Epoxy nanocomposites	Max. strength (MPa)	% change	Modulus (GPa)	% change	Strain %	% change
Epon 862 neat	102.4 ± 5.6	–	3.1 ± 0.3	–	$6.1 \pm .02$	–
MWCNT-UNMOD	152.1 ± 20.2	49	4.1 ± 0.2	32	4.7 ± 1.0	−23
MWCNT-COOH	151.1 ± 14.9	48	4.8 ± 0.6	55	4.7 ± 0.7	−23
MWCNT-F	136.1 ± 12.2	33	3.6 ± 0.0	16	4.6 ± 0.5	−25

3.3 *Flexural Characterization*

Functionalization not only improved the dispersion of MWCNTs in the polymer matrix but also enhanced the flexural properties as shown in Table 3. Brittle failure was observed in all systems. Increased flexural properties are attributed to increase in crosslinking, better dispersion, and improved interfacial interaction between the MWCNTs and the matrix. PNC becoming more brittle with the incorporation of CNTs is a common problem seen in most research. The strain to failure decreased by approximately 20% from 6.1% to 4.7% to 4.6% for the MWCNT-COOH and MWCNT-F nanocomposites, respectively. Compared to the MWCNT-UNMOD nanocomposites, mechanical properties for the MWCNT-COOH and MWCNT-F composites were lower, but still higher than that of the neat resin. This observed behavior for the MWCNT-F may be due to degradation of the sample during curing. The premature degradation is presumably catalyzed by hydrofluoric acid (HF), which evolves from the MWCNT-F during the curing process. HF etches the polymer matrix cracking its structure causing a loss in strength. As for the MWCNT-COOH nanocomposites, the disadvantage of shorter MWCNTs does not significantly delay the crack rate compared to the longer MWCNTs in the MWCNT-UNMOD nanocomposites. The shorter the MWCNTs, the lower the stress in the MWCNTs, corresponding to higher average stress in the matrix, resulting in faster crack growth rate. However, the MWCNT-COOH nanocomposites maintained the highest increase in modulus by 55%, simply because the composite is packed more densely with MWCNTS, compared to the others. Though, the results indicate that functionalized MWCNTs have more or less similar properties compared to unmodified MWCNTs, it is anticipated that with better dispersion methods and optimal MWCNTs content, these properties can certainly show improvement. In addition, other mechanical characterizations are currently being carried out.

3.4 *Morphology of Fracture Surface*

SEM was employed to determine the morphology and dispersion of the functionalized MWCNTs in the matrix. The fracture surface of the samples from three point bend test was investigated. Figure 3 illustrates scanning electron micrographs of samples after the flexure tests. From Fig. 3, it is obvious that the neat composite

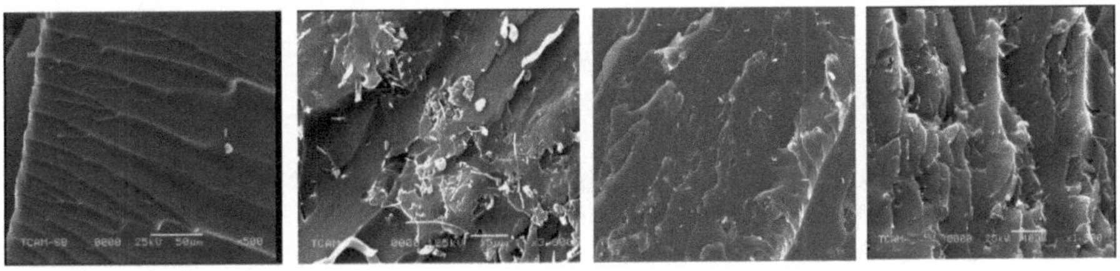

Fig. 3 Scanning electron micrographs of neat epoxy, epoxy with unmodified MWCNT, MWCNT-COOH, and MWCNT-F (*left* to *right*)

has a clean and smooth fracture surface. Incorporation of pristine MWCNT into the matrix increases the fracture surface roughness. These nanocomposites contain agglomeration of MWCNT that seem to be sitting on top of the matrix as opposed to being embedded inside the matrix. These results are signs of poor dispersion resulting from a lack of interfacial interaction and adhesion to the matrix. Incorporating functionalized MWCNTs increased the surface roughness to an even higher degree making the ridges, which dissipate energy during their formation, thereby promoting a higher resistance to material separation. In the MWCNT-COOH composites, a continuous and uniformed river pattern was formed at the fracture surface and the MWCNT-COOH were very well dispersed. In the MWCNT-F composites, matrix cracking leading to debonding in some areas and fiber pullout was evident. As mentioned earlier, the matrix cracking could be the consequences resulting from the damages created by the HF.

3.5 Conclusion

In this study, MWCNTs were functionalized through oxidation, and fluorination. The modified as well as unmodified MWCNTs were then dispersed in EPON 862 polymer at 1% loading by weight. Performance of the nanocomposites samples was compared with that of neat epoxy. It was found that chemical functionalization, in most cases, is a useful method to improve dispersion leading to increased properties. Curing kinetics studies show evidence of how these functionalized MWCNT affect the curing process by lowering activation energy, and increasing change in enthalpy. Despite variations in curing parameters degrees of conversion was relatively the same for all systems. Incorporation of functionalized MWCNTs into epoxy matrix improves the flexural strength and modulus of the neat composite. SEM investigation of the fracture surface showed evidence of improvement in the dispersion of MWCNTs in the matrix and the chemical interaction between them.

Acknowledgments Financial support through the National Science Foundation through PREM program (DMR-0611612) is gratefully acknowledged.

References

1. Yaping Z, Aibo Z, Jiaoxia Z, Rongchang N. Materials Science and Engineering A 2006; 435–436: 145–149.
2. Gojny FH, Schulte K. Composites Science and Technology 2004; 64:2303–2308.
3. Chen GX, Kim HS, Park B, Yoon JS. Polymer 2006; 47(13):4760–4767.
4. Zhu J, Peng H, Rodriguez-Macias F, Margrave J, Khabashesku V, Imam A, Lozano K, Barrera E. Advanced Functional Materials 2004; 14(7):643–648.
5. Wang J, Fang Z, Gu A, Xu L, Liu F. Applied Polymer Science 2005; 100:97–104.
6. Gojny FH, Wichmann MHG, Fiedler B, Kinloch IA, Bauhofer W, Windle AH, Schulte K. Polymer 2006; 47:2036–2045.
7. Breton Y, Desarmot G, Salvetat JP, Delpeux S, Sintruel C, Beguin F, and Bonnamy S, Carbon 2004; 42:1027–1030.
8. Eitan A. Chemical Materials 2003; 15:3198–3201.

A Study on Polarization-Electric Field Nonlinearity in Smart Composite Structures

V.L. Sateesh, C.S. Upadhyay, and C. Venkatesan

Abstract This paper presents the development of nonlinear coupled electro-thermo-elastic constitutive relations applicable for piezo materials. These nonlinear constitutive relations model the experimentally observed hysteresis and butterfly loops accurately and, can be easily incorporated in structural analysis of smart structures. Using this nonlinear formulation, static analysis of smart composite plates has been carried out to bring out the effect of polarization-electric field nonlinearity on the deformation of composite plates. Results of static analysis are correlated with the experimental data and it is shown that nonlinear formulation provides a better correlation with the experimental results than the linear case.

Keywords P-E hysteresis modeling · Smart composite plates · Static analysis

1 Introduction

With the development of piezo ceramic sensors and actuators of varying shapes and sizes for use in structural applications, the field of smart structures has emerged as an area of research of great importance [1]. The introduction of piezo materials in structural applications has created a necessity to have a relook on the traditional structural modeling and analysis. Under cyclic variation of applied electric field, piezo materials exhibit polarization – electric field (P-E) hysteretic losses. Mathematical modeling of hysteresis has been approached at two different levels, one at the microscopic level [2] and the other at macroscopic level [3–6] of the piezo material. Macroscopic models can be categorized as empirical models [3, 4] or thermodynamically consistent models [5, 6].

By the method of local state [7] an internal variable can be introduced at the macroscopic level to represent the micro level phenomenon. The physical processes

V.L. Sateesh, C.S. Upadhyay, and C. Venkatesan (✉)
Department of Aerospace Engineering, Indian Institute of Technology, Kanpur, India
e-mail: vlsateesh@gmail.com; shekhar@iitk.ac.in; cven@iitk.ac.in

B. Dattaguru et al. (eds.), *IUTAM Symposium on Multi-Functional Material Structures and Systems*, IUTAM Bookseries 19, DOI 10.1007/978-90-481-3771-8_16,

described in this manner will always be thermodynamically admissible provided Clausius-Duhem inequality is satisfied at every instant of evolution. Even though micromechanical models have been able to capture hysteresis and butterfly loops, they are computationally time consuming and cumbersome, particularly with regard to their application for structural analysis. Therefore it is necessary to develop a thermodynamically consistent model which can be easily integrated with traditional finite element formulation for the analysis of smart structures.

Resolution of these effects (sensing, actuation and hysteresis effects) requires analysis of a smart structure using coupled electro-thermo-elastic constitutive models [8, 9]. Three dimensional exact solutions of smart plates with simply supported boundary conditions for cylindrical bending have been given by Ray et al. [10]. Analytical solutions of laminated piezo electric plates have been given by Heyliger [11]. Simply supported mechanical boundary conditions are adopted in the analysis and electric potential is assumed to be set to zero (grounded) in all the four vertical edges. Crawley and Lazarus [12] have conducted experiments on smart cantilever plates containing segmented piezo patches on either side.

Ha et al. [13] used layerwise model with few higher order compatible modes in formulating cantilever plates. Segmented piezo patches have been considered in the analysis for dynamic and static cases. In static analysis, good correlation of longitudinal deflection is achieved with the experimental results given in [12], whereas correlation for lateral bending and twisting, with the experiments, is not satisfactory.

In all the above mentioned studies, linear constitutive models are used for the analysis. However it was shown by Tiersten [14] and Ahmad et al. [15] that the governing equilibrium equation and the constitutive relationships are inherently nonlinear in nature, due to the electric field polarization interaction. Based on this nonlinear formulation, Ahmad et al. [15] have analyzed smart beams by layerwise finite element model. They have concluded that the effect of nonlinearity in the deflection increases with the increase in actuation voltage. In the present analysis, an attempt is made to study the effect of P-E nonlinearity for the case of actuation in piezo electric plates with segmented and continuous piezo patches.

2 Hysteresis Modeling

In this section, hysteresis modeling is carried out by using consistent thermodynamic formulation and validated with the experimental hysteresis data.

2.1 Formulation

The governing equations have been developed by applying the conservation laws (mass, linear momentum, angular momentum). The conservation of energy (first law of thermodynamics) and the rate of entropy production (second law of thermodynamics) are used for deriving the constitutive relations for piezo materials. The development of the mathematical modeling of electro-thermo-elasticity follows the procedure given in [14, 15].

The free energy Ψ is expressed as a quadratic function of all state variables and quartic functions of strain and electric field.

$$\begin{aligned}\Psi = \frac{1}{2}\{ & C_{ijkl}\varepsilon_{ij}\varepsilon_{kl} - b_{ij}E_iE_j + C_\theta\theta^2 + \xi\Theta^2 - 2e_{ijk}E_i\varepsilon_{jk} + 2\alpha_{ij}\theta + 2\lambda_iE_i\theta \\ & - 2\Omega_{ij}\varepsilon_{ij}\Theta - 2\beta_iE_i\Theta + 2\zeta\Theta\theta - Q_{ijkl}E_iE_jE_kE_l - H_{ijklm}\varepsilon_{ij}E_kE_lE_m \\ & + f_{ijklmn}\varepsilon_{ij}\varepsilon_{kl}E_mE_n - g_{ijklmno}\varepsilon_{ij}\varepsilon_{kl}\varepsilon_{mn}E_o + X_{ijklmnop}\varepsilon_{ij}\varepsilon_{kl}\varepsilon_{mn}\varepsilon_{op}\}\end{aligned} \tag{1}$$

The set of constitutive relations between associated variables (force terms) and the state variables are given as:

$$\begin{aligned}\sigma_{ij} &= C_{ijkl}\varepsilon_{kl} - e_{ijk}E_k - H_{ijklm}E_kE_lE_m - \alpha_{ij}\theta - \Omega_{ij}\Theta - P_iE_j \\ &\quad + f_{ijklmn}\varepsilon_{kl}E_mE_n - g_{ijklmno}\varepsilon_{kl}\varepsilon_{mn}E_o + X_{ijklmnop}\varepsilon_{kl}\varepsilon_{mn}\varepsilon_{op} \qquad (2)\\ P_i &= b_{ij}E_j + Q_{ijkl}E_iE_kE_k + e_{ijk}\varepsilon_{jk} + H_{ijklm}\varepsilon_{jk}E_lE_m + \beta_i\Theta - \lambda_i\theta \\ &\quad - f_{ijklmn}\varepsilon_{jk}\varepsilon_{lm}E_n + g_{ijklmno}\varepsilon_{jk}\varepsilon_{lm}\varepsilon_{no} \qquad (3)\\ S &= \alpha_{ij}\varepsilon_{ij} + \lambda_kE_k + C_\theta\theta - \zeta\Theta \qquad (4)\\ \Theta &= -\left[\sum_{i=1,2,5\ldots}\left(\Gamma_i\frac{\partial\psi}{\partial\Theta}\right)^i + \sum_{j=2,4,6..}\Gamma_j\left(\Gamma_j\frac{\partial\psi}{\partial\Theta}\right)^{j-1}\left|\frac{\partial\psi}{\partial\Theta}\right|\right] \qquad (5)\end{aligned}$$

$$\text{where}\quad \frac{\partial\psi}{\partial\Theta} = \xi\Theta - \beta_iE_i - \Omega_{ij}\varepsilon_{ij} + \zeta\theta.$$

Where C_{ijkl} is the elastic constant, b_{kl} is the electric susceptibility; C_θ is thermal constant; e_{ijk} is the piezoelectric constant; α_{ij} is the thermo-elastic constant, λ_i is pyroelectric constant; ζ, β_i, Ω_{ij}, ξ and Γ_i and Γ_j are real positive constants; H_{ijklm} are Q_{ijkl} real constants. Detailed version of the derivation can be found in [16].

2.2 Validation of Constitutive Relations

The validity of the constitutive relations given in Eqs. (2)–(5), are verified by solving the problem of free piezo patch subjected to an external time varying electric field along z-direction. Since the piezo patch is treated as free crystal $\sigma_3 = 0$. Hence, strain in z-direction can be written as:

$$\varepsilon_{33} = \frac{[e_{333}E_3 + H_{33333}E_3E_3E_3 + \Omega_{33}\Theta + E_3P_3]}{C_{3333} + f_{333333}E_3E_3} + \frac{g_{3333333}\varepsilon_{33}E_3 - X_{33333333}\varepsilon_{33}\varepsilon_{33}\varepsilon_{33}}{C_{3333} + f_{333333}E_3E_3} \tag{6}$$

The material used is PZT-5H, for which the constants corresponding to the linear part of model are taken from the literature. The nonlinear and dissipative constants have been evaluated based on error minimization between experimental and theoretical *P-E* hysteresis loop and also by imposing the limiting conditions on maximum polarization (P_{max}), maximum electric field (E_{max}) and maximum strain (ε_{max}).

Assume that the piezo patch is acted on by an electric field having an amplitude 2 kV/mm and a frequency of 0.0165 Hz. Keeping the initial conditions $P = 0$, $\varepsilon = 0$ and $\Theta = 0$, at $t = 0$, the coupled nonlinear electro-elastic equations (Eqs. (2)–(6)) are solved iteratively. The P-E hysteresis loop and the corresponding ε-E variation are shown in Figs. 1 and 2, respectively. These results are generated by keeping terms upto seventh power in Eq. (5). It can be seen from Fig. 1 that the theoretically generated hysteresis loop correlates well with the experimental data [17], including both the initial variation and steady state response. Figure 2 shows the corresponding variation of strain with respect to electric field (butterfly loop) and the correlation is satisfactory.

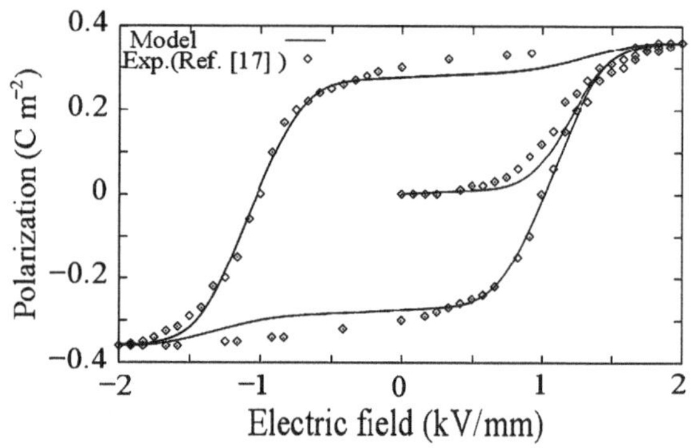

Fig. 1 Comparison of theoretical and experimental hysteresis loop

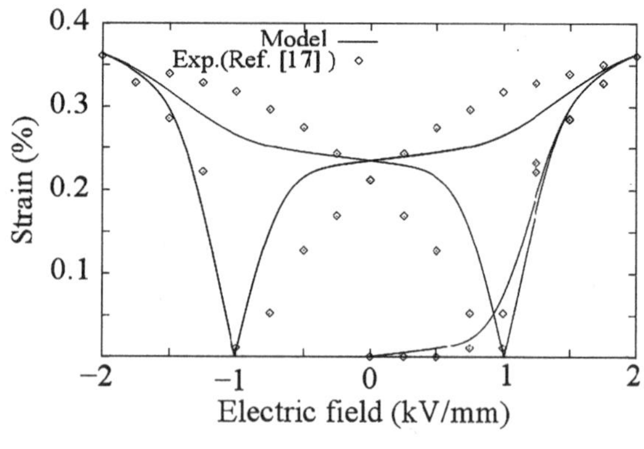

Fig. 2 Variation of strain with electric field (butterfly loop)

3 Analysis of Laminated Composite Plates

Linear and nonlinear static analyses of smart plates have been carried out with layer-by-layer finite element formulation. Laminated composite plate with segmented piezo patches is considered for the analysis.

3.1 Finite Element Formulation

Variational approach has been used for deriving finite element formulation. Virtual work done by any general electro-elastic medium in a domain V with a surface S is given by:

$$\int_{v}\left(\sigma_{ij,i}+P_iE_{j,i}\right)\delta_{uj}dv+\int_{v}D_{i,i}\delta\phi dv=0 \tag{7}$$

where δu_j and $\delta\varphi$ are the virtual displacement and virtual electric potential, respectively. $P_i E_{j,i}$ is the force term due to polarization electric field interaction [9]. Integrating by parts and using Gauss divergence theorem and substituting for stress as $\sigma_{ij} = \sigma^L{}_{ij} - P_i E_j$, the above expression can be written as:

$$\begin{aligned}&\int_{v}\sigma_{ij}^{L}\delta\varepsilon_{ij}dv-\int_{v}D_i\delta E_idv+\int_{v}\left(P_jE_{i,j}\right)\delta u_idv-\int_{v}P_iE_j\delta\frac{\partial u_i}{\partial x_j}dv\\&=\int_{s}\mathrm{T}_i\delta u_idS+\int_{s}Q\delta\phi dS\end{aligned} \tag{8}$$

where ε is the Green's strain, T is surface traction force, Q is the free electric charge per unit area applied on the surface S of the piezo actuator. Here, E_i is the electric field and can be written in terms of electric potential as $\frac{\partial\phi}{\partial x_i}$. In the piezoelectric domain, either the applied surface potential is given or the free charge on the surface Q is specified as zero, in the both cases $\int_S Q\delta\phi ds$ term is zero.

Piezo ceramic is considered as transversely isotropic along x_3 axis, which is also the poling direction. Here x_1, x_2 and x_3 axes are same as x, y and z respectively.

3.2 Deformation Models

The axis system of the problem is taken as Cartesian coordinate system. Deformations u, v and w are displacements along x, y and z axes respectively. The displacement field and electric potential in each element are represented as:

$$u_j\left(x,y,z\right)=\sum_{i=1}^{n_u}\sum_{l=1}^{n_\phi}N_i\left(x,y\right)\bar{u}_j^{il}\Phi_l\left(z\right)\text{ and }\phi_l\left(x,y,z\right)=\sum_{i=1}^{n_w}\sum_{l=1}^{n_\phi}N_i\left(x,y\right)\bar{\phi}_{il}\Phi_l\left(z\right) \tag{9}$$

where $\bar{u}_j^{il}$ are nodal displacements u, v and w for j = 1, 2 and 3 respectively, $N(x, y)$ represents inplane basis functions, Φ is the out of plane basis function. n_u and n_w are number of inplane basis functions used for u, v, w and ϕ respectively and n_ϕ is number of out of plane basis functions. Both inplane and out of plane shape functions have been obtained by Legendre polynomials. Cubic approximation has been taken for all the displacements and the electric potential. Triangular elements have been used in the inplane direction and line elements are used for out of plane direction. Substituting the displacement field Eq. (9) in Eq. (8) and using the constitutive relations, the final nonlinear equations can be written in matrix form as:

$$[K]\{q\} + ([K_1(\{q\})] + [K_2(\{q\})])\{q\} = \{F\} \tag{10}$$

where [K] is linear stiffness matrix, $[K_1(\{q\})]$ and $[K_2(\{q\})]$ are nonlinear stiffness matrix due to nonlinear force and extended stiffness matrices respectively, {q} is the nodal displacements ($\{q\} = \{u, v, w, \phi\}^T$) and {F} is the external load vector. Linear stiffness matrix has contribution from mechanical stiffness, electrical stiffness, and electromechanical coupling. In the linear analysis, nonlinear stiffness terms ($[K_1]$ and $[K_2]$) are excluded.

3.3 Linear and Nonlinear Analysis of Composite Plates with Segmented Piezo Patches

The composite plate has 15 piezo patches on each side of the plate, as shown in Fig. 3. The core material consists of graphite epoxy and is having 0.83 mm thickness

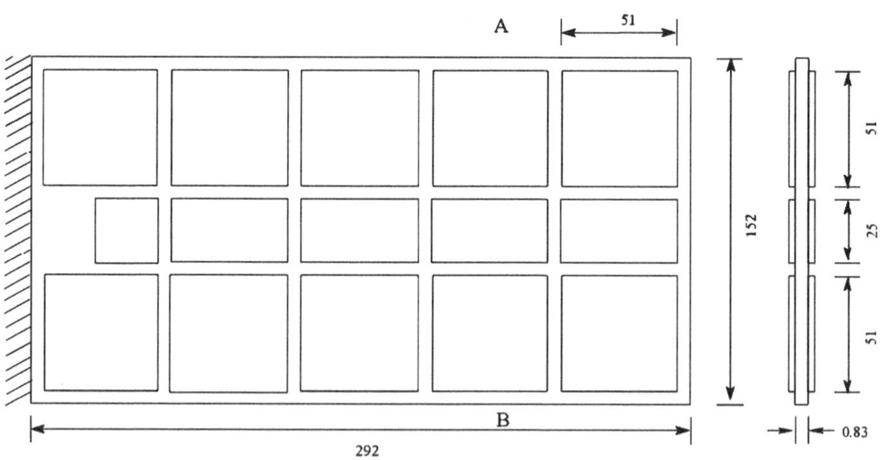

Fig. 3 Schematic representation of cantilever plate with segmented piezo patches (all dimensions are in millimetre)

and ply orientation is $[0/\pm 45^0]_s$ [12]. Piezo patches are made of PZT-G1195 and thickness of 0.25 mm. Properties of piezo and composite materials used for the analysis are given in Table 1. Fixed boundary condition is considered on the boundary $x = 0$ and other three boundaries are left free. The graphite-epoxy layers are modeled as orthotropic material and piezo is considered as transversely isotropic about z axis which is axis of poling. An electric field of amplitude 394 V/mm is applied with opposite polarity on either side of the plate on all patches. The longitudinal bending, transverse bending and twisting are calculated and nondimentinalzed as:

$$\begin{aligned} \text{Longitudinal} &= \mathrm{w}_2/\mathrm{L}_2 \\ \text{Transverse bending} &= \frac{1}{L_2}\left[w_2 - \frac{w_3 + w_1}{2}\right] \\ \text{Twisting} &= \frac{1}{L_2}\left[w_3 - w_1\right] \end{aligned} \tag{11}$$

where w_2 is the deflection along centerline of the plate in longitudinal direction; and w_1 and w_3 are the deflections at points A and B at the two edges of the plate, respectively. L_2 is the width of the plate.

Direct iteration method is used to solve the nonlinear set of equations (Eq. (10)). The resulting linear and nonlinear bending deflections are given in Fig. 4 along with the experimental results by Crawley and Lazarus[12]. From the figure it can be seen that the linear analysis results are able to capture the experimental data reasonably. The nonlinear results show better correlation with the experimental data. Polarization-electric field interaction nonlinear effects increase the longitudinal bending and twist considerably, whereas its effect on the transverse bending is less.

Table 1 Material properties of Graphite-epoxy and PZT actuator

Property	Graphite-epoxy	PZT-G1195
E_{ll}(GPa)	147.00	63.0
E_{tt}(GPa)	9.7	63.0
G_{lt}(GPa)	6.0	24.2
G_{tt}(GPa)	6.0	24.2
V_{lt}	0.3	0.3
V_{tt}	0.3	0.3
d_{31}(pm/V)	0.0	310.0
d_{32}(pm/V)	0.0	310.0
d_{33}(pm/V)	0.0	374.0
d_{24}(pm/V)	0.0	584.0
$\varepsilon_{11}(10^{-10}$F/m)	153.00	153.0
$\varepsilon_{22}(10^{-10}$F/m)	153.00	153.0
$\varepsilon_{33}(10^{-10}$F/m)	153.00	150.0

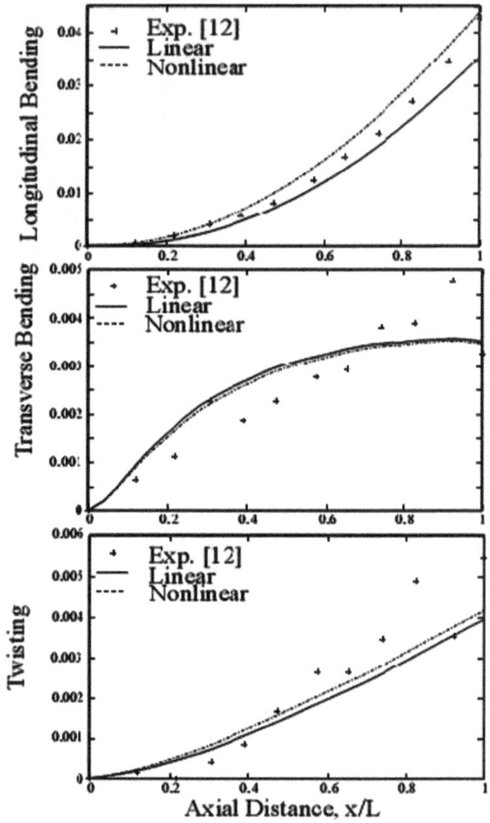

Fig. 4 Linear and nonlinear deformations of cantilever plate with piezo patches

4 Concluding Remarks

A set of nonlinear constitutive equations have been derived for piezo material to model the hysteresis effects by using consistent thermodynamic formulation. The nonlinear constitutive relations are shown to model the experimentally observed hysteresis and butterfly loops with reasonable accuracy.

Using these constitutive relations, linear and nonlinear static analyses are carried out by layer-by-layer finite element formulation. The nonlinear results are found to be well in agreement with the experimental data. It is observed that the polarization-electric field interaction nonlinear effect is of softening type.

References

1. Crawley, E. F., "Intelligent Structures for Aerospace: A Technological Overview and Assessment," *AIAA Journal*, Vol. 31, No. 8, 1994, pp. 1689–1699.
2. Hwang, S. C., Huber, J. E., McMeeking, R. M., and Fleck, N. A., "The Simulation of Switching in Polycrystalline Ferroelectric Ceramics," *Journal of Applied Physics*, Vol. 84, No. 3, 1998, pp. 1530–1540.

3. Chen, P. J., and Montgomery, S. T., "A Macroscopic Theory for the Existence of the Hysteresis and Butterfly Loop in Ferroelectricity," *Ferroelectrics*, Vol. 23, 1980, pp. 199–208.
4. Bossong, H., Lentzen, S., and Schmidt, R., "Experimental Investigation and Modeling of Piezoelectric Actuator Hysteresis for FE Analysis of Smart Structures," *Proceedings of Indo-German Discussion Meeting on Smart Materials and Structures*, Dept. of Aerospace Engineering, IIT Kanpur, India, 2005.
5. Bassiouny, E., Ghaled, A. F., and Maugin, G. A., "Thermodynamical Formulation for Coupled Electromechanical Hysteresis Effects-I. Basic Equations, and II. Poling of ceramics," *International Journal of Engineering Science*, Vol. 26, No. 12, 1988, pp. 1279–1306.
6. Bassiouny, E., Ghaleb, A. F., and Maugin, G. A., "Thermodynamical Formulation for Coupled Electromechanical Hysteresis Effects-III. Parameter Identification, and IV. Combined Electromechanical Loading," *International Journal of Engineering Science*, Vol. 27, No. 8, 1989, pp. 975–1000.
7. Lemaitre, J., and Chaboche, J. L., *Mechanics of Solid Materials*, Cambridge University Press, Cambridge, UK, 1995.
8. Tiersten, H. F., "Linear Piezoelectric Plate Vibration," *Plenum Press*, NY, 1994.
9. Ahmad, S. N., Upadhyay, C.S., and Venkatesan, C., "Electro-thermo-elastic Formulation for the Analysis of Smart Structures," *Smart Materials Structures*, Vol. 15, 2006, pp. 401–416.
10. Ray, M. C., and Samanta, B., "Exact Solutions for Static Analysis of Intelligent Structures Under Cylindrical Bending," *Computers and Structures*, Vol. 47, 1993, pp. 1031–1042.
11. Heyliger, P., "Exact Solution for Simply Supported Laminated Piezoelectric Plates," *Journal of Applied Mechanics*, Vol. 64, 1997, pp. 299–306.
12. Crawley, E. F., and Lazarus, K. B., "Induced Strain Actuation of Isotropic and Anisotropic Plates," *AIAA Journal*, Vol. 29, No. 6, 1991, pp. 944–951.
13. Ha, S. K., Keilers, C., and Chang, F. K., "Finite Element Analysis of Composite Structures Containing Distributed Piezoceramic Sensors and Actuators," *AIAA Journal*, Vol. 30, No. 3, 1992, pp. 772–780.
14. Tiersten, H. F., "On the Nonlinear Equations of thermo-electro-elasticity," *International Journal of Engineering Science*, Vol. 9, 1971, pp. 587–604.
15. Ahmad, S. N., Upadhyay, C. S. and Venkatesan, C., "Linear and Nonlinear Analysis of a Smart Beam Using General Electro-thermo-elastic Formulation," *AIAA Journal*, Vol. 42, No. 4, 2004, pp. 840–848.
16. Sateesh, V. L., Upadhyay, C. S. and Venkatesan, C., "Thermodynamic Modeling of Hyteresis Effects in Piezoceramics for Application to Smart Structures," *AIAA Journal*, Vol. 46, No. 1, 2008, pp. 280–284.
17. Kamlah, M., "Ferroelectric and Ferroelastic Piezoceramics Modeling of Electromechanical Hysteresis Phenomena," *Continuum Mechanics and Thermodynamics*, Vol. 13, 2001, pp. 219–269.

Multifunctional Components in Sodium Cooled Fast Reactor: Design and Development

Baldev Raj and P. Chellapandi

Abstract The important role of Sodium Cooled Fast Reactors (SFR) in ensuring sustainable supply of electricity over the long periods and SFR programme in India are highlighted. The salient features of SFRs and challenges in the design technology are addressed in this paper. The current status of 500 MWe Prototype Fast Breeder Reactor (PFBR), which is under construction at Kalpakkam is presented. Engineering of a few critical SFR components, which perform multiple functions are described. To illustrate a typical application of advanced numerical simulations, testing and evaluations methods to demonstrate the reliable operation, results of theoretical and experimental investigations carried out for the main vessel, the most critical component in PFBR, are highlighted.

Keywords Buckling of thin shells · Core disruptive accident · Design challenges Design validation Failure modes · Fuel clad · Grid plate · Main vessel · Manufacturing challenges · Seismic forces · Sodium cooled fast reactors · Subassembly · Top shield

1 Introduction

Sodium cooled Fast Reactors (SFRs) are preferred internationally for providing environmentally acceptable and sustainable energy over the years to come. Particularly, in the Indian context, SFRs are essential for effective utilization of limited natural uranium resource and also converting the abundant thorium available to ^{233}U, the fissile material for third stage of Indian Nuclear Energy Programme. SFR is an efficient tool for burning minor actinides, which contributes to waste minimization and also useful for production of isotopes like Cs^{137}, which have societal applications. Further, SFRs would provide critical liquid metal technology

B. Raj (✉) and P. Chellapandi
Indira Gandhi Centre for Atomic Research, Kalpakkam-603 102, India
e-mail: dirsec@igcar.gov.in; dmg@igcar.gov.in

B. Dattaguru et al. (eds.), *IUTAM Symposium on Multi-Functional Material Structures and Systems*, IUTAM Bookseries 19, DOI 10.1007/978-90-481-3771-8_17,

and high temperature design inputs for the future fusion and high temperature reactor systems. These apart, they can provide electricity at competitive costs over long periods. Hence, SFRs are the inevitable option for providing sustainable and energy systems, for India.

India started its fast reactor program with a research reactor called FBTR, which is a loop type 13/45 MWe, sodium cooled, mixed carbide fuelled reactor, operating since 1985. The experience gained in design, construction, commissioning and operation of FBTR as well as 390 r-y worldwide SFR operational experience, rich experience with oxide fuel, 35 years of focused R&D programme, systematic manufacturing technology development, peer reviews and synergism among DAE, R&D institutions and industries, have provided the necessary confidence to launch a 500 MWe capacity Prototype Fast Breeder Reactor (PFBR). The reactor construction, which was started in 2003, is scheduled to be commissioned by 2011. As a follow-up to PFBR, it is planned to construct four units of 500 MWe SFR reactors with improved economy and safety during 2012–2020. It has been realized that for the enhanced growth of fast reactors in the country, it is imperative to develop metallic fuelled SFRs, which promise a much higher breeding. Hence, a comprehensive programme on development of metal fuelled reactor and its fuel cycle has been undertaken with the aim of introducing them beyond 2020.

To realize the large scale deployment of SFRs, it is essential to make them competitive from economic and reliability considerations. One of the mechanisms to achieve this is to have minimum number of systems and components, which have capability of executing multiple functions, by employing novel/innovative engineering techniques. The multifunctional performance of each of these components should be demonstrated using numerical simulation, testing and evaluation techniques. A few critical multifunctional components and associated engineering aspects are addressed in this paper, with reference to PFBR. [1] provides more details.

2 Challenges and Achievements in Design and Manufacture of PFBR Components

Design of components operating in sodium at high temperature comprehensively addressing all the possible failure modes [2] as depicted in Fig. 1 for long reliable operation at temperatures around 820 K for the design life of 40 years, design of mechanisms and rotating equipment operating in sodium and argon cover gas space, handling the sodium leaks and sodium water reactions in the steam generators, seismic analysis of interconnected buildings resting on the common base raft, seismic design of thin walled vessels, pumps and absorber rod mechanisms and in-service inspection of reactor internals immersed in sodium are a few challenging issues addressed in the design. These issues have been successfully resolved through extensive numerical simulations with strong experimental investigations.

Development of large size bearings, inflatable seals, high temperature fission chambers, manufacture of large size thin walled vessels made of stainless steel with tight form tolerances, machining and assembly of grid plate and steam generators

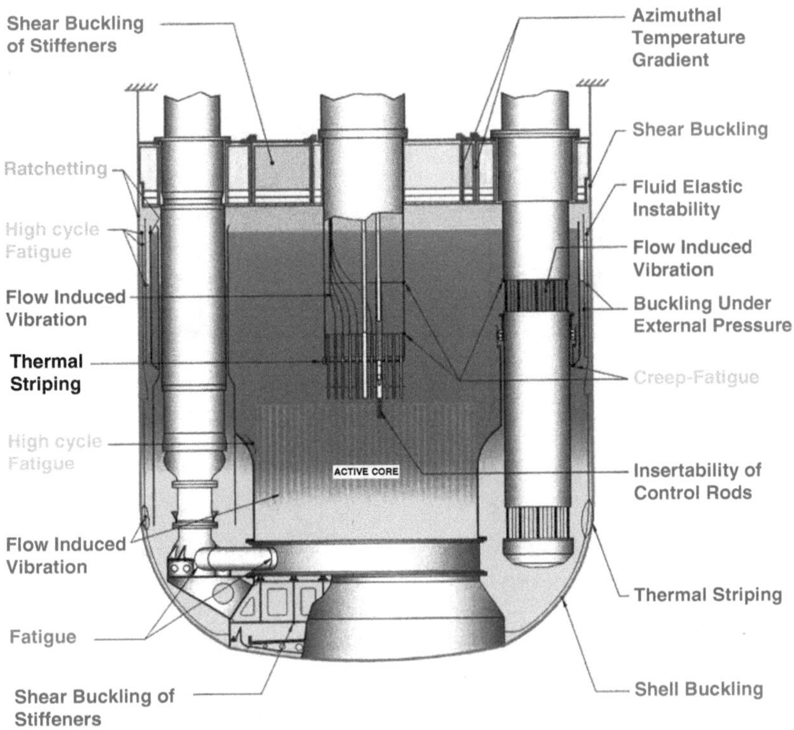

Fig. 1 Failure modes for structural design

with close tolerances, etc. are some of the challenging issues, which have been successfully resolved through detailed technology development exercise. In particular, for the large diameter thin vessels, the major manufacturing challenges are the stringent control on the manufacturing deviations, such as form tolerances ($<^1/_2$ thickness), verticality and horizontality ($<\pm 2$ mm) and high quality welds with low residual stress to be achieved without any heat treatment. Transportation and erection of the large dimensioned components respecting the stringent tolerance requirements are the challenging activities. The successful erection of safety vessel meeting the stringent geometrical tolerances (Fig. 2) is a fore runner giving confidence on the success of the project.

3 Important Multifunctional Components in FBR

There are many components that perform multiple functions. Among them, fuel clad and fuel subassembly, grid plate, main vessel, top shield and reactor vault are the critical ones which are addressed in the following paragraphs.

Fig. 2 Successful erection of safety vessel on the inner wall of reactor vault

3.1 Fuel Clad and Fuel Subassembly

Figure 3 presents the details of fuel clad pin arranged within fuel subassembly (FSA) and in addition, their functions are indicated in the figure. The fuel for SFR is manufactured in the form of pellets. Typical dimensions of pellet are 5.5 mm diameter and 8 mm height and size of clad tube is 6.6 mm outer diameter with 0.4 mm wall thickness. Each pellet has energy potential of 3,000 kWh of energy, but only 300 kWh is extracted due to structural material limitation. Hence, proper material choice is made with 20% cold worked D9 material (15% Cr-15% Ni with nano size TiC), which has better swelling resistance. Further, To achieve high extraction of heat, thin walled tube is chosen giving due considerations to manufacturing. For accommodating the fission gas release, a plenum is provided in the bottom. The triangular arrangement of tubes separated by thin wires wound helically on the tubes helps to achieve the required flow distribution. Many design features are incorporated in the FSA to achieve the intended functions: (i) concept of self standing support system helps to simplify the fuel handling scheme, (ii) providing a spacer pad at suitable locations so that the bowing of FSA would not result in positive reactivity addition, (iii) choosing hexagonal shape to achieve compactness.

3.2 Grid Plate

Figure 4 shows the details of grid plate and its functions. The grid plate (GP) is the safety class 1 component made of SS 316 LN. In order to ensure insertion of absorber rods on demand, it should have sufficient rigidity to remain flat under all operating condition and to achieve this, without increasing the weight, box type structure is opted, provided with sleeves to share the loads in a novel way. These sleeves are used for positioning the subassemblies. Apart from this, a rigid box type structure called core support structure is provided as a strong backup. The GP

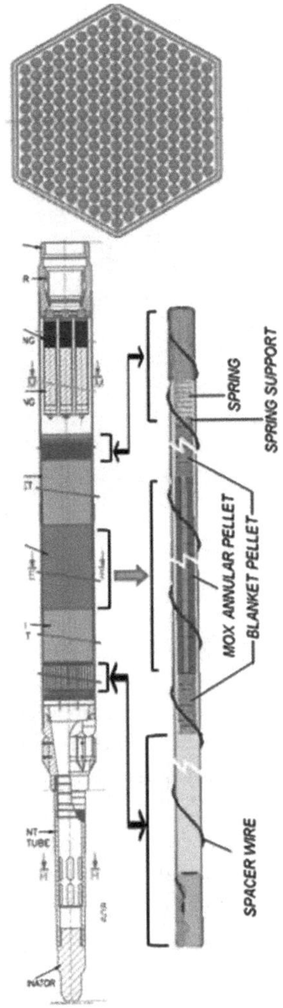

Fig. 3 ESA details

Functions
• Supports entire core consisting of 181 fuel, 120 blanket, 138 reflectors, 125 inner B4C, 12 absorber rod, 156 storage, 1026 shielding SAs.
• Acts as a plenum which receives sodium from two pumps through four primary pipes
• Distributes the sodium to various SAs meeting the flow zoning requirements.

Fig. 4 View of GP and primary sodium and their functions

Fig. 5 Hard facing of grid plate assembly

contains the coolant at a pressure of about eight bars to facilitate the smooth flow of coolant overcoming high pressure drop across core subassemblies. To prevent any unacceptable stresses, due to varying temperatures from operation to transients, the interfaces between the bottom plate and core support structure and top plate and inner vessel, are allowed to slide freely through hard facing of appropriate materials compatible to radiation environments. Machining of large diameter plates and shell assembly to the required tight tolerances, achieving sound and crack free hard facing on continuous, large diameter (diameter 6.7 m) annular tracks in bottom plate of grid plate and also in the inner surface of sleeves with nickel based cobalt free hard facing material, heat treatment of large austenitic stainless steel parts at 1,320 K with controlled rates of cooling and heating, assembly of a large number of parts (∼14,900) meeting the important requirements on verticality of sleeve assemblies (Ø0.1 mm) and handling without introducing any sagging due to self weight are the challenges in the manufacturing technology. Figure 5 shows the challenging stages of manufacture of GP.

3.3 *Main Vessel*

The geometrical details of main vessel and its associated functions are presented in Fig. 6. The main vessel is made of austenitic stainless steel type SS 316LN, which is very ductile material having high potential to absorb significant amount of mechanical energy under core disruptive accident (CDA) loading. Cylindrical vessel is chosen with specially shaped dished end that enhances buckling strength under combined pressure and line loads. Tight form tolerance ($\leq$0.2% R) is specified to enhance the overall buckling strength under normal and seismic loadings. The vessel has smooth geometry without any penetrations to eliminate stress concentrations. In order to avoid failure modes of thermal origin such as creep and embrittlement, the vessel is maintained at temperatures less than 420°C.

3.4 *Top Shield*

The top shield consists of roof slab, large rotatable and small rotatable plugs as shown along with their functions in Fig. 7. The plugs are to be rotated for the fuel

Functions
• Contains radioactive sodium and thus forms the primary containment. • Supports the core catcher, core support structure, grid plate along with four primary sodium pipes and inner vessel. • Absorbs mechanical energy released under CDA

Fig. 6 Main vessel and its functions

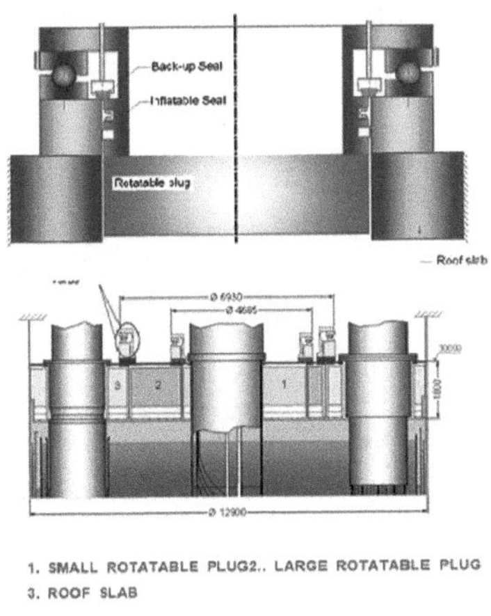

Functions
• Prevents direct contact of sodium with air in reactor containment building (RCB). • Supports important components, viz. pumps, heat exchangers, control plug and fuel handing systems. • Provides gamma radiation shielding, thermal shielding in top axial direction and facilitates fuel handing operations through rotatable plugs. • Absorbs energy release due to CDA

Fig. 7 Top shield, its components and functions

handling without any unacceptable leak of cover gas, containing radioactive elements. Combination of backup seal and inflatable seal arrangements are employed to achieve required leak tightness respectively during normal as well as fuel handling conditions. The box type concept filled with concrete is adopted to meet the rigidity and shielding requirements. In order to cool the top shield due to heat flux from sodium free surface through conduction, convection, radiation and condensation mode of sodium aerosols, required air flow is maintained through passages provided by cooling box sectors. In order to prevent buckling of bottom plate under sodium impact pressure due to CDA, spacer rods are embedded in the concrete with the optimized gap so that air flow passages are not penalized. In the domain of manufacturing, apart from machining of large diameter bearing, welding of carbon steel plates without laminar tearing is one of the critical problems during welding, is the challenging task.

4 R&D Towards Design Validation: A Case Study on Main Velsse

4.1 Investigation of Buckling Under Seismic Induced Forces

The stress and pressure distribution extracted from the seismic analysis performed in the first phase are used as input data for the subsequent buckling investigation. The analysis has been validated with shake table tests. Figure 8(a) shows 3D finite element mesh generated for the analysis. The integrated seismic forces acting at the mass center are indicated schematically (Fig. 8(b)), under which the upper cylindrical portion of the main vessel is subjected to shear buckling (Fig. 8(c)). Buckling analysis is performed taking into account the effects of plasticity, geometrical

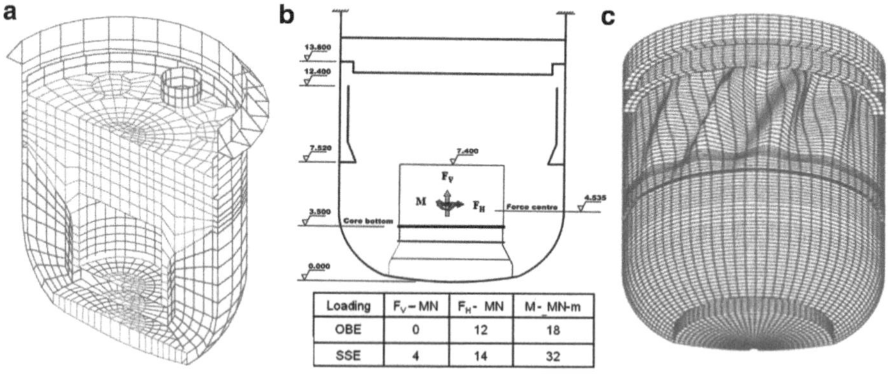

Loading	F_V – MN	F_H - MN	M - MN-m
OBE	0	12	18
SSE	4	14	32

Fig. 8 Buckling analysis of main vessel under seismic forces

Fig. 9 Buckled shells (1/18th model of main vessel) vessels under shear force

imperfections and temperatures. The effect of accumulated plastic deformation due to thermal ratcheting mechanism, in the vicinity of sodium free level on the buckling strength is also included [3]. Finally it is demonstrated that main vessel is free from risk of buckling. The details are given in [4]. By extensive testing and evaluation based on scaled down model studies, the buckling integrity of main vessel has been well established (Fig. 9).

4.2 Structural Integrity Assessment Under CDA

The transient response analysis under CDA calls for advanced numerical simulation techniques for the treatment of large distortions in the fluid domains, strong geometrical and material non-linearities in structures, fast transient fluid structure interaction, sodium slug impact on the top shield, shock wave propagation, interactions between the moving interfaces of various fluids and automatic rezoning of finite element mesh. Addressing all the complexities, an in-house computer code FUSTIN has been developed based on Arbitrary Lagrangean Eulerian (ALE) formulation for fluids and convected co-ordinate formulation for structures. Tests were conducted at Terminal Ballistic Research Laboratory, Chandigarh, to generate the test data towards validating the code. The performance code is found to be extremely satisfactory (Fig. 10). Subsequently, the structural integrity of main vessel is ensured. Towards this, the transient displacements and strains were predicted for the main vessel by sequentially introducing the reactor internals which provided good understanding of the sequence of events. Figure 11 shows the numerically predicted core bubble expansion at a few discrete intervals during CDA having the work potential of 100 MJ. The important sequence of events observed from the analysis without any internals are: the loading on the bottom portion of main vessel due to direct impact of shock/pressure waves, upward lifting of sodium slug compressing the cover gas space, bulging of upper portion of main vessel below the roof slab junction due to local pressure generated, subsequent to slug impact and downward motion of sodium slug compressing and distorting the core bubble. More details can be found in [5].

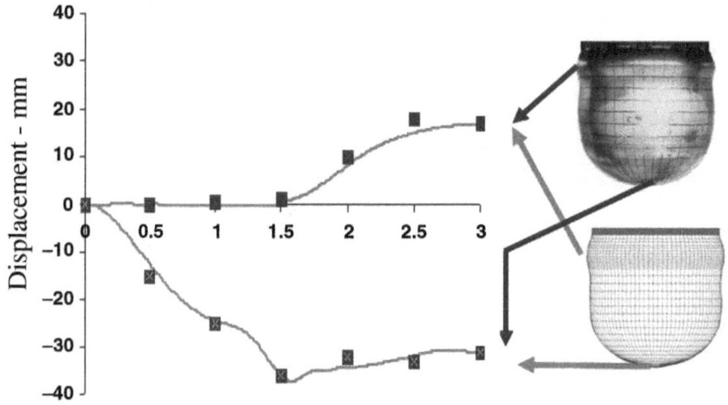

Fig. 10 Validation of FUSTIN Code for MV model

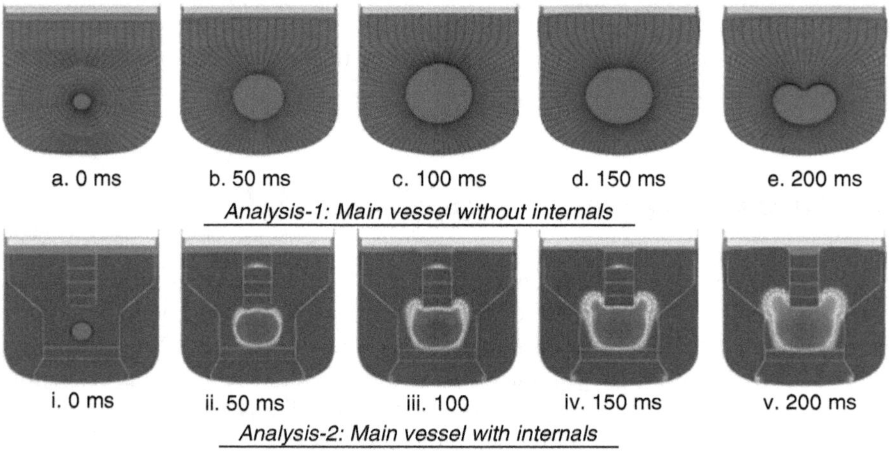

Fig. 11 Core bubble expansion behavior predicted by FUSTIN

5 Conclusion

The critical multifunctional components in a typical pool type SFR are described along with associated engineering aspects incorporated to accomplish the intended functions. Results of a few typical R&D activities completed to validate the structural design of main vessel are highlighted. It is clear from the presentations that SFRs have ever growing demand for the advanced R&D activities both in science and technology, which needs to be carried out through multi-institutional collaborations.

References

1. Baldev Raj (2008), Science and Technology of Fast Breeder Reactor Programme in India: Challenges and Achievements, Jai Krishna Memorial Lecture.
2. P. Chellapandi, et al. (2009), Investigation of Structural Mechanics Failure Modes in FBR, Pressure Vessels and Piping: Codes, Standards, Design and Analysis, Naros publishing House, New Delhi.
3. P. Chellapandi, et al. (2000), Theoretical and Experimental Investigations on Thermal Ratchetting of Main Vessel of an FBR, Tran. Indian Institute of Metals, 53: 3: 391–399.
4. P. Chellapandi, et al. (2008), Investigation on Buckling of FBR Vessels Under Seismic Loading with Fluid Structure Interactions, J N Eng Design, 238: 3208–3217.
5. P. Chellapandi, et al. (2000), Effect of Reactor Internals on Structural Integrity of PFBR Main Vessel under CDA, ASME, PVP, 403:161–172.

Section IV
Smart Sensors, Structural Health Monitoring

From Structural Mechanics to Inspection Processes: Getting Structural Health Monitoring into Application for Riveted Metallic Structures

C. Boller and M.R. Mofakhami

Abstract Structural health monitoring (SHM) has been around now in research for more than 2 decades. However not many SHM applications have been realised so far for various reasons. The paper addresses those reasons for the case of acoustic wave monitoring by discussing state-of-the-at in SHM, the issue of scattered acoustic waves around holes in plates with gradually enhancing the complexity of the structural component considered from an infinite to the finite plate and allowing a crack to start and to grow from the hole. This sequence intends to exemplify in the end a riveted structure and a way on how increased structural complexity enhances the complexity of an acoustic signal sent through these structures for the purpose of monitoring. This is discussed on the basis of numeric and experimental results regarding the size of cracks to be monitored and what strategy of monitoring might be used best such that a reliable monitoring result is obtained for a riveted structure. Finally a procedure is explained that can be applied such that structural components, and in this case riveted structures, can be identified on an aircraft for which SHM is a promising solution in terms of aircraft operability.

Keywords Acoustic waves · Fatigue · Riveted metallic structures · Structural health monitoring

1 Introduction

Structural health monitoring (SHM) has now been around for more than 2 decades, at least as an acronym for something which as its ‘ingredients’ has been around for an even much longer period of time. It may be difficult to give a definition of what SHM exactly is but what it could roughly be could be the following:

C. Boller
Saarland University, Gemany
e-mail: christian.boller@izfp.fraunhofer.de

M.R. Mofakhami (✉)
Bombardier Aero, Canada
e-mail: Mohammad.Reza.Mofakhami@aero.bombardier.com

B. Dattaguru et al. (eds.), *IUTAM Symposium on Multi-Functional Material Structures and Systems*, IUTAM Bookseries 19, DOI 10.1007/978-90-481-3771-8_18,

SHM is the integration of sensing and possibly also actuation devices into a structure to allow the loading and damaging conditions to be recorded, analysed, localised and predicted in a way that non-destructive testing (NDT) becomes an integral part of the structure. SHM is ***not*** *an additional NDT application, nor quantitative NDT (QNDT), a fatigue analysis, vibration analysis, signal processing, material science or any other single technological area but rather the lateral integration of all those in terms of a structure's life cycle management.*

SHM has been triggered in a variety of industries but where it has been possibly discussed most is in aerospace. Although there is no 'birth' date of SHM in aerospace truly defined, a rather logical one would be the de Havilland Comet crashes in 1954 and the consequences resulting from this. A consequence of this was loads monitoring to be integrated at least into military aircraft using accelerometers, strain gauges and flight parameters. Another major step triggering SHM was the Aloha Airlines accident in 1988 where damage monitoring raised awareness. Although much awareness has been generated and a lot of dedicated research work in SHM done, this has now at least been summarised in documents such as [1]. A valid question raised from all this may however be: Why is there so little SHM in aerospace today? Some of the reasons for this may include:

- Lack of SHM technology readiness level and reliability
- SHM potentials not sufficiently quantified
- Manufacturers only consider SHM for new aircraft (design issue)
- Aircraft operators not aware of the SHM potentials, reluctant to release their operational conditions (operational issue), and afraid of getting more sensors into aircraft (reliability issue)
- Airworthiness authorities not very well integrated into the technology development and integration process
- Data integration issues not sufficiently solved (vertical integration)
- Insufficient appreciation of activities in neighbouring fields (i.e. engine health management, avionics built in test equipment, helicopters health and usage management).

Some further argumentation for those reasons will be given in the following.

2 State-of-the-Art in Riveted Joint Monitoring

It is difficult to say which of an aircraft's components is relevant for SHM if this has not been analysed before. However, as a consequence of the Aloha Airlines accident, monitoring of riveted lap joints looks relevant. A riveted lap joint panel taken from a real Airbus aircraft has therefore been tested recently in a fatigue test (Fig. 1, where different commercially available SHM technologies including acoustic emission, acousto-ultrasonics (AU) and phased array were validated. From those three SHM technologies the Smart Suitcase and Smart Layer from Acellent Technologies [2] scored slightly better than the others. Figure 1 shows the Smart Layer placed under

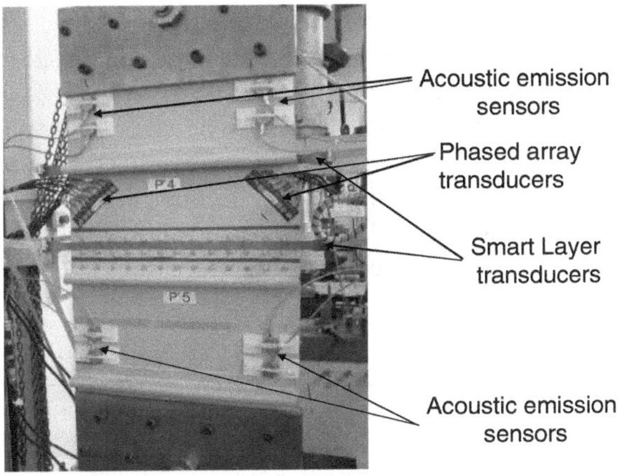

Fig. 1 Riveted aircraft fuselage lap joint fatigue test setup for SHM technology comparison

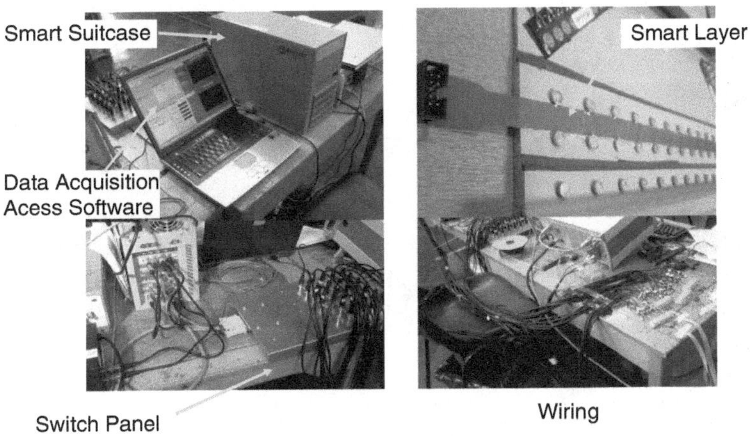

Fig. 2 Snapshots of test setup with Acellent Smart Suitcase and Smart Layer AU equipment

the stringer (actuator) and between the rivet lines (sensor) respectively. Figure 2 shows some snapshots of the 'infrastructure' that had to be established. The upper left hand side shows the Smart Suitcase as a PC rack containing a signal generation and acquisition unit combined with an amplifier. The Smart Suitcase is linked to a laptop, which serves as a graphical unit interface and on which the data processing and visualisation software ACESS is run. The upper right hand figure shows the Smart Layer bonded between the rivet lines with a connecting plug at each end to serve seven of the piezoelectric transducers on each half of the layer's strip. All of this is what has been provided from the commercial side.

The lower two pictures on Fig. 2 show what all had to be done manually on site, which mainly consisted of configuring and building a switch box to serve all the different transducers (Fig. 2 lower left) and getting the transducers and plugs wired (Fig. 2 lower right). Hence from that situation it can be concluded that SHM is still more on a laboratory than on an industrial scale so far.

The monitoring strategy to be applied has been based on a direct path monitoring approach (Fig. 3). Here only the wave package arriving first (direct path) is analysed with all other wave packages resulting from reflections being left out (Fig. 4). It is worth noting that the fist wave package shown in Fig. 4 does not result from an acoustic signal but rather from an electromagnetic crosstalk.

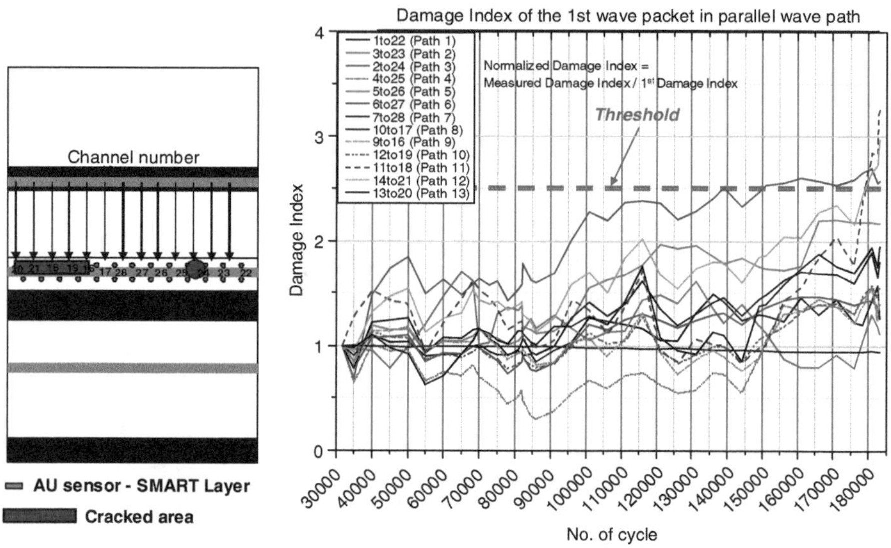

Fig. 3 Riveted panel and direct signal paths considered (*left*) and resulting damage indices for the different paths (*right*)

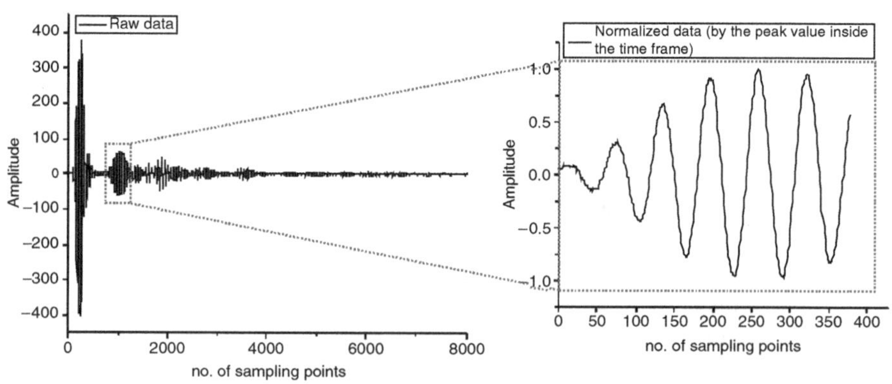

Fig. 4 Recorded time domain sensor signal (*left*) with direct path signal windowed (*right*)

Figure 3 (right) shows the results obtained for each direct path in terms of a damage index, which was simply defined as the signal amplitude where the amplitude of each of the reference signals (signal at the onset for the undamaged condition) was normalised to one. A broad spread of these signals along the fatigue test is seen, the reasons for this being various, such as transducer adhesion, strain transfer and possibly others. While the results gradually diverge there are some of the paths where the damage index suddenly increases considerably towards the end of the fatigue test. A threshold value was therefore introduced, that was defined above the scattering of the different path signals. Since cracks appearing on the surface of the specimen only happened at a very late stage, not many opportunities for comparing the results obtained from SHM with those obtained from classical NDT were provided. However, analysing the fracture surfaces around the different rivets for the broken panel provided a solution (Fig. 5) to measure cracks that had not appeared on the panel's surface before final fracture. Figure 5 shows a crack starting from a rivet and the various crack lengths measured visually on the fracture surface versus the damage index determined with the SHM procedure. Two groups separated by the threshold value set, as well as a crack length of around 5 mm can be identified, where the crack length has been mentioned already before [3]. Above this value crack lengths seem to be clearly monitored by SHM while below a cluster of rather more indifferent results is obtained. It can be concluded that if a crack in a riveted panel structure and possibly even others was to be monitored by means of SHM it would have to be significantly above 5 mm.

The variations in damage index progression for the different direct paths can again be seen in Fig. 5 (left). Some of the reasons might be understood when looking at the cross section of the lap joint shown in Fig. 6.

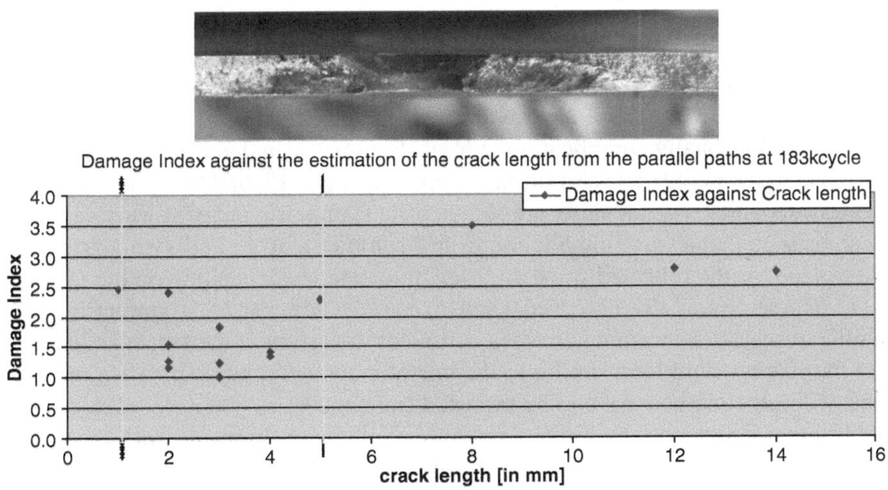

Fig. 5 Fracture surface around countersunk rivet holes after fatigue test (*top*) and SHM based damage index versus measured crack lengths at different rivet holes (*bottom*)

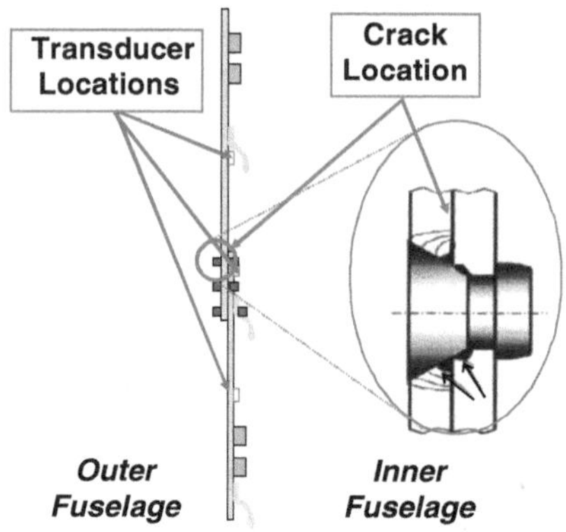

Fig. 6 Schematic view of the cross section of a riveted lap joint with counter sunk rivets

3 Acoustic Signals Reflected from Differently Shaped Boundaries

The geometry of a lap joint can be complex. This is further enhanced once a crack emerges. Furthermore the transducer layers would be placed on the inside of an aircraft fuselage hence a signal to be transmitted from one Smart Layer to another with a crack to occur on the inside of the outer panel would at least have to cross a lap interface once enhancing an acoustic signal's attenuation.

Considering cracked riveted lap joints and trying to understand the various phenomena ongoing while an acoustic signal would be sent through, may lead to gradually analyse a structure starting from an infinite plate over a finite plate with orthogonal edges, tapering the edges, introducing a hole, adding a notch to the hole, filling the hole with a rivet, adding a crack into the hole, doubling the structure to a lap joint, etc. Some of those cases are described in the following.

Solutions for waves in an infinite plate and for a circular hole in a finite plate can be found variously [4]. However, looking at a tapered edge resulting from a countersunk rivet additionally complicates the topic. Symmetric and anti-symmetric waves were sent into tapered plates for different tapering angles [5] and the results obtained numerically and experimentally are shown for a symmetric incident wave in Fig. 7. It can be seen that a wave modulation from a symmetric to an anti-symmetric wave takes place for the reflected wave at different intensities depending on the tapering angle. The same was also observed when inversely sending an anti-symmetric wave. Similar observations were made when cutting off a portion of the tapered edge such that the shape would look similar to the one of a countersunk rivet. Wave modulations will therefore also have to be expected from cracks.

Different lengths of cracks starting from a hole were simulated with the results being plotted in terms of the scattered waves around the holes as shown in Fig. 8. It becomes obvious that transducers need to be no more than 150 mm away from the crack such that it can be reliably monitored.

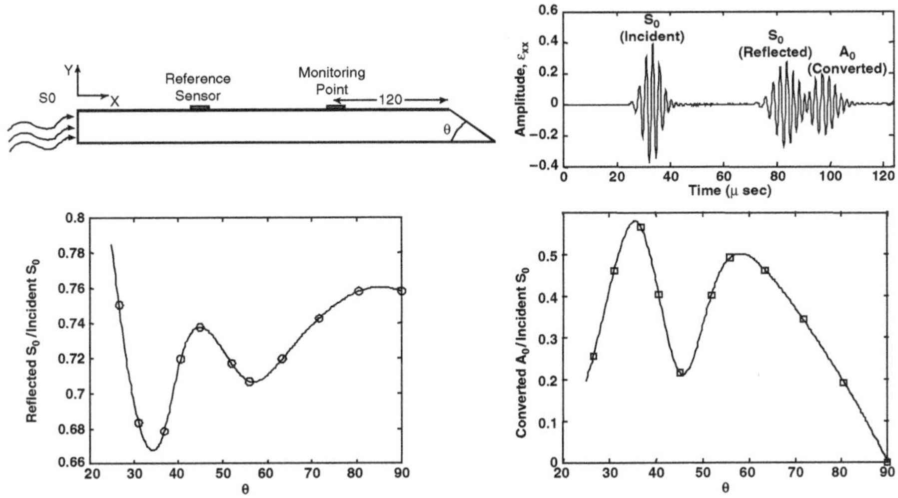

Fig. 7 Wave modulation of a reflected symmetric wave at a tapered edge: reflected symmetric wave (*lower left*) and anti-symmetric wave (*lower right*)

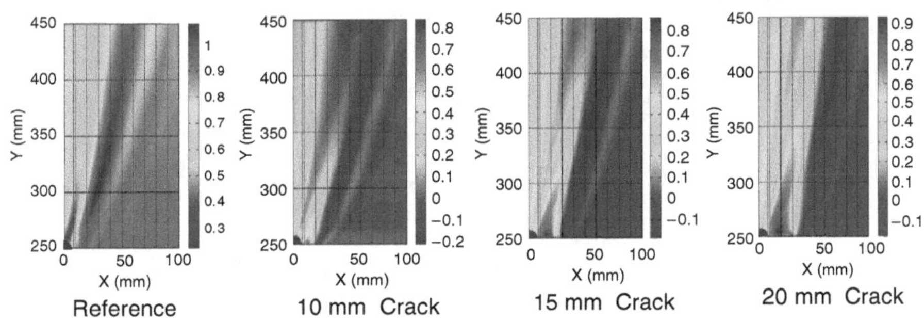

Fig. 8 Distribution of scattered waves behind a hole (quarter section) for different crack lengths

Knowledge of the distribution of scattered waves around holes can further help to establish monitoring strategies. For a plate with three holes run in a fatigue test and monitored with a set of piezoelectric transducers a crack was observed to start to grow to the inside of the plate from what is shown as the upper hole on Fig. 9 [7]. By using a variety of actuator-sensor paths going through the cracked area allowed crack propagation to be identified through a correlation of the different paths. Additional attention has to be placed as to what degree a crack might be open or not with regard to the loading condition, as this might influence the transmission of the acoustic signal as well [7].

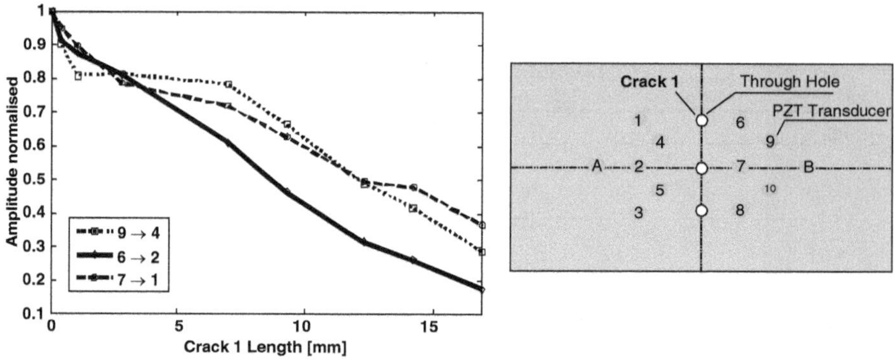

Fig. 9 Crack monitoring strategy for a three hole plate cracked specimen under fatigue loading

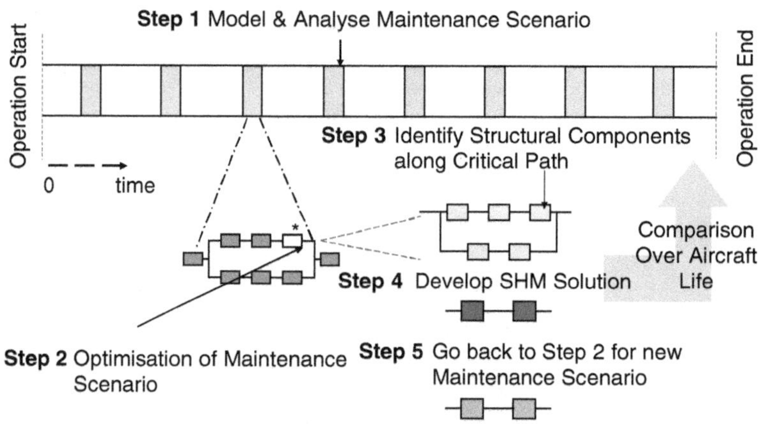

Fig. 10 Approach for identification of aircraft components with SHM potential along an aircraft maintenance process

4 Finding Components with Adequate SHM Potential

Application of SHM applies where it provides an economic benefit. With aircraft where asset costs are high, operability becomes crucial. Any hour gained is therefore easily worth a good four digit value in Euros per hour. Aircraft operability could be enhanced by reducing the maintenance time without compromising safety and reliability. SHM could help by automating the inspection process. A way on how to identify which components would positively contribute to enhance the inspection process can be by simply simulating the aircraft maintenance process and analysing the critical path. Any components being along the critical path would be those considered for SHM first, where the resulting potential could again be obtained from simulation. Figure 10 shows a five step approach that can be iteratively applied and which has been described in further detail in [8, 9].

5 Conclusions

It can be concluded that SHM still does not possess the maturity levels required for industrial applications. There is still work to be done to provide the reliability levels required and to identify components with SHM relevance. It can be assumed that riveted lap joints would be one type of metallic structures with high SHM relevance. Riveted lap joints are sufficiently complex to be well understood in terms of SHM. A systematic approach from a mechanics point of view therefore needs to be established, where limitations of SHM can be recognised. So far cracks of a size larger than 5 mm might be reliably monitored with SHM using AU with transducers to be placed 50–150 mm away from the crack location. Reliability and precision of the results obtained might also depend on the monitoring strategy to be used, where more needs to be done to get the appropriate monitoring strategies qualified. Finally SHM critical components need to be identified through critical path analysis of the maintenance process and an aircraft's potential quantified in terms of operability gained, by feeding the SHM solution back into the maintenance process simulation.

Acknowledgements Much of the work presented here has been funded under the European Commission Framework Programme 6 which is highly acknowledged together with the contributions from Yiu Sum Leung and Hrshi Kapoor while performing research at the University of Sheffield.

References

1. Boller C, Chang F-K, Fujino Y (Eds.) (2009) Encyclopedia of Structural Health Monitoring. Wiley
2. www.acellent.com
3. Staszewski W J, Boller C, Tomlinson G R (2003) Health Monitoring for Aircraft Structures. Wiley
4. Worden K (2001) Rayleigh and Lamb waves – Basic principles. Strain, Vol. 37, p167–172
5. Mofakhami R M, Boller C (2008) Lamb Wave Interactions with Non-symmetric Features at Structural Boundaries. Manuscript submitted to Zeitschrift für Angewandte Mathematik und Mechnik (ZAMM)
6. Mofakhami R M, Boller C (2008) Scattering of Lamb waves from edge cracks in circular holes for understanding damage detection in more complex structures. Proc. of the 2nd Asia-Pacific Workshop on SHM, Melbourne/Australia
7. Mofakhami R M, Boller C (2008) Effect of the Senor Location on Damage Detection in a Plate like Structure Using Acousto-Ultrasonic Technique. International Workshop on Smart Materials and Structures (Cansmart), Montreal/Canada
8. Boller C, Kapoor H and Goh W T (2007) Structural Health Monitoring Potential Determination Based on Maintenance Process Analysis. SPIE Smart Materials and Structures Symposium, San Diego/CA, USA, SPIE Vol. 6527
9. Kapoor H, Goh W T, Boller C (2008) Optimisation and Simulation Approach for the Determination of SHM Potentials on Commercial Aircraft. Proceedings of 4th European Workshop on Structural Health Monitoring; DEStech Publ. Inc.

Shaped Modal Sensors for Uncertain Dynamical Systems

Michael I. Friswell and Sondipon Adhikari

Abstract The idea of using modal sensors and actuators for beam and plate type structures has been a subject of intense interest for many years. Using modal sensors in active control reduces problems of spillover, where high frequency unmodelled modes affect the stability of the closed loop system. This paper is concerned with distributed sensors made of piezoelectric material to measure the response of beam and plate structures. The design of modal sensors for beam structures is well established. For example, a modal sensor for a beam type structure may be obtained by varying the sensor width along the length of the beam. If the sensor covers the whole beam the shape of the sensor may be derived using the mode shape orthogonality property. Friswell considered modal sensors that cover only part of the beam, segmented modal sensors for multiple modes, and the effect of geometric tolerances during manufacture on the quality of the sensors. Friswell parameterized the width of the sensor for beam structures using the finite element method and uses the underlying shape functions to approximate the transducer shape. For plate structures constant thickness sensors are difficult to design, although methods that parameterize sensor boundary, or using topology optimization, have been suggested. Currently modal sensors and actuators are designed using deterministic models of the structure, and assuming that they may be manufactured perfectly. The performance of the transducers is critically dependent on these assumptions and this paper investigates the effect of modelling uncertainty and manufacturing errors. The origin of these uncertainties and their form are discussed, and their effect is modeled and propagated through the system to determine the errors induced on the modal outputs. Simulated examples are used to demonstrate the issues raised.

Keywords Beam · PVDF · Piezoelectric · Shaped sensor · Stochastic

M.I. Friswell and S. Adhikari
School of Engineering, Swansea University, Singleton Park, Swansea SA2 8PP, UK
e-mail: m.i.friswell@swansea.ac.uk; s.adhikari@swansea.ac.uk

B. Dattaguru et al. (eds.), *IUTAM Symposium on Multi-Functional Material Structures and Systems*, IUTAM Bookseries 19, DOI 10.1007/978-90-481-3771-8_19,

1 Introduction

The consideration of uncertainties in numerical models to obtain the probabilistic descriptions of vibration response and robustness of vibration control with respect to uncertainties is becoming more desirable for industrial scale finite element models. Such uncertainties include, but are not limited to (a) parametric uncertainty (e.g., uncertainty in geometric parameters, friction coefficient, strength of the materials involved); (b) model uncertainty (arising from the lack of scientific knowledge about the model which is unknown a-priori); (c) experimental error (uncertain and unknown errors percolate into the model when they are calibrated against experimental results) and (d) computational uncertainty (e.g., machine precision, error tolerance and the so called ‘h’ and ‘p’ refinements in finite element analysis). These uncertainties must be assessed and managed for robust vibration control.

In this paper we focus our attention on dynamical systems with parametric uncertainties. In the parametric approach, the uncertainties associated with the distributed system parameters, such as Young’s modulus, mass density, Poisson’s ratio and geometric parameters are quantified using statistical methods. Problems of structural dynamics in which the uncertainty in specifying mass, damping and stiffness of the structure is modeled within the framework of random fields can be treated using the stochastic finite element method [10, 14, 16, 17].

The stochastic finite element model of the structure can be represented in the form

$$\mathbf{M}\ddot{\mathbf{q}}(\theta) + \mathbf{D}(\theta)\dot{\mathbf{q}} + \mathbf{K}(\theta)\mathbf{q} = \mathbf{B}\mathbf{u} \tag{1}$$

$$\mathbf{y} = \mathbf{C}\mathbf{q} \tag{2}$$

Here $\mathbf{M}(\theta)$, $\mathbf{D}(\theta)$ and $\mathbf{K}(\theta)$ are the random mass, damping and stiffness matrices based on the degrees of freedom, $\mathbf{q}$. The inputs to the structure, $\mathbf{u}$, are applied via a matrix $\mathbf{B}$ which determines the location and gain of the actuators (or the actuator shape for distributed actuators). Similarly the outputs, $\mathbf{y}$, are obtained via the output matrix $\mathbf{C}$ which is determined by the sensor shape. The notation θ is used to denote the random nature of the system matrices.

This paper develops a method to design shaped modal sensors for Euler-Bernoulli beams with uncertain properties. It will be shown that due to the uncertain nature of the system matrices, several mathematical choices are possible for optimal design. Here a simple and physically realistic approach is developed and the resulting formulation is applied to a beam with bending stiffness and mass per unit length modeled as stationary normal random field with a given mean value and exponentially decaying auto-correlation function.

2 Defining Shaped Sensors for Beam Structures

The shape of a transducer is a continuous function. However this function needs to be parameterised to enable the optimisation of the sensor shape. Using the shape functions of the underlying finite element model is a convenient approach to approximate the width of the piezoelectric material. In this way modal transducers may be designed for arbitrary beam type structures. Furthermore modal transducers that only cover part of a structure may be designed. Most of the development will concern sensors, although actuators may be dealt with in a similar way.

Suppose a single polyvinylidene fluoride (PVDF) film sensor is placed on the beam with a shape defined by a variable width $f(\xi)$, where ξ denotes the length along the beam element. Incorporated into $f(\xi)$ is both the physical width of the sensor, and also the polarisation profile of the material. The central feature of the proposed method is to approximate this width using the shape functions of the underlying finite element model. Thus the sensor width within element number e is approximated as

$$f_e(\xi)\left[N_{e1}(\xi)\ N_{e2}(\xi)\ N_{e3}(\xi)\ N_{e4}(\xi)\right]\begin{Bmatrix} f_{e1} \\ f_{e2} \\ f_{e3} \\ f_{e4} \end{Bmatrix}, \quad 0 \leq \xi \leq \ell_e \tag{3}$$

where ℓ_e is the length of the element, $N_{ej}, j = 1, 2, 3, 4$ are the shape functions for an Euler-Bernoulli beam, and the constants f_{ei} must be determined. This approximation has the advantage that the width and slope of the sensor are continuous at the nodes of the finite element model. The output (voltage or charge) from the part of the sensor with element number e is

$$y_e(t) = K_s \int_0^{\ell_e} f_e(\xi) \frac{\partial^2 w_e(\xi, t)}{\partial^2 \xi} d\xi \tag{4}$$

where the constant K_s is determined by the properties of the piezoelectric material and w_e is the translational displacement of the beam [13, 15]. Approximating this displacement by the shape functions the sensor output for the element can be expressed as

$$y_e = \begin{Bmatrix} f_{e1} \\ f_{e2} \\ f_{e3} \\ f_{e4} \end{Bmatrix}^\top \mathbf{C}_e \begin{Bmatrix} w_{e1} \\ w_{e2} \\ w_{e3} \\ w_{e4} \end{Bmatrix} \tag{5}$$

where the (i, j)th element of the matrix $\mathbf{C}_e$ is

$$\mathbf{C}_{e_{ij}} = K_s \int_0^{\ell_e} N_{ei}(\xi) N''_{ej}(\xi)\, d\xi \tag{6}$$

The sensor output, y, is the sum of the contributions of the elements given by,

$$y = \sum_e y_e = \mathbf{f}^\top \mathbf{C}_s \mathbf{q}. \tag{7}$$

Here the element matrices have been assembled into the global matrix $\mathbf{C}_s$, in the usual way. The element nodal displacements, w_{ei}, have been incorporated into the global displacement vector $\mathbf{q}$, and the sensor nodal *widths* f_{ei} have been assembled into a global vector $\mathbf{f}$. However, the sensor nodal widths at the clamped or pinned boundary conditions are not set to zero, whereas the corresponding displacements are set to zero. Thus in general $\mathbf{C}_s$ is a rectangular matrix. Comparing Eqs. (2) and (7), it is clear that

$$\mathbf{C} = \mathbf{f}^\top \mathbf{C}_s. \tag{8}$$

3 Modal Sensors for the Baseline System

In this section we review the derivation of the modal sensors for the baseline (deterministic) system. Proportional damping will be assumed so that the mode shapes of Eq. (1) are real, and equal to the mode shapes of the undamped system. For light damping this approximation will introduce small errors. The mode shapes, $\boldsymbol{\Phi}$, are assumed to be normalized arbitrarily so that the modal mass is

$$\boldsymbol{\Phi}^\top \mathbf{M} \boldsymbol{\Phi} = \mathbf{M}_m. \tag{9}$$

Applying the transformation to modal co-ordinates, $\mathbf{q} = \boldsymbol{\Phi}\mathbf{p}$, to Eq. (1), gives

$$\ddot{\mathbf{p}} + 2\mathbf{Z}\Omega\dot{\mathbf{p}} + \Omega^2\mathbf{p} = \mathbf{M}_m^{-1}\boldsymbol{\Phi}^\top \mathbf{B}\mathbf{u} \tag{10}$$

$$\mathbf{y} = \mathbf{C}\boldsymbol{\Phi}\mathbf{p} = \mathbf{C}_\mathbf{p}\mathbf{p} \tag{11}$$

where

$$\Omega = \text{diag}\left[\omega_1, \omega_2, \ldots, \omega_n\right], \quad \mathbf{Z} = \text{diag}\left[\zeta_1, \zeta_2, \ldots, \zeta_n\right] \tag{12}$$

are diagonal matrices of the natural frequencies and modal damping ratios and

$$\boldsymbol{\Phi} = \left[\phi_1, \phi_2, \ldots, \phi_n\right] \tag{13}$$

is a $n \times n$ matrix of mode shapes. Thus, using Eq. (8),

$$\mathbf{C}_\mathbf{p} = \mathbf{C}\boldsymbol{\Phi} = \mathbf{f}^\top \mathbf{C}_s \boldsymbol{\Phi}. \tag{14}$$

The modal sensor design problem is then to determine the sensor shape, defined by $\mathbf{f}$, to give the required modal output gain matrix, $\mathbf{C}_\mathbf{p}$. Clearly the modes must be scaled in some consistent manner, since otherwise for a particular set of output gains the modal output gain matrix will change depending on the mode scaling. This is particularly important when more than one mode is measured. Usually the number of elements describing the sensor shape is large and so Eq. (14) will be underdetermined. In this case the pseudo inverse solution will produce the minimum norm solution.

An alternative is to minimize transducer curvature, while ensuring zero sensitivity to unwanted modes [8]. The minimum curvature ensures that the transducer may be manufactured as easily as possible. Thus, we wish to minimize

$$J_c(\mathbf{f}) = \sum_e \int_0^{\ell_e} f_e''(\xi)^2 \, \mathrm{d}\xi = \sum_e \begin{Bmatrix} f_{e1} \\ f_{e2} \\ f_{e3} \\ f_{e4} \end{Bmatrix}^{\top} \mathbf{H}_e \begin{Bmatrix} f_{e1} \\ f_{e2} \\ f_{e3} \\ f_{e4} \end{Bmatrix} \tag{15}$$

where

$$\mathbf{H}_{eij} = \int_0^{\ell_e} N_{ei}''(\xi) \, N_{ej}''(\xi) \, \mathrm{d}\xi. \tag{16}$$

$\mathbf{H}_e$ looks like the element stiffness matrix with a unit flexural rigidity. Assembling the contributions from all of the elements gives

$$J_c(\mathbf{f}) = \mathbf{f}^{\top}\mathbf{H}\mathbf{f} \tag{17}$$

where $\mathbf{H}$ contains the element matrices, $\mathbf{H}_e$, and is symmetric. The sensor design problem then requires that J_c is minimized, subject to the constraints given by Eq. (14). This problem may be solved using Lagrange multipliers [7, 8], as the solution of

$$\begin{bmatrix} 2\mathbf{H} & \mathbf{C}_s\boldsymbol{\Phi} \\ \boldsymbol{\Phi}^{\top}\mathbf{C}_s^{\top} & \mathbf{0} \end{bmatrix} \begin{Bmatrix} \mathbf{f} \\ \boldsymbol{\lambda} \end{Bmatrix} = \begin{Bmatrix} \mathbf{0} \\ \mathbf{C}_{\mathrm{p}}^{\top} \end{Bmatrix} \tag{18}$$

where $\boldsymbol{\lambda}$ is the vector of Lagrange multipliers.

4 Modal Sensors for Uncertain Systems

The essential requirement for modal sensors is to enforce the constraints given by Eq. (14) for the modes of interest. In most practical cases the number of constraints is fewer than the number of parameters determining the sensor shape and hence other metrics may be optimized, for example the sensor curvature outlined above. The development thus far has only considered a deterministic system. In reality the physical system will be uncertain, meaning that the mode shapes will be stochastic. One possibility is to use the freedom available in designing the sensor shape to ensure robustness to the system uncertainty for the modal sensors.

Consider the terms in Eq. (14), repeated here for convenience,

$$\mathbf{C}_{\mathrm{p}} = \mathbf{f}^{\top}\mathbf{C}_s\boldsymbol{\Phi}(\theta). \tag{19}$$

Note that the effect of system uncertainty is reflected by the random nature of the modal matrix $\boldsymbol{\Phi}(\theta)$.

$\mathbf{C}_s$ is determined from the element shape functions and is therefore fixed for a given mesh and will not change with the system parameters. Clearly the mode shapes of interest, $\boldsymbol{\Phi}$, will vary with the uncertain parameters. The vector $\mathbf{f}$ determines the shape of the sensor, and is obtained from the system optimization. Here we will assume this vector is deterministic, although of course implementing a required sensor shape in practice will be subject to manufacturing errors not considered in this paper. Thus for a given sensor the system uncertainty will produce a stochastic modal output vector $\mathbf{C_p}$. Suppose the desired modal output vector is denoted $\mathbf{C}_{\mathbf{p}d}$. Then to ensure the correct modal response we will enforce the constraint

$$\mathrm{E}\left[\mathbf{C_p}\right] = \mathbf{C}_{\mathbf{p}d} = \mathbf{f}^\top \mathbf{C}_s \mathrm{E}\left[\boldsymbol{\Phi}(\theta)\right] \tag{20}$$

where E [ ] denotes the expected value. Assuming that the mode shapes have mean $\boldsymbol{\Phi}_0$, we have

$$\mathbf{C}_{\mathbf{p}d} = \mathbf{f}^\top \mathbf{C}_s \boldsymbol{\Phi}_0 \tag{21}$$

To ensure robustness we will minimize the sum of the variances of the modal outputs given by

$$J_s = \mathrm{E}\left[\left(\mathbf{C_p} - \mathbf{C}_{\mathbf{p}d}\right)\left(\mathbf{C_p} - \mathbf{C}_{\mathbf{p}d}\right)^\top\right]. \tag{22}$$

The above choice is clearly not unique. Another alternative could be to generate the full covariance matrix between different modal outputs and perhaps minimize some (matrix) norm of the full covariance matrix $\left(\mathbf{C_p} - \mathbf{C}_{\mathbf{p}d}\right) \otimes \left(\mathbf{C_p} - \mathbf{C}_{\mathbf{p}d}\right)$ (where $\otimes$ denotes the Kroneker product). This would involve the solution of a very large n^2 dimensional matrix optimisation problem. Recalling that the mode shapes have mean $\boldsymbol{\Phi}_0$, the required optimization is to minimize

$$J_s = \mathbf{f}^\top \mathbf{C}_s \mathrm{E}\left[(\boldsymbol{\Phi} - \boldsymbol{\Phi}_0)(\boldsymbol{\Phi} - \boldsymbol{\Phi}_0)^\top\right] \mathbf{C}_s^\top \mathbf{f} \tag{23}$$

with the constraint

$$\mathbf{C}_{\mathbf{p}d} = \mathbf{f}^\top \mathbf{C}_s \boldsymbol{\Phi}_0. \tag{24}$$

This optimization is equivalent to minimizing the sensor curvature, discussed in the previous section, with

$$\mathbf{H} = \mathbf{C}_s \mathrm{E}\left[(\boldsymbol{\Phi} - \boldsymbol{\Phi}_0)(\boldsymbol{\Phi} - \boldsymbol{\Phi}_0)^\top\right] \mathbf{C}_s^\top. \tag{25}$$

The calculation of the above quantity requires the calculation of second-order statistical properties of the mode shapes and is discussed in the next section.

5 Modal Statistics for Uncertain Dynamical Systems

The estimation of the means of the mode shapes and the $\mathbf{H}$ matrix in Eq. (25) requires the solution of a random eigenvalue problem for linear stochastic dynamical systems. Specifically, in this section we propose explicit expressions to obtain $\mathrm{E}\left[\boldsymbol{\Phi}(\theta)\right]$ and elements of the matrix

$$\mathbf{W} = \mathrm{E}\left[(\boldsymbol{\Phi}(\theta) - \boldsymbol{\Phi}_0) (\boldsymbol{\Phi}(\theta) - \boldsymbol{\Phi}_0)^{\mathrm{T}} \right] \tag{26}$$

for linear dynamical systems with uncertain properties modeled by random fields.

Suppose $H(\mathbf{r}, \theta)$ is a random field with a covariance function $C_H(\mathbf{r}_1, \mathbf{r}_2)$ defined in a space Ω. Since the covariance function is finite, symmetric and positive definite it can be represented by a spectral decomposition. Using this spectral decomposition, the random process $H(\mathbf{r}, \theta)$ can be expressed in a generalized Fourier type series as

$$H(\mathbf{r}, \theta) = H_0(\mathbf{r}) + \sum_{i=1}^{\infty} \sqrt{\lambda_i} \xi_i(\theta) \varphi_i(\mathbf{r}) \tag{27}$$

where $\xi_i(\theta)$ are uncorrelated random variables, λ_i and $\varphi_i(\mathbf{r})$ are eigenvalues and eigenfunctions satisfying the integral equation

$$\int_{\Omega} C_H(\mathbf{r}_1, \mathbf{r}_2) \varphi_i(\mathbf{r}_1) \mathrm{d}\mathbf{r}_1 = \lambda_i \varphi_i(\mathbf{r}_2), \quad \forall\; i = 1, 2, \ldots \tag{28}$$

The spectral decomposition in Eq. (27) is known as the Karhunen-Loève expansion. The series in Eq. (27) can be ordered in a decreasing series so that it can be truncated after a finite number of terms with a desired accuracy. We refer the books [10, 18] and references therein for further discussions on Karhunen-Loève expansion. In this paper one dimensional systems (beam elements) are considered. Moreover, Gaussian random fields with exponentially decaying autocorrelation functions are considered for the stiffness and density of the system.

Utilizing the series expansion of the random fields describing the uncertain parameter of the system and dicretisation of the displacement fields, the equation of motion can be expressed as Eq. (1). Due to the presence of uncertainty, the n dimensional matrices $\mathbf{M}(\theta)$, $\mathbf{C}(\theta)$ and $\mathbf{K}(\theta)$ become random matrices. These random matrices can be expressed as

$$\mathbf{K}(\theta) = \mathbf{K}_0 + \boldsymbol{\Delta}\mathbf{K}(\theta), \quad \mathbf{M}(\theta) = \mathbf{M}_0 + \boldsymbol{\Delta}\mathbf{M}(\theta) \quad \text{and} \quad \mathbf{C}(\theta) = \mathbf{C}_0 + \boldsymbol{\Delta}\mathbf{C}(\theta) \tag{29}$$

where, for example

$$\boldsymbol{\Delta}\mathbf{K}(\theta) = \sum_{j=1}^{N_{\mathrm{K}}} \xi_{\mathrm{K}_j}(\theta) \sqrt{\lambda_{\mathrm{K}_j}} \mathbf{K}_j, \tag{30}$$

In the above expression $\xi_{\mathrm{K}_j}(\theta)$ are sets of uncorrelated random variables and N_{K} denote its numbers. The number of random variables depend on the truncation of the infinite series (27) and are functions of the respective correlation lengths of the underlying random fields. The deterministic matrices $\mathbf{K}_j$, $\mathbf{M}_j$ and $\mathbf{C}_j$ are symmetric and non-negative definite. These matrices can be obtained by assembling respective element matrices which in turn depend on the displacement shape functions and the eigenvectors corresponding to the eigenvalues $\sqrt{\lambda_{\mathrm{K}_j}}$, $\sqrt{\lambda_{\mathrm{M}_j}}$ and $\sqrt{\lambda_{\mathrm{C}_j}}$ respectively. We refer to the book by Ghanem and Spanos [10] for further details.

One needs to solve *random eigenvalue problems* to obtain statistical properties of the eigenvalues and eigenvectors appearing in stochastic dynamical systems. Several studies have been conducted on this topic since the mid-sixties. The paper by Boyce [4], the book by Scheidt and Purkert [19] are useful sources of information on early works in this area of research. The literature on random eigenvalue problems arising in engineering systems is dominated by the mean-centered perturbation methods [1–5, 11, 12]. These methods work well when the uncertainties are small and the parameter distribution is Gaussian. Here, first-order perturbation approach is proposed to calculate the second-order statistics appearing in Eq. (25).

The random eigenvalue problem corresponding to an undamped stochastic system can be expressed as

$$[\mathbf{K}_0 + \boldsymbol{\Delta}\mathbf{K}(\theta)]\,\boldsymbol{\phi}_j = \omega_j^2\,[\mathbf{M}_0 + \boldsymbol{\Delta}\mathbf{M}(\theta)]\,\boldsymbol{\phi}_j. \tag{31}$$

Using the first-order perturbation method, each eigenvector can be expressed as

$$\boldsymbol{\phi}_j \approx \boldsymbol{\phi}_{0_j} + \sum_{k=1}^{N_{\mathrm{K}}} \frac{\partial \boldsymbol{\phi}_j}{\partial \xi_{\mathrm{K}_k}} \xi_{\mathrm{K}_k}(\theta) + \sum_{k=1}^{N_{\mathrm{M}}} \frac{\partial \boldsymbol{\phi}_j}{\partial \xi_{\mathrm{M}_k}} \xi_{\mathrm{M}_k}(\theta). \tag{32}$$

The derivative of the eigenvectors can be obtained using the expression derived by Fox and Kapoor [6]. Equation (32) can be can be expressed express

$$\boldsymbol{\phi}_j \approx \boldsymbol{\phi}_{0_j} + \mathbf{A}^{(j)}\boldsymbol{\xi}(\theta); \tag{33}$$

where the $n \times (N_{\mathrm{K}} + N_{\mathrm{M}})$ coefficient matrix $\mathbf{A}^{(j)}$ and the $(N_{\mathrm{K}} + N_{\mathrm{M}})$ dimensional random vector $\boldsymbol{\xi}_t$ are defined as

$$\mathbf{A}^{(j)} = \left[\frac{\partial \boldsymbol{\phi}_j}{\partial \xi_{\mathrm{K}_k}}, k = 1, 2, \ldots, N_{\mathrm{K}} \middle| \frac{\partial \boldsymbol{\phi}_j}{\partial \xi_{\mathrm{M}_k}}, k = 1, 2, \ldots, N_{\mathrm{M}}\right] \quad \text{and} \quad \boldsymbol{\xi} = \begin{bmatrix} \xi_{\mathrm{K}_k} \\ \xi_{\mathrm{M}_k} \end{bmatrix}. \tag{34}$$

Recalling that all of the random variables are uncorrelated, after elementary matrix manipulations we can show that

$$W_{kl} = \sum_{j=1}^{n} \sum_{s=1}^{N_{\mathrm{K}}+N_{\mathrm{M}}} A_{ks}^{(j)} A_{ls}^{(j)}. \tag{35}$$

The elements of $A_{kr}^{(j)}$ are defined by Eq. (34).

6 Numerical Example

A clamped-clamped beam example [8, 9] will be used to demonstrate the design of modal sensors. The baseline model of the steel beam is assumed to be of 1.5 m in length with cross-section 20×5 mm. Bending in the more flexible plane is modeled by using 15 finite elements. Only the first nine modes are considered important and damping is assumed to be 1% in all modes. The first 12 natural frequencies for the baseline beam are 11.81, 32.57, 63.86, 105.6, 157.8, 220.7, 294.3, 379.0, 474.9, 582.7, 702.6, 834.9 Hz. The force input is applied node 7 for the sensor design. A sensor is designed by considering only the first nine modes of the beam. The sensor gain constant is assumed to be unity, $K_s = 1$, since it is most important to compute the sensor shape, rather than the calibration constant.

It is assumed that the bending stiffness $EI(x)$ and mass per unit length $\rho A(x)$ are random fields of the form

$$EI(x) = EI_0 \left(1 + \epsilon_{EI} f_1(x)\right) \tag{36}$$

$$\text{and} \quad \rho A(x) = \rho A_0 \left(1 + \epsilon_{\rho A} f_2(x)\right) \tag{37}$$

Here EI_0 and ρA_0 are the corresponding properties of the baseline model. The strength parameters are assumed to be $\epsilon_{EI} = 0.05$, and $\epsilon_{\rho A} = 0.1$. The random fields $f_1(x)$ and $f_2(x)$ are assumed to be delta-correlated homogenous Gaussian random fields. A 1,000-sample Monte Carlo simulation is performed to obtain the FRFs and modal statistics. The following two cases are considered:

- Case 1: sensor to be designed to excite only the first mode, with a peak in the receptance of 0.01 m/N.
- Case 2: sensor to be designed to excite only the third mode, with a peak in the receptance of 0.01 m/N.

For each case, design for the deterministic and uncertain systems were considered.

Figure 1(a) shows the shape of the sensor designed to excite only the first mode of the underlying deterministic system, with a peak in the receptance of 0.01 m/N. The shape was obtained by minimising the curvature, and the nodal width values

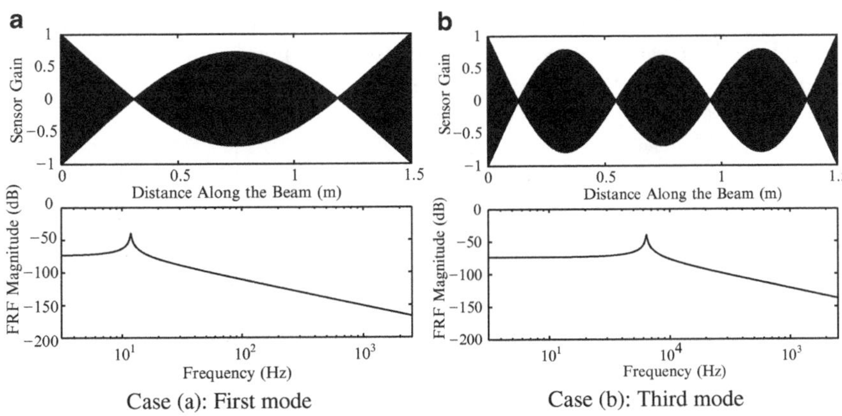

Case (a): First mode Case (b): Third mode

Fig. 1 The distributed sensor shape to measure specified modes for the baseline system and the associated receptance

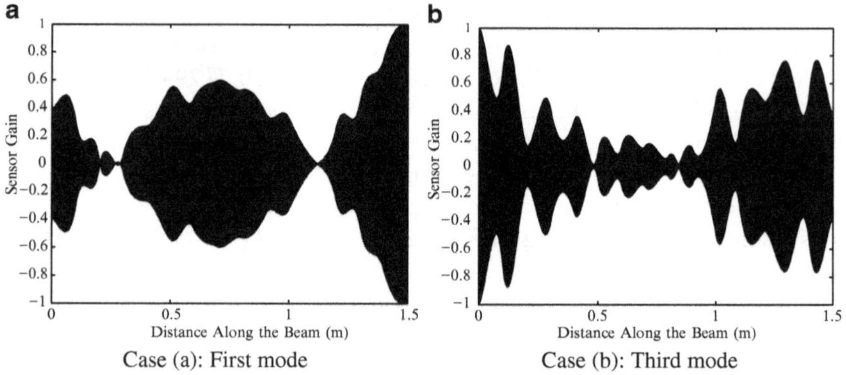

Fig. 2 The distributed sensor shape to measure specified modes for the stochastic system

used, together with the element shape functions, to produce a continuous sensor width function. The shape is very similar to the curvature of the first mode of the deterministic beam. Figure 1(b) shows the shape of the sensor designed to excite only the third mode of the underlying deterministic system, with a peak in the receptance of 0.01 m/N. Notice that the shape is very similar to the curvature of the third mode of the deterministic beam.

Now we turn our attention to the stochastic system. The shapes of the sensors are obtained using the formulation outlined in Section 4. The results for both the cases are shown in Fig. 2. Unlike the deterministic case, the designs are now much more complex and also the intuitive nature of the sensor shapes are less obvious. Our numerical studies (results not shown here) indicate that the sensor shapes are sensitive to the amount of uncertainty in the system. It should be noted that the sensor shapes obtained here are specific to the design criteria given by Eq. (22) for the stochastic system. Due to the sensitivity of the shapes, it is highly likely that the sensor shapes would be different if a different criteria is selected. The results obtained here lead us to conclude that the sensor shapes for stochastic systems could be significantly different from the corresponding deterministic system.

7 Conclusion

Uncertainties in the system need to be taken into account for the robust design of sensors and actuators for engineering dynamical systems. This paper has considered the problem of designing modal actuators and sensors using a discrete approximation to the equations of motion for linear stochastic systems. Transducer shapes are represented by utilizing the underlying finite element shape functions. This allows the actuators and sensors to be designed by using a discrete approximation and the shape recovered by using the shape functions. Optimal shape design has been coupled with the stochastic finite element method to consider parametric uncertainty described using random fields. It was shown that only the eigenvector statistics are needed to obtain the optimal shape. A first-order perturbation based approach is proposed to obtain the second-order covariance of the modal matrix in closed-form. Numerical results show that the shape of the sensors of the stochastic system differs significantly from the corresponding deterministic system. The analytical and numerical results are derived for beam structures. Work is currently underway to extend the results to complex systems.

Acknowledgements SA gratefully acknowledges the support of UK Engineering and Physical Sciences Research Council (EPSRC) through the award of an Advanced Research Fellowship and The Leverhulme Trust for the award of the Philip Leverhulme Prize.

References

1. Adhikari, S.: Random eigenvalue problems revisited. Sādhanā – Proceedings of the Indian Academy of Engineering Sciences **31**(4), 293–314 (2006). (Special Issue on Probabilistic Structural Dynamics and Earthquake Engineering)
2. Adhikari, S.: Joint statistics of natural frequencies of stochastic dynamic systems. Computational Mechanics **40**(4), 739–752 (2007)
3. Adhikari, S., Friswell, M.I.: Random matrix eigenvalue problems in structural dynamics. International Journal for Numerical Methods in Engineering **69**(3), 562–591 (2007)
4. Boyce, W.E.: Random Eigenvalue Problems. Probabilistic methods in applied mathematics. Academic Press, New York (1968)
5. Collins, J.D., Thomson, W.T.: The eigenvalue problem for structural systems with statistical properties. AIAA Journal **7**(4), 642–648 (1969)
6. Fox, R.L., Kapoor, M.P.: Rates of change of eigenvalues and eigenvectors. AIAA Journal **6**(12), 2426–2429 (1968)
7. Friswell, M.I.: Partial and segmented modal sensors for beam structures. Journal of Vibration and Control **5**, 619–637 (1999)
8. Friswell, M.I.: On the design of modal actuators and sensors. Journal of Sound and Vibration **241**, 361–372 (2001)
9. Gawronski, W.: Modal actuators and sensors. Journal of Sound and Vibration **229**, 1013–1022 (2000)
10. Ghanem, R., Spanos, P.: Stochastic Finite Elements: A Spectral Approach. Springer-Verlag, New York, USA (1991)
11. Hart, G.C.: Eigenvalue uncertainties in stressed structure. Journal of Engineering Mechanics, ASCE **99**(EM3), 481–494 (1973)
12. Hasselman, T.K., Hart, G.C.: Modal analysis of random structural system. Journal of Engineering Mechanics, ASCE **98**(EM3), 561–579 (1972)
13. Hsu, C.Y., Lin, C.C., Gaul, L.: Vibration and sound radiation controls of beams using layered modal sensors and actuators. Smart Materials and Structures **7**, 446–455 (1998)
14. Kleiber, M., Hien, T.D.: The Stochastic Finite Element Method. John Wiley, Chichester (1992)
15. Lee, C.K., Moon, F.C.: Modal sensors/actuators. Journal of Applied Mechanics **57**, 434–441 (1990)
16. Manohar, C.S., Adhikari, S.: Dynamic stiffness of randomly parametered beams. Probabilistic Engineering Mechanics **13**(1), 39–51 (1998)
17. Manohar, C.S., Adhikari, S.: Statistical analysis of vibration energy flow in randomly parametered trusses. Journal of Sound and Vibration **217**(1), 43–74 (1998)
18. Papoulis, A., Pillai, S.U.: Probability, Random Variables and Stochastic Processes, fourth edn. McGraw-Hill, Boston, USA (2002)
19. Scheidt, J.V., Purkert, W.: Random Eigenvalue Problems. North Holland, New York (1983)

Sensor Failure Detection Using Interaction Matrix Formulation

Zhiling Li, Satish Nagarajaiah, and B.H. Koh

Abstract A novel sensor failure detection method is developed in this paper. Sensor failure considered in this paper can be any type of measurement error that is different from the true structural response. The sensors are divided into two groups, sensors that correctly measure the structural responses, are termed as reference sensors, and sensors that may fail to correctly measure the structural responses, are termed as uncertain sensors henceforth. A sensor error function, one for each uncertain sensor, is formulated to detect the corresponding uncertain sensor failure in real-time, using the measurements from reference sensors and the uncertain sensor being monitored. The sensor error function is derived using the indirect and direct approaches. In the indirect approach, the error function is obtained from the state space model in combination with the inverse model and interaction matrix formulation. The input term is eliminated from the error function by applying inverse model and the interaction matrix is applied to eliminate the state and all uncertain sensor measurement terms excepted the examined uncertain sensor from the error function. In the direct approach, the coefficients of the error function can be directly calculated from the healthy measurement data from the examined uncertain sensor and all reference sensors without having to know the state-space model of the system. Thus the need to know the state-space model of the plant can be bypassed. The sensor failure detection formulations are investigated numerically using a four degree-of-freedom spring-mass-damper system and experimentally using a 4 m long NASA 8-bay truss structure. It is shown by means of numerical and experimental results that the

---

Z. Li
Departmental of Civil & Environmental Engineering, Rice University, Houston, TX 77005, USA
e-mail: li@rice.edu

S. Nagarajaiah (✉)
Departmental of Civil & Environmental Engineering and Mechanical Engineering & Material Science, Rice University, Houston, TX 77005, USA
e-mail: nagaraja@rice.edu

B.H. Koh
Departmental of Civil & Envi. Engineering and Mechanical Engineering & Material Science, Rice University, Houston, TX 77005, USA
e-mail: bkoh@dongguk.edu

B. Dattaguru et al. (eds.), *IUTAM Symposium on Multi-Functional Material Structures and Systems*, IUTAM Bookseries 19, DOI 10.1007/978-90-481-3771-8_20,

developed sensor failure formulations correctly detect the instants of sensor failure and can be implemented in real structural systems for sensor failure detection.

Keywords Experimental validation · Input error function · Interaction matrix formulation · Inverse model · Mathematical formulation · Sensor failure detection · State space systems

1 Introduction

Sensor failure is defined as the measured output that is different from the structural response. It is critical to detect sensor failure on real-time basis for vibration control and structural health monitoring. Various Fault Detection and Isolation (FDI) techniques for sensors and actuators failure detection have been discussed over the past decades [1, 2]. The two major sensor failure detection methods are approaches that exploit (1) physical redundancy, and (2) analytical redundancy. Physical redundancy uses more than one physical sensor to measure the responses from the same degree of freedom. Analytical Redundancy (AR) method, which implies the inherent redundancy contained in the static and dynamic relationship among the system inputs and measured outputs [1], has been applied broadly both in adaptive control and fault tolerant systems. The method developed in this study belongs to one of the AR-based methods.

Kammer and co-worker [5,6,9] have applied inverse model to estimate the remote sensor responses and estimate the input forces from the structural responses. To estimate the remote sensor responses [5], the input terms are first eliminated from the state-space model by introducing inverse model and only measured sensor responses and remote sensor responses are retained. Then the Markov parameters corresponding to the inverse model are obtained by a transformation matrix, which is either obtained indirectly from the state-space model or directly from measured data.

Phan et al. [3, 4, 8] have shown that the interaction matrix can be applied to eliminate state variables so that only input-output measurement data are needed to represent the system's dynamics. Previous work [7] have applied the concept of interaction matrix to build actuator failure detection function, which eliminates the influence of other actuators on the actuator failure function except the examined one. The actuator failure function will be non-zero in case of failure in the corresponding actuator.

In this paper, interaction matrix [10–17] formulation and inverse model are combined to develop a new algorithm to detect and isolate sensor failure in real-time. In this formulation, sensors are separated into two groups as needed by inverse model technique (see Section 2 for detail discussion). Sensors in the first group are assumed to correctly measure the structural responses and are termed as reference sensors henceforth. Sensors in the second group that need to be monitored are termed as uncertain sensors henceforth. A sensor error function, one for each uncertain sensor, is developed to monitor the failure of the uncertain sensor in real-time. Non-zero value in the sensor error function indicates the time instant of failure of the examined uncertain sensor and it is not influenced by the measurement of other uncertain sensors.

2 Indirect Method

Consider an n-th order, r-input, q-output controllable and observable linear time invariant discrete time model in state-space system

$$\begin{aligned} x(k+1) &= Ax(k) + Bu(k) \\ y(k) &= Cx(k) + Du(k) \end{aligned} \tag{1}$$

The output in Eq. (1) can be separated into two groups $y_s(k)$ and $y_d(k)$, where $y_s(k)$ is the measurement from reference sensors and $y_d(k)$ is the measurement from uncertain sensors. Rewrite Eq. (1) for the measurement $y_d(k)$

$$\begin{aligned} x(k+1) &= Ax(k) + Bu(k) \\ y_d(k) &= C_d x(k) + D_d u(k) \end{aligned} \tag{2}$$

Where $y_d(k)$ is an n_d (number of uncertain sensors) dimensional vector, C_d is an output influence matrix for the uncertain sensors, and D_d is an $n_d \times r$ direct transmission matrix from the input to the uncertain sensors. The direct transmission matrix is assumed to be full column rank here. For the case D_d is not full column rank, the method proposed by Steltzner and Kammer [9] can be adopted. Solving the output $y_d(k)$ in Eq. (2) for input $u(k)$

$$u(k) = -D_d^+ C_d x(k) + D_d^+ y_d(k) \tag{3}$$

Here D_d^+ is the pseudo inverse of D_d and $D_d^+ = (D_d^T D_d)^+ D_d^T$. In order for the pseudo-inverse to exist, the number of uncertain sensors n_d must be greater than or equal to the number of inputs r. Substituting Eq. (3) into the state equation in Eq. (2) produces

$$x(k+1) = \left(A - BD_d^+ C_d\right) x(k) + BD_d^+ y_d(k) \tag{4}$$

By defining $\overline{A} = A - BD_d^+ C_d$, $\overline{B} = BD_d^+$, $\overline{C}_d = -D_d^+ C_d$, $\overline{D}_d = D_d^+$, Eq. (3) and Eq. (4) can be rewritten as

$$\begin{aligned} x(k+1) &= \overline{A}x(k) + \overline{B}y_d(k) \\ u(k) &= \overline{C}_d x(k) + \overline{D}_d y_d(k) \end{aligned} \tag{5}$$

The inputs and outputs shown in Eq. (5) have been switched, which is an inverse model of Eq. (2). The state and output equations for the measurements of reference sensors can be expressed as

$$\begin{aligned} x(k+1) &= Ax(k) + Bu(k) \\ y_s(k) &= C_s x(k) + D_s u(k) \end{aligned} \tag{6}$$

$y_s(k)$ is an n_s (number of reference sensors) dimensional measurement vector, C_s is an $n_s \times n$ output influence matrix for the reference sensors, and $D_s(k)$ is an $n_s \times r$ direct transmission matrix from the inputs to the reference sensors. Substituting Eq. (4) into Eq. (6) produces

$$\begin{aligned} x(k+1) &= \overline{A}x(k) + \overline{B}y_d(k) \\ y_s(k) &= \left(C_s + D_s\overline{C}_d\right)x(k) + D_s\overline{D}_d y_d(k) \end{aligned} \tag{7}$$

Defining $\overline{C}_s = C_s + D_s\overline{C}_d$ and $\overline{D}_s = D_s\overline{D}_d$, Eq. (7) can be expressed as

$$\begin{aligned} x(k+1) &= \overline{A}x(k) + \overline{B}u(k) \\ y_s(k) &= \overline{C}_s x(k) + \overline{D}_s u(k) \end{aligned} \tag{8}$$

Equation (8) represents an inverse model that the measurements from uncertain sensors are treated as inputs and the measurements from the reference sensors are treated as outputs. The input terms that exert forces to the structure have been completely eliminated from the state-space equation. Assuming that sensors measure the responses of independent degree of freedoms, each sensor will not be collocated with the other sensors; then $\overline{D}_s$ term will be a null matrix. Equation (8) can be rewritten as

$$\begin{aligned} x(k+1) &= \overline{A}x(k) + \overline{B}y_d(k) \\ y_s(k) &= \overline{C}_s x(k) \end{aligned} \tag{9}$$

By repeated substitution for some $p > 0$ and regrouping the output from each uncertain sensor as one term

$$\begin{aligned} x(k+p) &= \overline{A}^p x(k) + \sum_{j=1}^{n_d} \overline{\Gamma}_j \overline{y}_{pd}^j(k) + \sum_{j=1}^{n_d} \overline{B}_j y_d^j(k+p-1) \\ y_{ps}(k) &= \overline{O}x(k) + \sum_{j=1}^{n_d} \overline{T}_j y_{pd}^j(k) \end{aligned} \tag{10}$$

An interaction matrix M_i is introduced by adding and subtracting the product $M_i y_{ps}(k)$ to Eq. (10) as follows

$$\begin{aligned} x(k+p) &= \overline{A}^p x(k) + \sum_{j=1}^{n_d} \overline{\Gamma}_j y_{pd}^j(k) + \sum_{j=1}^{n_d} \overline{B}_j y_d^j(k+p-1) \\ &\quad - M_i y_{ps}(k) + M_i y_{ps}(k) \\ &= \left(\overline{A}^p + M_i\overline{O}\right)x(k) + \sum_{j=1}^{n_d} \left(\overline{\Gamma}_j + M_i\overline{T}_j\right)\overline{y}_{pd}^j(k) \\ &\quad - M_i y_{ps}(k) + \sum_{j=1}^{n_d} \overline{B}_j y_d^j(k+p-1) \\ y(k+p) &= \left(\overline{C}_s\overline{A}^p + \overline{C}_s M_i\overline{O}\right)x(k) + \sum_{j=1}^{n_d} \left(\overline{C}_s\overline{\Gamma}_j + \overline{C}_s M_i\overline{T}_j\right) y_{pd}^j(k) \\ &\quad - \overline{C}_s M_i y_{ps}(k) + \sum_{j=1}^{n_d} \overline{C}_s\overline{B}_j y_d^j(k+p-1) \end{aligned} \tag{11}$$

By imposing the condition in Eq. (12), the coefficient matrices of $x(k)$ and $\overline{y}_{pd}^j(k)$ for $j \neq i$ in Eq. (11) vanish simultaneously except for the i-th uncertain sensor

$$\begin{aligned}\overline{C}_s\overline{A}^p + \overline{C}_s M_i \overline{O} &= 0\\ \overline{C}_s\overline{\Gamma}_j + \overline{C}_s M_i \overline{T}_j &= 0 \quad for\ j \neq i\end{aligned} \tag{12}$$

If the number of independent reference sensors n_s is greater than or equal to the number of uncertain sensors n_d, Eq. (12) will be solvable. Equation (11) can be rewritten here with the constraint conditions in Eq. (12) applied

$$\begin{aligned}y(k+p) = &\left(\overline{C}_s\overline{\Gamma}_i + \overline{C}_s M_i \overline{T}_i\right) y^i_{pd}(k) - \overline{C}_s M_i y_{ps}(k)\\ &+ \sum_{j=1}^{n_d} \overline{C}_s \overline{B}_j y^j_d(k+p-1)\end{aligned} \tag{13}$$

Up to this point, there still exists other uncertain sensor output y^j_d for $j \neq i$ that contributes to the relationship between examined uncertain sensor and all reference sensors in Eq. (13). The coefficient vector of y^j_d for $j \neq i$ can be eliminated by pre-multiplying Eq. (13) with a row vector that is orthogonal to all remaining column vectors $\overline{C}_s\overline{B}_j$, for $j \neq i$

$$N_i^T\left(\overline{C}_s\overline{B}_j\right) = 0 \quad \forall j \neq i \tag{14}$$

Because $n_s \geq n_d$, such a vector N_i^T can be easily found. Pre-multiplying Eq. (13) with N_i^T produces a scalar equation that involves outputs from all reference sensors and outputs from the i-th uncertain sensor only.

$$\begin{aligned}N_i^T y(k+p) = &N_i^T\left(\overline{C}_s\overline{\Gamma}_i + \overline{C}_s M_i \overline{T}_i\right) y^i_{pd}(k) - N_i^T\overline{C}_s M_i y_{ps}(k)\\ &+ N_i^T\overline{C}_s\overline{B}_i y^i_d(k+p-1)\end{aligned} \tag{15}$$

In Eq. (15), the outputs from other uncertain sensors have been eliminated except the monitored one. If $\overline{y}^i_d(k)$ represents the measurement at the i-th uncertain sensor point at k-th steps and $\overline{y}^{*i}_d(k)$ is the measurement error of the sensor at k-th step, then $y^i_d(k) = \overline{y}^i_d(k) - \overline{y}^{*i}_d(k)$. Substituting it into Eq. (15)

$$\begin{aligned}N_i^T y(k+p) = &N_i^T\left(\overline{C}_s\overline{\Gamma}_i + \overline{C}_s M_i \overline{T}_i\right)\left(\overline{y}^i_{pd}(k) - \overline{y}^{*i}_{pd}(k)\right) - N_i^T\overline{C}_s M_i y_{ps}(k)\\ &+ N_i^T\overline{C}_s\overline{B}_i\left(\overline{y}^i_d(k+p-1) - \overline{y}^{*i}_d(k+p-1)\right)\end{aligned} \tag{16}$$

By defining the error function as

$$e_i(k+p) = N_i^T\left(\overline{C}_s\overline{\Gamma}_i + \overline{C}_s M_i \overline{T}_i\right)\overline{y}^{*i}_{pd}(k) - N_i^T\overline{C}_s\overline{B}_i\overline{y}^{*i}_d(k+p-1) \tag{17}$$

The final form of the error function is

$$\begin{aligned}e_i(k+p) = &\overline{N}_i^T y_s(k+p) - N_i^T\left(\overline{C}_s\overline{\Gamma}_i + \overline{C}_s M_i \overline{T}_i\right)\overline{y}^i_{pd}(k)\\ &+ N_i^T\overline{C}_s M_i y_{ps}(k) - N_i^T\overline{C}_s\overline{B}_i\overline{y}^i_d(k+p-1)\end{aligned} \tag{18}$$

Failure of the i-th uncertain sensor in $\overline{y}^i_d(k)$ can be detected by the error function in Eq. (18). When there are no measurement errors for the i-th uncertain sensor from k-th to $k+p$-th steps, $e_i(k+p)$will be zero. When there are measurement errors from k-th to $k+p$-th steps, the error function will normally be different from zero. In the above procedure, there are two conditions that need to be satisfied for the error function to be valid (1)$n_d > r$ to ensure the pseudo-inverse of D_d to exist uniquely and (2) $n_s > n_d$ to ensure $\overline{C}_s M_i$ exists. Since the coefficients of the sensor error functions need to be calculated from the state-space matrices A, B, C, and D. This method is termed as indirect method.

3 Direct Method

The error function for the i-th examined sensor is found to assume the general form

$$\begin{aligned} e_i(k) = \alpha^i_0 y_s(k) + \alpha^i_1 y_s(k-1) + \cdots + \alpha^i_p y_s(k-p) \\ +\beta^i_1 \overline{y}^i_d(k-1) + \cdots + \beta^i_p \overline{y}^i_d(k-p) \end{aligned} \tag{19}$$

Having justified the existence and the structure of the error function of the i-th examined uncertain sensor, the coefficients of the error function can also be calculated from uncertain sensor and reference sensor measurement only without having to identify the state-space model of the system first. Thus the need to know the state-space model of the plant can be bypassed. This method is called direct method. This strategy should turn out to be particularly advantageously in practice because it bypasses the intermediate system identification step and all the error incurred during such identification.

When the examined actuator operates perfectly, the right hand side of Eq. (19) will be zero and the coefficients of the error function cannot be fixed unless they are 'normalized' in some way, say for example by keeping the value of β^i_p to be one. By doing so, Eq. (19) can be rewritten as

$$\begin{aligned} -\overline{y}^i_d(k-p) = \alpha^i_0 y_s(k) + \alpha^i_1 y_s(k-1) + \cdots + \alpha^i_p y_s(k-p) \\ +\beta^i_0 \overline{y}^i_d(k-1) + \beta^i_1 \overline{y}^i_d(k-1) + \cdots + \beta^i_{p-1} \overline{y}^i_d(k-p+1) \end{aligned} \tag{20}$$

From a set of sufficiently rich and long input-output data

$$\{\overline{y}^i_d(k-p), \overline{y}^i_d(k-p+1), \ldots, \overline{y}^i_d(k+l)\}, \{y_s(k-p), y_s(k-p+1), \ldots, y_s(k+l)\} \tag{21}$$

Input-output relationship can be written as

$$U_i = P_i V_i \tag{22}$$

Thus the coefficients of error function of the i-th examined actuator can be obtained by

$$P_i = U_i(V_i)^+ = U_i V_i^T \left(V_i V_i^T\right)^+ \tag{23}$$

Where the $()^+$ sign denotes the pseudo-inverse, which should be computed via singular value decomposition.

4 Simulation

A four degree-of-freedom (DOF) spring-mass-damper system is first used to demonstrate the sensor failure detection algorithms, as shown in Fig. 1. Four accelerometers are attached from m_1 to m_4; those sensors are separated into two groups. The two sensors attached to m_3 and m_4 are assumed to be uncertain sensors. The other two sensors attached to m_1 and m_2 are assumed to be reference sensors. Actuator 1 and 2 are driven by independent band-limited white noise excitation inputs. Since inputs have been eliminated from the sensor error function and only the measurements from the reference sensors and uncertain sensors are required to build the sensor error function, as shown in Eq. (19). For this system, $m_1 = m_2 = m_3 = m_4 = 1$ kg, $k_1 = k_2 = k_3 = k_4 = 100$ KN/m, and $c_i = 0.05\,m_i + 0.001 k_i$ N s/m. In order to simulate sensor failure, the measurements from sensor attached to m_3 (uncertain sensor 1) are set to sine response instead of actual structural responses during 2–8 s and the sensor attached to m_4 (uncertain sensor 2) gives out uniform zero measurements between 3 and 6 and half of the structural response between 8 and 10 s. The structural responses and sensor measurements are shown in Fig. 2(a) and (b). The error functions of sensor 1 and 2 by indirect method for noise free case are shown in Fig. 3. From Fig. 3, it is shown that the sensor error functions show non-zero error signal when measurement errors occur. Otherwise, the error function shows zero values. Also each sensor failure detection function uniquely indicates the corresponding sensor's failure and is not influenced by the other uncertain sensors. Figure 4 shows the sensor error functions by indirect method with 2% standard deviation noise-corrupted data for all the reference sensors. It can be seen that the

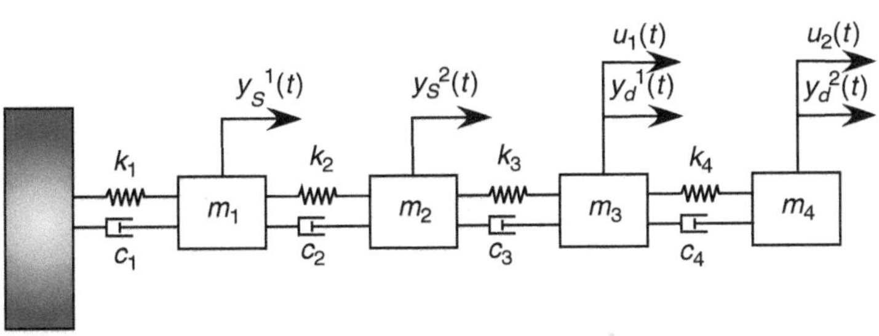

Fig. 1 A 2-input, 4-output, 4-DOF spring-mass-damper system

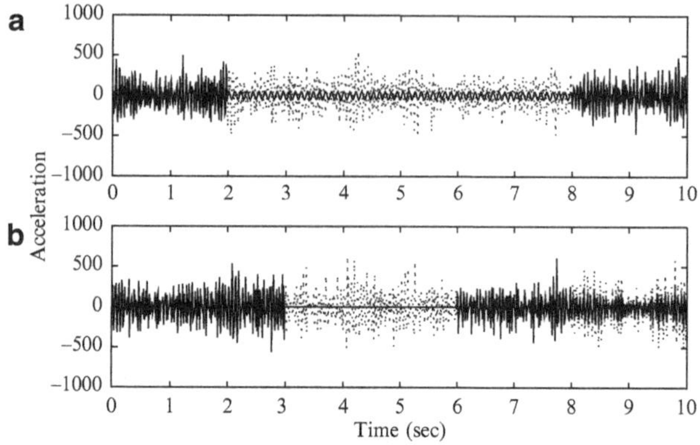

Fig. 2 The measurements [Solid] and structural responses [Dashed] at uncertain sensors 1 and 2 of a 4-DOFs spring-mass-damper system: (**a**) uncertain sensor 1, and (**b**) uncertain sensor 2

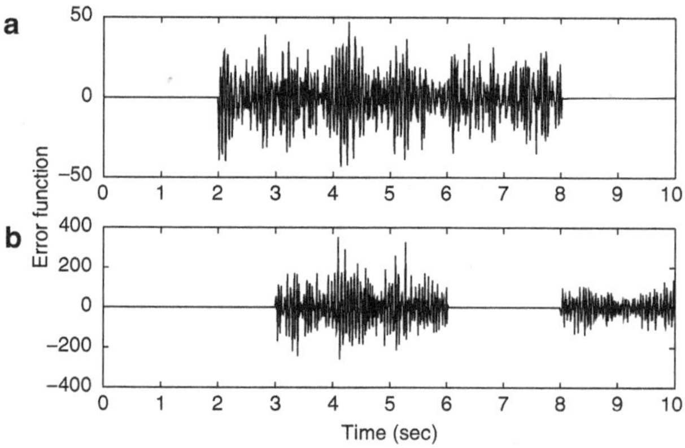

Fig. 3 Error function for uncertain sensors 1 and 2 of a 4-DOFs spring-mass-damper system by indirect method: (**a**) uncertain sensor 1, and (**b**) uncertain sensor 2

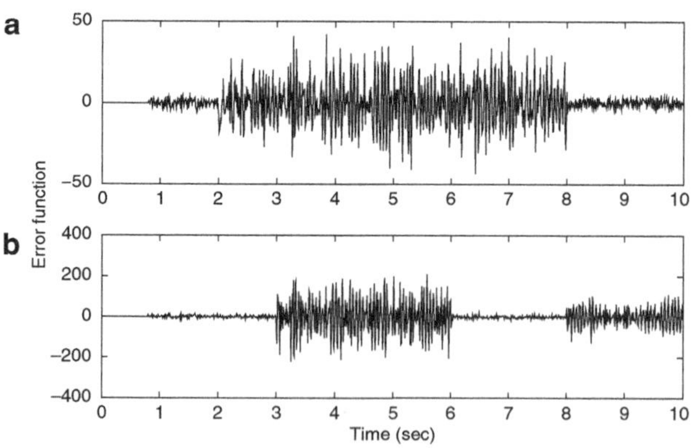

Fig. 4 Error function for uncertain sensor 1 and 2 of a 4-DOFs spring-mass-damper system with 2 % standard deviation measurement noise at all reference sensors by indirect method: (**a**) uncertain sensor 1, and (**b**) uncertain sensor 2

error function can still identify the failure of each uncertain sensor in real-time. It is noted that the corruption of measurement noise on reference sensors will make the error functions non-zero even when the corresponding uncertain sensor correctly measures the structural response. However, in the current simulation, the amplitude of the error function produced by the measurement noise is much less than that of the failure case. For $p = 8$, The error functions by direct method for noise free case and 2% standard deviation measurement noise at all reference sensors case are shown in Figs. 5 and 6. From the figures, it is evident that the error function by direct method can also isolate and detect the failure of the examined uncertain sensor in real-time.

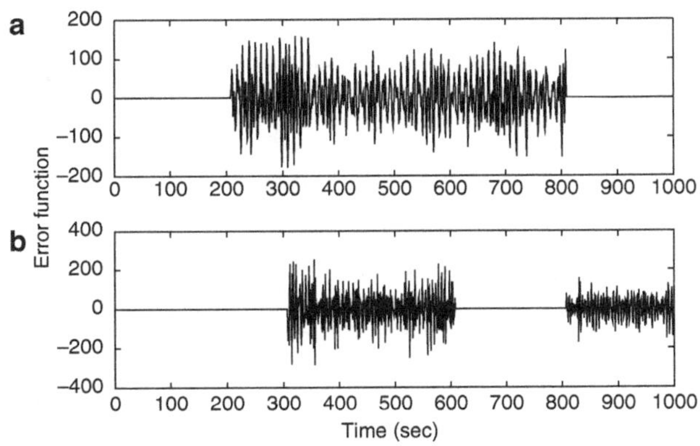

Fig. 5 Error function for uncertain sensors 1 and 2 of a 4-DOFs spring-mass-damper system by direct method: (**a**) uncertain sensor 1, and (**b**) uncertain sensor 2

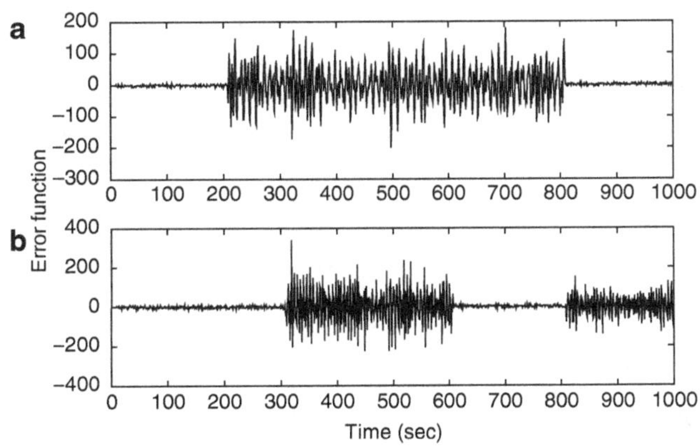

Fig. 6 Error function for uncertain sensors 1 and 2 of a 4-DOFs spring-mass-damper system by direct method: (**a**) uncertain sensor 1, and (**b**) uncertain sensor 2

5 Experimental Verification

The NASA 8-bay truss structure at Rice University, which consists of 36 nodes and 109 aluminum members, is used to verify the aforementioned sensor failure detection algorithm. The plan view of the truss are shown in Fig. 7. Picture of the NASA 8-bay truss can be found in previous paper [7].The members are made up of hollow tubes fitted with aluminum bolt ends which can be screwed into the nodes. One end of the truss is mounted to a steel frame, which is fixed to the ground. The total truss length is four meters. Two hangers are vertically connected to the truss node at the fourth and eighth bays (counted from the fixed end) to restrain vertical movement. The experimentally identified fundamental frequency of the truss is 12.45 Hz (fundamental period of 0.08 s). Two electromagnetic shakers are attached to the fourth bay and eighth bay of the truss structure, respectively, by stingers, to produce horizontal excitation. Four accelerometers are mounted on the nodes of the truss from fifth to the eighth bay to measure the horizontal responses of the truss at those nodes, as shown in Fig. 7. Total simulation time of the experiment is 80 s at a sampling rate of 1,000 Hz. The accelerometers mounted on the third and fourth bays are assumed to be the reference sensors that measure the correct structural response. These sensors are used to monitor the condition of the uncertain sensors mounted on the first and second bays. Uncertain sensor mounted on the first bay (sensor 2) was disconnected from the signal conditioner during 30–40 s to simulate one type of sensor failure – the sensor totally fails to measure the structural response and produces zero measurement. Between 50 and 60 s the amplitude of the output from uncertain sensor 2 was reduced by half using the built in function in the Simulink toolbox to simulate the amplitude reduction type of sensor failure. Uncertain sensor mounted on the second bay (sensor 1) was disconnected from the signal conditioner from 35 to 45 s and the amplitude of outputs was reduced by forty percent from 55 to 65 s. The first 1.5 s of the data, which is nearly 19 times of the fundamental period of the truss structure, was used to calculate the coefficients of the error function for each uncertain sensor. The error function for uncertain sensor 1 and uncertain sensor 2 by direct method for different values of p are shown from Figs. 8 and 9. In these figures,

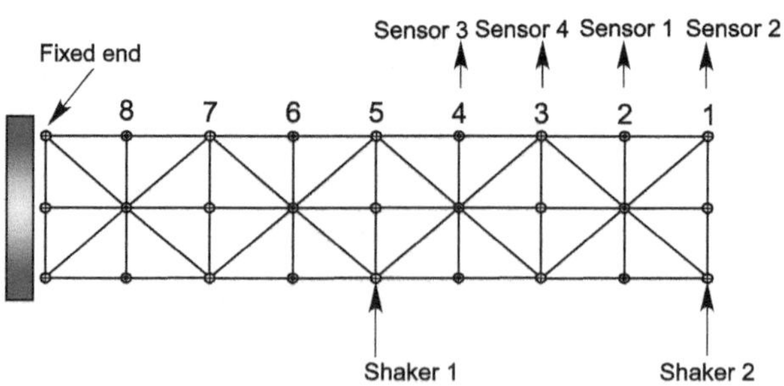

Fig. 7 Plan view of four meter long NASA eight bay truss structure

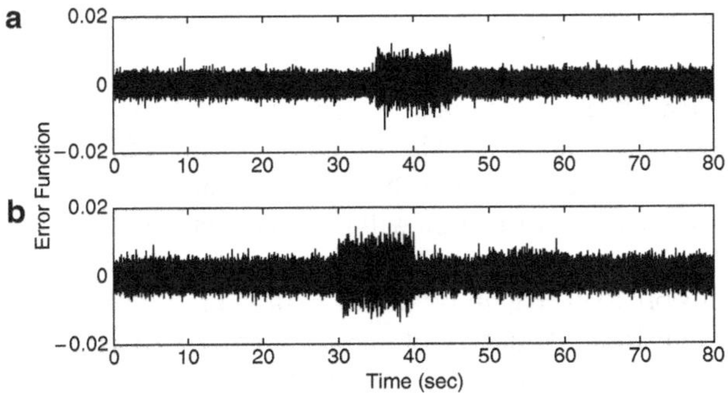

Fig. 8 Error function for uncertain sensors 1 and 2, calculated by direct method 1 with p = 40: (**a**) uncertain sensor 1, and (**b**) uncertain sensor 2

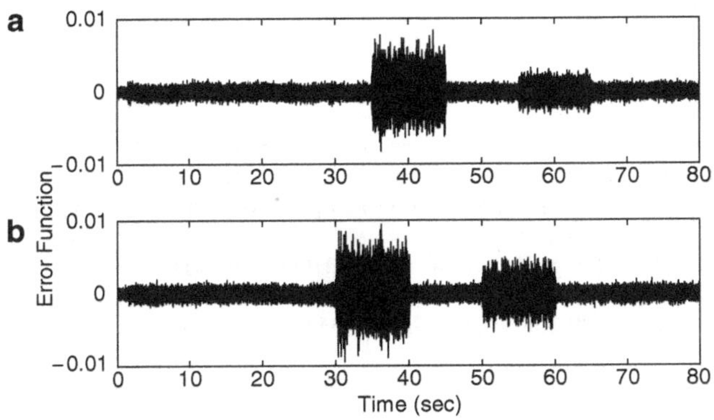

Fig. 9 Error function for uncertain sensors 1 and 2, calculated by direct method 2 with p = 150: (**a**) uncertain sensor 1, and (**b**) uncertain sensor 2

the norm of the coefficients vector P_i for the error functions are kept to one for the direct method. From these figures, it is evident that the error functions by direct methods can successfully detect and isolate the failure of the examined uncertain sensor in real time. Due to the existence of measurement noise at the reference sensors, the error functions for uncertain sensors also show non-zero value even when the uncertain sensor does not fail. In the presence of measurement noise, for different direct methods, there exists certain value of p that producing distinct non-zeros in the error function due to the examined uncertain sensor failure when compared to the error function due to the measurement noise. For example, for the cases studied for direct method, for $p = 150$, the error function due to uncertain sensor is most distinct from the error signal due to measurement noise.

6 Conclusion

The analytical and experimental studies demonstrate the performance of indirect and direct sensor failure detection algorithm. An error function, one for each examined uncertain sensor, can monitor the working status of the examined uncertain sensor in real-time. The error function will produce non-zero values when the examined uncertain sensor fails to correctly measure the structural response. Numerical simulation results prove the capability of the both indirect and direct methods in detecting and isolating multiple uncertain sensor failure. Experimental results from the eight bay NASA truss show that the direct method can successfully isolate and detect the examined uncertain sensor failure in real-time, in the presence of reasonable experimental measurement noise.

References

1. Frank P M (1990) Automatica 26(3):459–474
2. Gertler J (1991) Proc of IFAC, IMACS safeprocess symp, Baden Baden
3. Goodzeit N E, Phan M Q (2000) J of guidance, ctrl and dyn 23(2):251–259
4. Lim R K, Phan M Q, Longman R W (1998) Dept of Mech and Aero Engrg technical report, No. 3046, Princeton University
5. Kammer D C (1997) J of guidance, ctrl and dyn 20(3):501–508
6. Kammer D C, Steltzner A D (2001) J of vibration and acoustics 123(2):230–237
7. Koh B H, Li Z, Dharap P, Nagarajaiah S, Phan M Q (2005) J of guidance, ctrl and dyn, AIAA, 28(5):895–901
8. Phan M Q, Lim R K, Longman R W (1998) Dept of Mech and Aero Engrg technical report, No. 3044, Princeton University
9. Steltzner A D, Kammer D C $17th$ Intrl Modal Analysis Conf, Kissimmee, FL, SEM, 954–960.
10. Chen B, Nagarajaiah S (2008) J of Smart Structs and Sys 4(6):779–794
11. Chen B, Nagarajaiah S (2008) J of Smart Matrls and Structs 17 (3), Article No. 035019
12. Chen B, Nagarajaiah S (2007) J of guidance, ctrl and dyn, AIAA, 30(6):18311835
13. Koh B H, Nagarajaiah S, Phan M Q (2008) J Mec Sci and Tech 22(01):103–112
14. Li Z, Koh B H, Nagarajaiah S (2007) J of Engrg Mechanics, ASCE, 133(11):1222–1228
15. Dharap P, Koh B H, Nagarajaiah S (2006) J of Intell Materl Sys and Structs 17(6):469–481
16. Koh B H, Li Z, Dharap P, Nagarajaiah S, Phan M Q (2005) J of guidance, ctrl and dyn, AIAA 28(5):895–901
17. Koh B H, Dharap P, Nagarajaiah S, Phan M Q (2005) J of American Insti of Aero and Astro, AIAA 43(8):1808–1814

Resonant MEMS Sensors

Rudra Pratap

Abstract In this paper, we discuss various issues related to MEMS resonator design, in particular, design for high *quality factor* (Q). Our work focuses on ways to attain high-Q, especially in the context of non-vacuum packaged devices. We discuss the effect of fluid damping on the resonator response and point out two different routes to achieve high-Q by taking the fluid flow to equivalent low pressure regimes, and hence low damping, even without changing the actual fluid pressure. This is achieved by designing the resonator for high Knudsen number via dimensional reduction of the device or by operating the resonator in higher modes of oscillation. We also discuss the relative importance of various design parameters in the context of the resonator application. The paper includes basic introduction to MEMS resonators written with readers in mind who may not necessarily be familiar with MEMS devices.

Keywords Fluid damping · MEMS resonator · Models · Q-factor · Resonant frequency

1 MEMS Resonator Structures

Microelectromechanical Systems (MEMS) are gradually developing into a technology platform for changing the world of sensors and actuators quite dramatically. One of the important areas that is currently at the forefront of research is that of MEMS resonant devices. While some well known applications of resonance principle at the macroscale are being successfully mapped onto MEMS technology, some new and interesting applications are also emerging at micro and nanoscales. Among the new applications, vibratory MEMS gyroscopes and resonant mass detectors are currently under intense development.

One of the key issues in development of resonant devices at micro and nanoscales is the understanding and control of damping. Devices such as gyroscopes and mass

R. Pratap (✉)
Department of Mechanical Engineering, Indian Institute of Science, Benguluru, India
e-mail: pratap@mecheng.iisc.ernet.in

B. Dattaguru et al. (eds.), *IUTAM Symposium on Multi-Functional Material Structures and Systems*, IUTAM Bookseries 19, DOI 10.1007/978-90-481-3771-8_21,

detectors require high Q. If a silicon based MEMS resonator can be operated under vacuum packed conditions, Qs of the order of several thousand, say 100,000 is not hard to achieve. It is, however, not always desirable to have high vacuum. It is also harder to maintain high vacuum over long periods of time. On the other hand, the need of high-Q MEMS resonators is growing steadily. With continuous improvement in resonator design and fabrication, it seems likely that MEMS resonators may replace the dominance of quartz oscillators in electronic circuits for clock applications. Also, there are several applications of MEMS oscillators in RF devices.

Resonant MEMS devices are largely made up of simple mechanical structures that vibrate in response to some actuation, typically a periodic electrostatic force. The main elements used in the design of these devices are a mechanical resonator structure that provides inertia and elastic restoring force for basic oscillations, air or gas around the structure that provides external damping, and an actuation mechanism that sets the structure in motion. On this basic structure one can build fairly sophisticated system with controls for frequency stability or frequency tuning, etc.

1.1 Resonator Building Blocks: Beams, Plates and Membranes

The most common mechanical structures used in resonators are beams, plates and membranes. A large variety of resonator structures can be obtained by an appropriate combination of these simple mechanical structures. These structures are used as building blocks not only because they are simple mechanical structures but also because they confirm to the essentially planar structural architecture used in most MEMS and NEMS devices. The constraint of planarity is mostly because of the limitations of microfabrication technologies that depend largely on deposition, patterning and etching of planar films of suitable materials. This constraint results in resonator structures that are particularly easy to model, design and fabricate. In particular, these structures can be easily modeled by lumped parameters which, in turn, helps tremendously in creating complete system level models consisting of several electrical components.

All mechanical resonators, simple or complex, share the basic DNA of elasticity and inertia. The mother of all resonators is the single degree of freedom spring-mass model that captures the essence of oscillatory systems and resonance. It is, therefore, natural that we try to model all mechanical resonators with spring-mass models. Irrespective of how complex the structure is, if its oscillatory motion of interest can be produced by thinking of an equivalent spring (representing the relevant stiffness of the structure) and a mass (representing, again, the relevant inertia), the problem of modelling the resonator reduces to well studied and well known responses of a spring-mass system. The only new things that we must worry about are the phenomena that are peculiar to MEMS and NEMS devices, e.g., the effect of electrostatic actuation on the frequency, the damping in the system due to squeeze films, etc. The excitement in research on these resonators comes largely because of the various effects showing up at these scales that we are either not familiar with or have not dealt with enough to develop the right intuition.

Table 1 Lumped mechanical parameters for simple resonant structures commonly used in MEMS resonant sensors. Here, E is the Young's modulus, ρ is the material density, ℓ is the length of the beam and edge length of the square membrane and plate, b is the width of the beam, h is the thickness of each structure

Fixed–fixed beam		Square membrane		Square plate	
k_{eq}	m_{eq}	k_{eq}	m_{eq}	k_{eq}	m_{eq}
$\frac{16Ebh^3}{\ell^3}$	$0.5m$	$12.3T$	$0.16\rho h\ell^2$	$205\frac{Eh^3}{12(1-\nu^2)\ell^2}$	$0.16\rho h\ell^2$

Central to the design of MEMS resonators is the idea of resonant frequency which, for a single degree of freedom (SDOF) spring-mass system, is given by $\omega = \sqrt{k/m}$, where k is the spring stiffness and m is the mass. For a given mechanical structure, if the mode of vibration is chosen as the first fundamental mode, it is usually not very hard to model the structure with a SDOF spring-mass system by computing the equivalent k and m. For example, for a cantilever beam, one can find k from the relationship between the free end deflection $\delta = \frac{F\ell^3}{3EI}$ under a point load F applied at the free end, where ℓ is the length of the beam, E is Young's modulus, and I is the sectional moment of inertia. Since $k = F/\delta$, we find that the equivalent spring stiffness for a cantilever beam in its simplest form of deformation is $k_{eq} = \frac{3EI}{\ell^3}$. The determination of equivalent mass requires a bit more work since the kinetic energy in a vibrating cantilever is spatially distributed. By assuming an approximate mode shape of the vibrating cantilever (say, with half of a cosine wave), one can find the total maximum kinetic energy of the beam by integration, equate it with the maximum kinetic energy of an equivalent spring-mass system, and thus find the equivalent mass (see Senturia [14], Sec. 10.5). It turns out that $m_{eq} = 0.23m_{beam}$, or approximately just one fourth of the beam mass counts for the spring-mass model. Thus computing the first resonant frequency of a cantilever beam under pure bending motion is almost trivial. The cantilever beam resonator happens to be one of the most popular MEMS and NEMS resonators. In fact, the field of MEMS resonators started with the seminal paper by Nathanson et al. [7] with the cantilever resonator used in a resonant gate transistor. Now, there are numerous designs and applications of MEMS resonators. The equivalent spring stiffness and mass of other simple structures used in MEMS resonant sensors are listed in Table 1. Even for complex resonator structures, one can always create a simple lumped parameter model that can be used for studying the response of the resonator. For such structures, one usually resorts to numerical techniques such as the finite element method (FEM) to compute frequencies and estimate lumped parameters.

1.2 Scaling Resonant Frequency

The cantilever beam resonator can be used to see the effect of scaling on the resonant frequency of MEMS and NEMS devices. Following the simple expressions for k_{eq} and m_{eq}, we have

$$\omega = \sqrt{\left(\frac{k_{\text{eq}}}{m_{\text{eq}}}\right)} \approx \sqrt{\left(\frac{\frac{Ewt^3}{\ell^3}}{\frac{1}{4}\rho wt\ell}\right)} = \sqrt{\frac{E}{\rho}}\frac{t}{\ell^2}$$

where w is the width of the beam and t is its thickness. From this expression it is clear that $\omega \sim \frac{t}{\ell^2} = \frac{L}{L^2} = L^{-1}$. Thus the natural or resonant frequency of the beam scales as L^{-1}. It is this scaling that makes MEMS and NEMS structures very attractive for high frequency resonators. In addition, the biggest advantage, perhaps, comes from the low damping at these scales that results in very high Q resonators.

2 Resonator Response and the Effect of Q

One of the key issues in the design and realization of resonators is the quality factor or Q. The Q of a device is a measure of energy dissipation during its oscillations. It is defined as the ratio of the energy available in one cycle of oscillation to the energy lost during that cycle of oscillation. Since the energy lost depends on the damping present in the system, the computation of Q usually requires a careful modelling of damping in the system. This turns out to be perhaps one of the hardest things in resonator design. Damping is just not amenable to exact analysis and hence our effort must go into as good an estimate of damping as we can possibly make. Fortunately, it is not very hard to measure damping in resonators from their dynamic response curves.

The response of a spring-mass-dashpot system as shown in Fig. 1, subjected to a harmonic force $F(t) = F_0 \sin \omega t$ is governed by the equation of motion:

$$m\ddot{x} + c\dot{x} + kx = F(t)$$

and the response is given by,

$$x(t) = \frac{F_0/k}{\sqrt{(1-r^2)^2 + (2\xi r)^2}}$$

where $r = \omega/\omega_n$, $\omega_n = \sqrt{k/m}$ is the natural frequency of the system, and $\xi = c/2\sqrt{km}$ is the damping ratio. Usually, the response is plotted as a normalized quantity $x(t)/x_{\text{st}}$ where $x_{\text{st}} = F_0/k$ is the static displacement of the mass due to the constant force F_0. The response function, normalized or not, peaks at $r = 1$, the condition of resonance and the normalized amplitude then is determined solely by the damping ratio ξ. These things are very well known in the theory of oscillations and can be found in any elementary book on vibrations or linear differential equations.

The quality factor Q is defined as the ratio of peak response frequency to the frequency bandwidth over which the response drops to half of its peak power (or, in terms of the amplitude, $1/\sqrt{2}$ of the peak amplitude). This drop in power

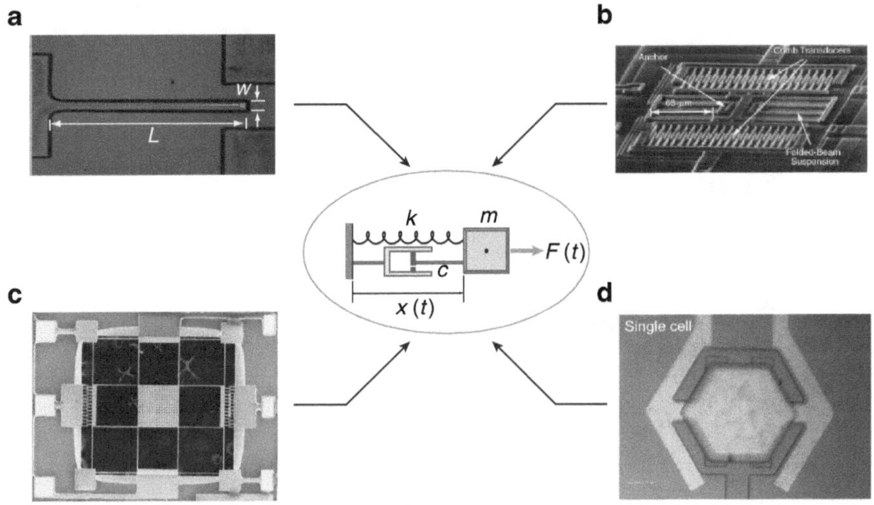

Fig. 1 Different MEMS resonators, with varying degrees of structural complexities, can be modeled as a SDOF spring-mass-dashpot system if the lumped parameters k, m and c are estimated accurately. Here, four MEMS resonant devices are shown (**a**) a simple cantilever beam [9], (**b**) a comb-drive folded beam (taken from Nguyen and Howe [8]), (**c**) a rate grade gyroscope made of a proof mass suspended by multiple beams, and (**d**) a hexagonal membrane used in a capacitive micromachined ultrasound transducer (cMUT)

corresponds to 3 dB on the decibel scale. Thus, if ω_{max} refers to the frequency corresponding to the peak response and ω_1 and ω_2 happen to be the frequencies corresponding to half power points, then

$$Q = \frac{\omega_{max}}{\omega_2 - \omega_1} = \frac{\omega_{max}}{\Delta\omega}$$

where $\Delta\omega = \omega_2 - \omega_1$ is the frequency bandwidth over which the response drops down by 3dB from the peak response. Using the expression for $x(t)$, it is easy to find the peak response frequency as well as the bandwidth corresponding to $x_{max}/\sqrt{2}$ and find that $Q \approx \frac{1}{2\xi}$. Substituting in the expression for $x(t)$, we have,

$$\frac{x(t)}{x_{st}} = \frac{1}{\sqrt{(1 - r^2)^2 + (r/Q)^2}} \qquad (1)$$

which clearly shows the effect of Q on the response of the system. It is also evident from this equation that exactly at resonance, i.e., at $r = 1$, the normalized response is exactly equal to Q. Thus the gain of a resonator at resonance depends solely on Q. This is why the Q of a resonator is probably its most important characteristic. Unfortunately, determining Q theoretically, especially for MEMS resonators is also probably the hardest task. The difficulty arises mostly from the inaccuracies in modelling various mechanisms of losses in MEMS devices, especially if they

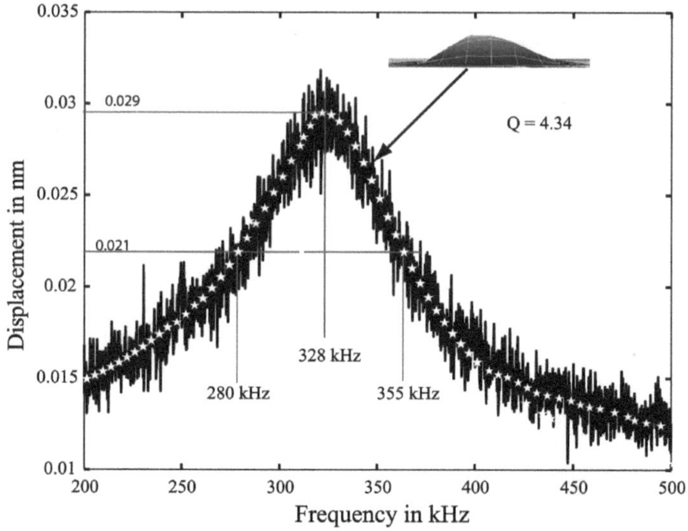

Fig. 2 The Q of a MEMS plate resonator is found here from the averaged FRF of the plate obtained experimentally by plate motion measurement using a scanning laser vibrometer, MSA 400

vibrate in air or some surrounding gas. In typical silicon MEMS resonators, if they are not vacuum packaged, the fluid damping, known as squeeze film damping [2,14], dominates all other loss mechanisms. While significant advances have been made in modelling squeeze film damping [1, 12], the predictions made by these models at the design stage rarely match Qs of fabricated devices. Fortunately, the experimental measurement of Q are not so hard and can be done at various levels of fluid pressure ranging from continuum flow conditions to rarefied, molecular flow conditions [11].

Typically, for experimental determination of Q, one excites the MEMS resonator (for example, electrostatically by applying a DC bias with a small component of AC) and obtains the frequency response function (FRF) either by direct measurements (e.g., with a scanning laser vibrometer) or indirect electrical measurements. One example of such measurement is shown in Fig. 2 where the Q of a vibrating MEMS plate resonator is determined from the averaged response curve.

3 The Quest for High Q and the Control of Damping

For resonant sensors, sharpness of the resonant peak is one of the most desirable attributes. This sharpness is achieved by designing the resonator for high Q. MEMS resonators can have extremely high Q, as much as several hundreds of thousand. Two factors that produce very high Q MEMS resonators are (i) their small size, and (ii) extremely low damping in silicon structures operating under vacuum conditions. Under very high vacuum, the only significant damping comes from internal, thermoelastic damping [13], baring the losses from degeneracies in supports, etc. Since thermoelastic damping is a few orders of magnitude smaller than air damping, it is not very difficult to get high Q resonators in the absence of air or gas damping. The challenge, however, is to design high Q resonators even in the presence of air damping. There are several applications where it is not desirable to have the MEMS resonant sensors work in vacuum. Ultrasound transducers and resonant mass detectors are examples of such sensors. It turns out that MEMS resonators can be designed to have high Q even in the presence of air or gas damping.

It is now well established that the Q of a resonator depends on the pressure of the air surrounding the resonator, especially in the squeeze film region [6, 11]. With reducing pressure, the air flow becomes rarefied and at very low pressures it reduces to molecular flow. The rarefied flow is characterized by the nondimensional number Kn, the Knudsen Number, which is the ratio of the mean free path λ of air molecules at a given pressure to the characteristic flow length h. A very low value of Kn (<0.01) signifies continuum flow and a high value (>10) signifies molecular flow [11]. In between, the flow goes through slip and transition regimes. The squeeze film damping can be computed through all these regions by solving the modified Reynolds equation using effective viscosity that accounts for rarefaction [16]. It is intuitive that as Kn increases, damping due to the squeeze film decreases.

What is more instructive, however, is to think of increasing the value of $\mathrm{Kn} = \frac{\lambda}{h}$ in various ways in order to reduce squeeze film damping. Reducing pressure increases λ and thus increases Kn. The other possibility is to reduce h by either (i) reducing air-gaps (in nanometers), or (ii) reducing the geometric dimensions of the resonator in order to have smaller h. NEMS resonators employ this route to achieve very high Q even at atmospheric pressure. Li et al. [4] show a 20-fold increase in Q by reducing the dimensions of a 100 nm thick cantilever resonator from $10 \times 2\ \mu$m to $0.6 \times 0.4\ \mu$m. Nano resonators, however, are much harder to make and are not suitable for many applications.

One of the non-intuitive solutions is to use higher modes of oscillation for increasing Q. In the case of MEMS cantilever resonators, we have found that changing the mode of oscillation from the first mode to third mode results in a 15-fold increase in Q at atmospheric pressure [9] (see Table 2). This increase is due to the pressure redistribution in the squeeze film domain in higher modes that results in lower losses. The three modes of the MEMS cantilever resonator are shown in Fig. 3. The resonator itself is shown in Fig. 1(a).

Table 2 Experimentally obtained values of Q from the frequency response shown in Fig. 3 in the first three modes of the MEMS cantilever resonator shown in Fig. 1(a)

Mode no.	ω_n (kHz)	Q
1	43	1.2
2	245	7.6
3	690	18.5

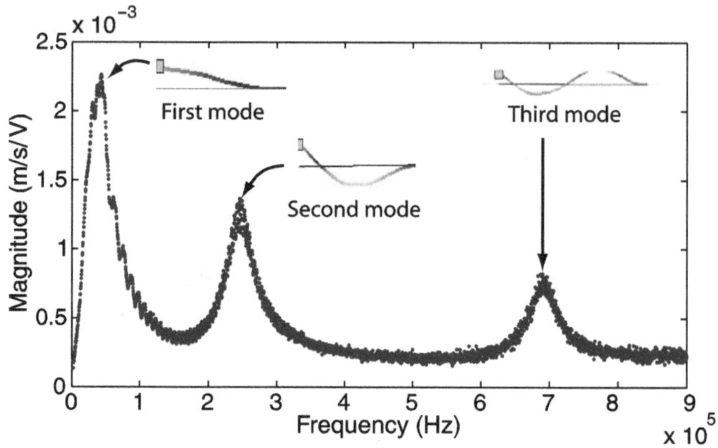

Fig. 3 The MEMS cantilever resonator shown in Fig. 1(a), with dimensions: length = 350 μm, width = 22 μm and thickness = 4 μm, is excited in the first three modes and its response measured with a scanning laser vibrometer. The Qs corresponding to the three modes are found to be $Q_1 = 1.2$, $Q_2 = 7.6$, and $Q_3 = 18.5$. See Pandey and Pratap [9] for details

4 Frequency Tuning and Stability

So far, we have only discussed the intrinsic resonant frequency of a MEMS structure and the response of the structure near resonance. Most MEMS resonant sensors use electrostatic forcing for actuation. This forcing is usually produced by applying a voltage $V(t) = V_{DC} + V_{AC} \sin \omega t$ between the resonator structure and an electrode on the substrate located at a distance, say d_0. In the parallel plate configuration of the resonator and the electrode (e.g., Fig. 1(a) and (d)), the electrostatic force is given by $F_e = \frac{\varepsilon_0 A V^2}{(d_0 - y)^2}$ where ε_0 is the permittivity of the dielectric in the gap (e.g., air), A is the area of overlap forming a parallel plate capacitor between the electrode and the resonator structure, $V = V(t)$ is the applied voltage, and $y = y(t)$ is the vertical displacement of the resonator structure at instant t during its motion. We point out that for most resonator structures (beams, plates, etc.), $y(t)$ is actually $y(x,t)$ or $y(x,z,t)$ (that is, the vertical displacement is distributed non-uniformly over the spatial domain). The electrostatic force is the applied external force $F(t)$ shown in Fig. 1. Since $F_e = f(y, V)$, a simple binomial expansion of F_e leads to a term that is linear in y and thus acts as a linear spring. This is an *electrostatic spring* with a spring stiffness proportional to V^2. This spring modifies the mechanical stiffness of the resonator structure and thus changes the resonant frequency of the resonator. For example, the changed resonant frequency of a cantilever beam resonator is approximately given by [7],

$$\omega_r \approx \omega_n \left(1 - \frac{4}{27} \frac{V_{DC}^2}{V_{\text{pull-in}}^2}\right)$$

where $V_{\text{pull-in}}$ is the *pull-in* voltage at which the electrostatic force overcomes the mechanical restoring force and the resonator structure collapses to the bottom electrode. The point here is that the resonant frequency of the structure changes due to the applied electrostatic force and is dynamic in nature. Thus, the absolute stability of the resonant frequency is destroyed. In order to maintain the resonant frequency within tight limits, one needs to employ feedback controls for appropriate voltage regulation. Of course, the upside is that we get certain amount of frequency tuning due to this electrostatic forcing.

Apart from the electrostatic spring, there is another source of induced spring effect from the squeeze film damping. If the resonator is expected to oscillate in ambient pressure with a very small gap between the resonator structure and the substrate, then the compressibility of the squeeze film gives rise to an air spring that adds to the mechanical stiffness of the resonator structure and shifts the resonant frequency [6, 12]. This is specially significant for high frequency (in MHz range) resonators. Even in the case of low frequency (in kHz range) resonators, if the boundaries of the air cavity between the oscillating structure and the substrate are flow-constrained, there is non-negligible spring effect from the squeeze film [10]. In the limiting case of a completely sealed cavity, as in the case of cMUTs (see Fig. 1(d)), the squeeze film effect reduces to only spring forces (no damping) and stiffens the structure to increase its resonant frequency.

5 Applications

Apart from the obvious applications of MEMS resonators in RF circuits, MEMS resonant structures find applications in rate grade gyroscopes [5], ultrasound transducers, biomolecule mass detection [3, 4], clock-oscillators [15], etc. In all these applications, one of the central issues is the Q of the resonator. In addition, resonators used in clock applications require high frequency stability. Most of these applications, with the possible exception of the vibratory gyroscopes, can be addressed with fairly simple resonator structures. Simpler structures can be designed quite well using lumped parameter models. As the complexity in structures increases, the accuracy of lumped parameter models starts to drop. However, one can always resort to more detailed and computationally involved continuous models for such structures. The applications and complexity of MEMS resonant sensors are continuously growing and new areas are opening up, especially due to the ease of integrability of a large number of these devices in a single array.

6 Conclusions

MEMS resonant sensors are made up of simple structures that can be easily modelled as spring-mass-dashpot systems using lumped parameters. MEMS resonators can be designed for high Q applications. The control of Q requires a good understanding of energy dissipation at micro and nano scales. Reducing the size of the structure, packaging them at low pressures and using higher modes of vibration are all possible means of increasing the Q of these devices. The ease of integration of such devices with electronics and the possibility of using an array of them on a single chip is likely to open up new areas of resonator applications.

Acknowledgement I wish to thank Suhas Mohite, Ashok Pandey, and Venkatesh for their contributions, and DST, Government of India, for the financial support.

References

1. Bao, M. and Yang, H.: Squeeze film air damping in MEMS. Sensors and Actuators A., **136**, 3–27 (2007)
2. Christen, M.: Air and gas damping of quartz tuning forks. Sensors and Actuators. **4**, 555–564 (1983)
3. Ilic, B., Yang, Y., and Craighead, H. G.: Virus detection using nanoelectromechanical devices. Appl. Phys. Lett., **85**, 2604–2606 (2004)
4. Li, M., Tang, H. X., and Roukes, M. L.: Ultra-sensitive NEMS based cantilevers for sensing, scanned probe and very high-frequency applications. Nat. Nanotechnol., **2**, 114–120 (2007)
5. Mohite S. S., Patil, N., and Pratap R.: Design, Modelling and Simulation of Vibratory Micromachined Gyroscopes. J. Phy.: Conf. Series **34**, 757–763 (2006)
6. Mohite S. S., Sonti, V. R., and Pratap R.: A Compact Squeeze-film Model including Inertia, Compressibility and Rarefaction Effects for Perforated 3D MEMS Structures. J. Microelectromech. Systems **17**, 709–723 (2008)
7. Nathanson, H., Newell, W. E., Wickstrom, A., and Davis Jr., J. R.: The Resonant Gate Transistor. IEEE Trans. Electron Devices **ED-14**, 117–133 (1967)
8. Nguyen, C. T.-C. and Howe, R. T.: An Integrated CMOS Micromechanical Resonator High-Q Oscillator. IEEE J. Solid-State Circuits. **34**, 440–455 (1999)
9. Pandey A. K. and Pratap, R.: Effect of Flexural Modes on Squeeze Film Damping In MEMS Cantilever Resonators. J. Micromech. Microengg. **17**, 2475–2484 (2007)
10. Pandey A. K., Pratap, R., and Chau, F. S.: Influence of Boundary Conditions on the Dynamic Characteristics of Squeeze Films in MEMS Devices. J. Microelectromech. Systems **16**, 893–903 (2007)
11. Pandey A. K., Pratap, R., and Chau, F. S.: Effect of Pressure on Fluid Damping In MEMS Torsional Resonators with Flow Ranging from Continuum to Molecular Regime. Experimental Mech. **48**, 91–106 (2008)
12. Pratap, R., Mohite, S., and Pandey, A. S.: Squeeze Film Effects in MEMS Devices. J. Indian Inst. of Science, **87**, 75–94 (2007)
13. Rozhart, T. V.: The effect of thermoelastic internal friction on the Q of micromachined silicon resonators. In IEEE Solid-State Sensor and Actutaor Workshop Tech. Dig., Hilton Head, SC, June 4–7, 13–16 (1990)
14. Senturia, S.: Microsystem Design, Springer Science, NY (2001)
15. Sunderesan, K., Ho, P. S., Pourkamali, S., and Ayazi, F.: A two-chip 4-MHz, microelectromechanical reference oscillator. ISCAS **6**, 5461–5464 (2005)
16. Veijola, T.: Compact models for squeeze film dampers with inertial and rarefied gas effects. J. Micromech. Microengg. **14**, 1109–1118 (2004)

Section V
Applications

Compressive Behavior of Fibre Reinforced Honeycomb Cores

S. Rao, S. Banerjee, K. Jayaraman, and D. Bhattacharyya

Abstract Honeycomb core sandwich panels have found extensive applications particularly in the aerospace and naval industries. In view of the recent interest in alternative, yet strong and lightweight materials, honeycomb cores are manufactured from sisal fibre reinforced polypropylene (PP) composites and the out-of-plane compressive behaviour of these cores is investigated. The cell wall material is modeled as a linear elastic, orthotropic plate/lamina and also as a linear elastic, quasi-isotropic material. The failure criteria for the reinforced honeycombs are theoretically developed. Failure maps that can be used for the optimal design of such honeycombs are constructed for a wide range of honeycomb densities. The results indicate a significant improvement in the load carrying capacity of the honeycomb cores after fibre reinforcement.

Keywords Compressive behaviour and modelling · Natural fibre · Reinforced honeycomb · Sandwich panel

1 Introduction

Honeycomb sandwich panels are being widely used in weight sensitive structural applications where high flexural rigidity is required. They are formed by bonding thin face sheets on either side of a low density honeycomb core. However, due to the high production costs, their application has been somewhat limited to aerospace and naval industries. To overcome this, low cost natural fibre reinforced thermoplastics are now being used in the manufacturing of core materials for sandwich panels.

S. Rao (✉), S. Banerjee, and K. Jayaraman
Centre for Advanced Composite Materials, Department of Mechanical Engineering,
The University of Auckland, Auckland Mail Centre 1142, New Zealand
e-mail: srao006@aucklanduni.ac.nz

D. Bhattacharyya
Centre for Advanced Composite Materials and Department of Mechanical Engineering, University of Auckland, Private Bag 92019, Auckland, New Zealand

B. Dattaguru et al. (eds.), *IUTAM Symposium on Multi-Functional Material Structures and Systems*, IUTAM Bookseries 19, DOI 10.1007/978-90-481-3771-8_22,

Honeycomb cores are commonly loaded in the out-of-plane direction as they exhibit excellent mechanical properties when loaded in that direction. Hence, the out-of-plane compressive behavior of honeycombs is of great importance. Research in this area is primarily concentrated in developing the relationship between the mechanical properties and the geometrical parameters of honeycombs. Extensive reviews of the mechanical properties of honeycomb materials can be found in the work of Gibson and Ashby [1]. The crushing behavior of metallic honeycombs under compressive loading was studied by Wierzbicki [2] which was a modification of McFarland's work [3]. Zhang and Ashby [4] developed expressions for the failure loads of honeycombs under transverse compression and shear loading, that agree well with the experimental data for aramid paper honeycombs. In a recent study, Banerjee et al. [5] have developed a general methodology for optimizing the specific out-of-plane shear strength of reinforced honeycomb cores.

The current work focuses on recycled sisal fibre reinforced polypropylene (PP) honeycombs. The manufacturing issues of these novel reinforced honeycombs are discussed first. The compressive behavior of honeycombs when subjected to out-of-plane compressive loading is investigated next. The failure criteria for reinforced honeycombs are theoretically developed and failure maps are constructed for a wide range of honeycomb densities. The improvement in the load carrying capacity of reinforced honeycombs as compared to that of unreinforced honeycombs is explained quantitatively. Experimental data of the manufactured honeycombs are compared with the theoretical predictions.

2 Honeycomb Core Manufacturing

The honeycombs were manufactured from extruded sisal-polypropylene composite sheets. The composite is made of sisal fibres of lengths 1–3 mm and aspect ratio ~30. Its tensile modulus is 4–9 GPa and tensile strength is 500–800 MPa. The base matrix PP has a tensile modulus of 0.9–1.2 GPa and a tensile strength of 33 MPa.

The sheets were manufactured from recycled sisal-PP pellets in a twin screw extruder through a die with 300 × 2.5 mm rectangular cross-section and were calendered to 1.5 mm thickness, with a fibre volume fraction of ~0.24. The extrusion process aligned the fibres more or less in the flow direction, making the material mildly orthotropic in nature. The extruded sheets were thermoformed between matched-dies to obtain profiled panels; these formed panels were assembled and bonded with adhesives to obtain hexagonal honeycombs, Fig. 1. The honeycombs were manufactured in such a way that the fibres in the cell walls were aligned in the loading direction, so that it can produce best performance under compressive loading. Tests were performed on the composite sheets (cell wall material) and honeycomb cores as per ASTM standards. The mechanical properties of the sheet material and the honeycomb cores are shown in Table 1.

Fig. 1 Honeycomb core sandwich panel manufactured from sisal-PP composite

Table 1 Mechanical properties of the composite sheet material and the honeycomb cores

Material property		Value	Test standard
Tensile strength	Longitudinal σ_{11}	36.40 MPa	ASTM D 638
	Transverse σ_{22}	21.40 MPa	ASTM D 638
Tensile modulus	Longitudinal E_{11}	3.87 GPa	ASTM D 638
	Transverse E_{22}	2.17 GPa	ASTM D 638
Poisson's ratio	Major υ_{12}	0.40	ASTM D 638
	Minor υ_{21}	0.20	ASTM D 638
Shear modulus G_{12}		2.87 GPa	ASTM D 4255
Shear modulus $G_{13} = G_{23}$		157.48 MPa	ASTM D 732
Sheet compressive strength		71.20 MPa	Modified ASTM D 695
Sheet compressive modulus		3.50 GPa	Modified ASTM D 695
Core compressive strength		8.73 MPa	ASTM C 365
Core compressive modulus		268.9 MPa	ASTM C 365
Sheet density		$960\,kg/m^3$	–
Core density		$156\,kg/m^3$	–

3 Out-of-Plane Compressive Behavior of Reinforced Honeycombs

A typical hexagonal cell and the associated geometrical parameters are shown in Fig. 2(a), where, h/ℓ, θ, t_1/ℓ and t_2/t_1 are the non-dimensional parameters that define the geometry of a hexagonal cell [1]. Figure 2(b) shows a unit cell made of three cell walls of half length connected at a node. H is the height of a cell, perpendicular to the plane of the paper. The unit cell shown in Fig. 2(b) has an

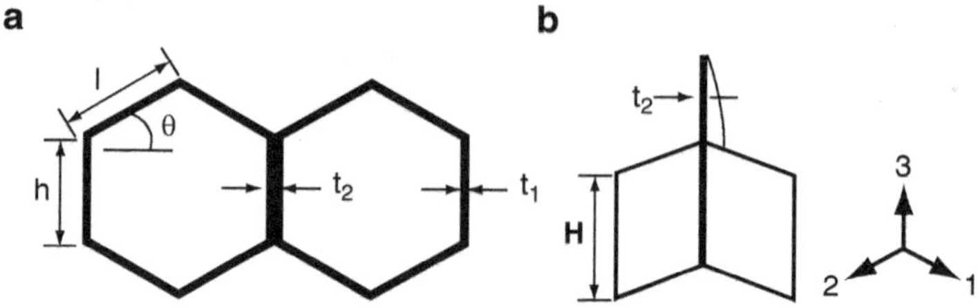

Fig. 2 Typical hexagonal cells showing (**a**) the associated geometrical parameters and (**b**) unit cell

area of $(h + l \sin\theta)l \cos\theta$, considering the periodicity and the symmetry of the honeycomb structure. The relative density ϕ of a low density hexagonal honeycomb can be approximately expressed as [5]

$$\phi = \frac{\rho^*}{\rho} = \frac{t_1}{l} \frac{(h/l)(t_2/t_1) + 2}{2(h/l + \sin\theta)\cos\theta} \tag{1}$$

where, ρ^* and ρ are the densities of the honeycomb and the cell wall material, respectively.

The possible failure mechanisms for thermoplastic honeycombs subjected to out-of-plane compressive force can be identified as elastic buckling of the cell walls, fracture of the cell wall material and de-bonding of the double thickness cell walls.

Because of the manufacturing process involved, the cell wall material is mildly orthotropic. In addition, the principal material direction (stiffer direction) is oriented in the depth direction of the honeycomb, i.e. three direction, perpendicular to the paper. Therefore, the cell wall material is modeled here as a linear elastic specially orthotropic plate/lamina under plane stress condition and also as a quasi-isotropic material, neglecting the mild orthotropy (as the degree of orthotropy ≈ 2). The failure loads are evaluated for the compressive loading applied on the unit cell in the three direction. For uniform compression of the cell walls, each wall carries an equal amount of compressive stress σ_{33}. Considering the force equilibrium in the three direction, the relationship between the external stress $\sigma_{33}{}^*$ and σ_{33} can be expressed as

$$\sigma_{33}^*(h + l \sin\theta)l \cos\theta = \sigma_{33} t_2 \frac{h}{2} + 2\sigma_{33} t_1 \frac{l}{2} \tag{2}$$

and by using (1),

$$\sigma_{33}^* = \sigma_{33}\phi \tag{3}$$

is obtained in the same form as in [4]. Therefore, the compressive stress in the cell wall is inversely proportional to the relative density of the honeycomb. The failure criteria is now developed for a honeycomb made of regular hexagonal cells with $h/\ell = 1$, $\theta = 30°$, $t_2/t_1 = 2$, $t_1 = t$ and hence, $\phi = \frac{8}{3\sqrt{3}}\frac{t}{l}$.

The elastic buckling load of the cell walls under compressive loading is now calculated, applying relevant boundary conditions. If all the edges are assumed to be simply supported, the lower bound of the critical buckling load is obtained, whereas the assumption of fixed edges predicts the upper bound. In reality, as cell walls are restrained by their neighbors at the edges and by the skins at the top and bottom, the edges between the core and skins may be considered fixed, but the fixity in the remaining edges is somewhere in between the simply supported and fixed. Hence, in this work, all the edges are assumed to be simply supported and thus, the lower bound of the critical buckling load is calculated. The lowest buckling load P_{cr} for a single specially orthotropic lamina (per unit width) under compression, with no extension-bending coupling, can be expressed as [6]

$$P_{cr} = \left(\frac{\pi^2 D_{22}}{l^2}\right)\left[\left(\frac{D_{11}}{D_{22}}\right)(l/H)^2 + 2\left(\frac{D_{12}+2D_{66}}{D_{22}}\right) + (H/l)^2\right] \quad (4)$$

where, $D_{ij} = Q_{ij}t^3/12$ are the typical bending stiffnesses of the cell wall, expressed in terms of the these stiffnesses Q_{ij} of the cell walls and i, j = 1, 2, 6.

Cell wall buckling is governed by the bending of the cell wall and hence, the buckling stress is proportional to $(t/\ell)^3$. Therefore, although the cell walls carry equal stresses, the buckling load for the double thickness (2t) cell wall is eight times higher than that of the inclined member (t). Hence, the inclined members buckle first and eventually, the double thickness cell walls buckle due to a loss of restraint at the edges. Therefore, neglecting post buckling, the critical applied stress for the honeycomb can be calculated considering the buckling of the inclined members, from Eqs. (3) and (4),

$$\begin{aligned}\sigma^*_{cr33} &= (\pi^2 Q_{22})\frac{2}{9\sqrt{3}}\frac{t^3}{l^3}\left[\left(\frac{D_{11}}{D_{22}}\right)(l/H)^2 + 2\left(\frac{D_{12}+2D_{66}}{D_{22}}\right) + (H/l)^2\right] \\ &= (\pi^2 Q_{22})\frac{9}{256}\phi^3\left[\left(\frac{D_{11}}{D_{22}}\right)(l/H)^2 + 2\left(\frac{D_{12}+2D_{66}}{D_{22}}\right) + (H/l)^2\right] \quad (5)\end{aligned}$$

If the cell wall is assumed as a quasi-isotropic material with $E = E_{11}$ and $\upsilon = \upsilon_{12}$, then using Euler's formula for elastic buckling load of a plate (per unit width) under compression [7], $P_{cr} = K\pi^2Et^3/12\ell^2(1-\upsilon^2)$, the critical buckling load for the honeycomb can be expressed using (3),

$$\sigma^*_{cr33} = \frac{2}{9\sqrt{3}}\frac{K\pi^2 E}{1-\upsilon^2}\frac{t^3}{l^3} = \frac{9}{256}\frac{K\pi^2 E}{1-\upsilon^2}\phi^3 \quad (6)$$

where the factor K represents the end conditions of the cell wall and also depends on the ratio of H/ℓ.

When the compressive stress in the cell walls reach the fracture stress of the cell wall material $\sigma_{33\,\max}$, cell walls fracture and cause the honeycomb failure. The corresponding external load is the critical fracture load for the honeycomb core and is given by

$$\sigma^*_{33} = \sigma_{33\,\max}\phi \quad (7)$$

4 Results and Discussion

The failure maps for reinforced honeycombs under compressive loading are shown in Fig. 3. The plot shows the maximum load carrying capacity of honeycombs as a function of the relative density and t/ℓ ratio, based on orthotropic and quasi-isotropic assumptions. For convenience, all the stress values are normalized with respect to E_{11}. The critical buckling load varies with the core depth and cell wall length ratio (refer (5) and (6)). The buckling load is plotted here for a core depth of 25 mm as a reference for which the laboratory experiments were performed. The value of K is taken as 4.02 for the isotropic case [7]. The ultimate compressive strength of the honeycomb is calculated using (7) on the basis of the average experimental value of compressive strength of the cell wall material which is 71.2 MPa, refer Table 1. The map shows that the buckling load prediction based on the orthotropic assumption is higher than that based on the quasi-isotropic assumption. The failure modes will now be described with reference to the orthotropic case and that is applicable for the quasi-isotropic case as well.

Figure 3(a) shows that when the honeycomb relative density is lower than the intersection point (P) of the two curves, honeycomb failure is governed by cell wall buckling. The density corresponding to point P indicates the occurrence of simultaneous buckling and fracture of the cell walls and indicates the balanced relative density for this material. With any further increase in the relative density beyond this point, the dominant failure mode changes from buckling to fracture of cell walls. The transition load at which the two modes of failure occur simultaneously, can be calculated by equating $\sigma_{33}{}^*$ from (5) and (7), and eliminating ϕ. In the present example, the load index is calculated as 1.5×10^{-3}, which corresponds to 5.87 MPa; the associated relative density is about 0.08.

The honeycomb load capacities are plotted in Fig. 3(b) for various t/ℓ ratios. For the critical load index 7.78×10^{-4}, t/ℓ ratio is 0.042 for the relative density of 0.065. Cell wall buckling is governed by the bending of the cell wall and hence, is proportional to $(t/\ell)^3$, refer Eq. (6). On the other hand, fracture resistance factor is t/ℓ, Eq. (7). Hence, with increasing honeycomb density ϕ and hence, t/ℓ ratio, effective buckling resistance increases at a much higher rate than the resistance to

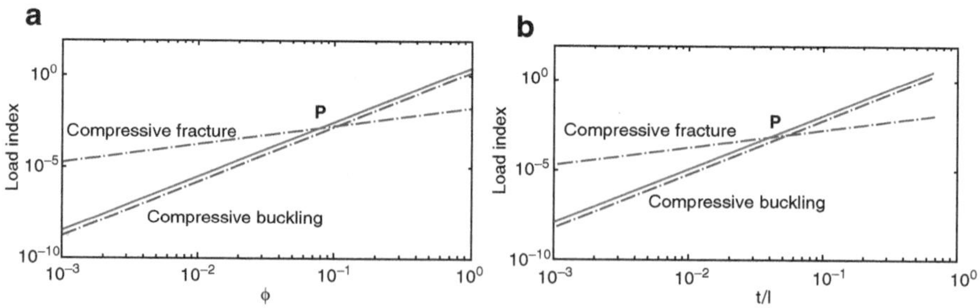

Fig. 3 Failure maps for reinforced honeycomb: (**a**) load index with respect to relative density ϕ and (**b**) load index with respect to (t/l) ratio. Legend: — indicates orthotropic case and – indicates quasi-isotropic case

fracture. As a result, cell walls become more prone to failure by fracture. Therefore, a change in the failure mode is observed when the relative density approaches a certain critical value (point P), as shown in Fig. 3. The failure map can also be used for designing the honeycomb density and geometrical parameters for a certain prescribed loading.

4.1 *Quantitative Comparisons of Strength Between the Reinforced and Un-reinforced Honeycombs*

For the *buckling* mode of failure, comparison of the expressions for the reinforced and un-reinforced cases, Eq. (6), show that for the same relative density, the ratio of the load carrying capacity of the reinforced PP honeycomb to that of the PP honeycomb is given by the ratio of the Young's modulus of the respective cell wall material, E^r/E, where superscript 'r' denotes the reinforced case. The cell wall buckling resistance improves with the increase in the Young's modulus of the cell wall material, and this, in turn, improves the load capacity of reinforced PP honeycomb. However, the addition of reinforcements would increase or decrease the density of the cell wall material depending upon whether the reinforcing fibre is of high or low density as compared to the base material itself. If a low density fibre is used, then the density of the composite is reduced and it offers an added advantage.

From Eq. (6), the ratio of the load carrying capacity of the reinforced honeycomb and the base honeycomb of same densities is in the ratio of its specific stiffnesses (E/ρ) of the respective cell wall materials, multiplied by $(\rho/\rho^r)^2$. In this work, E^r/E is 387%, and the increase in density is from 900 to 960 kg/m^3, by ~6.67%. Hence, the overall improvement in the load carrying capacity of the reinforced honeycomb to that of the PP honeycomb of the same density is about $3.87 \times 0.82 = 3.17$, i.e. 317%. Thus showing that, the increase in specific stiffness of cell wall material would result in an enhanced load carrying capacity of the honeycomb. On the other hand, for the same loading, relative density ratio of the reinforced and unreinforced honeycomb can be obtained using Eq. (6) as $\sqrt[3]{E/E^r} = 0.64$, i.e. the relative density reduction is about 36%. The reduction is less now because the relative density is proportional to the cube root of the respective moduli. The overall reduction in honeycomb density after taking into account the effect of the reinforcement (ρ/ρ^r) is about $0.64 \times 1.07 = 0.68$, or 32%.

For the *fracture* mode of failure, as seen from Eq. (7), improvement in the load carrying capacity of the reinforced honeycomb as compared to the un-reinforced case having the same relative density is in the ratio $\sigma^r_{33\,max}/\sigma_{33\,max}$. If the increase in density of the cell wall material is taken into account (same for both failure mechanisms), net improvement in strength for the same honeycomb density is in the ratio of the specific strengths $(\sigma_{33\,max}/\rho)$ of the corresponding cell wall materials. In the current work, the ratio is $\sigma^r_{33\,max}/\sigma_{33\,max} = 2.16$ and therefore, improvement in strength is 216% for the reinforced honeycomb. Taking into account of the effect of density, load carrying capacity of the reinforced honeycomb improves by $2.16 \times 0.94 = 2.02$, i.e. 202%.

4.2 Comparison of Theory with Experimental Result

The average compressive strength of these honeycombs was measured as 8.73 MPa; corresponding relative density is 0.16. Theoretical calculations indicate that the dominant failure mode of these honeycombs is cell wall fracture with the value of σ_{33}^{*} as 11.39 MPa. Observation of the real honeycomb specimens indicates that cell wall buckling is the dominant mode of failure. Figure 3 indicates that the buckling strengths of the honeycomb based on the orthotropic and quasi-isotropic assumptions to be 44.45 and 26.39 MPa, respectively. Therefore, both the assumptions indicate higher buckling load as compared to the experimental value, with the quasi-isotropic assumption predicting lower load of the two. In the buckling load calculation, shear deformation of the cell wall is neglected. As the t/ℓ ratio of the cell wall is about six, including the shear deformation of the cell wall would reduce the buckling load. In addition, the cell wall material can have elastic-plastic buckling, that can cause buckling at a lower load. Any imperfection/damage in real honeycomb can also reduce the buckling load considerably. Hence, the elastic buckling load calculated here can be considered as the upper bound. With the reduction in buckling load, the intersection point (P) would move further right and thus honeycombs with a wide range of densities would fail by buckling instead of fracture. Hence, the correlation with the experiments would improve, especially for higher densities. In future, the cell wall will be modeled as an elastic-plastic material for an improved prediction of the buckling load. Further research is in progress in this area.

5 Concluding Remarks

With an aim to produce low cost yet stiff and strong core materials, honeycombs were manufactured by a matched die forming process using recycled sisal fibre reinforced PP sheets. The out-of-plane compressive behaviour of these honeycomb cores was investigated and the failure loads were evaluated. Failure maps that can be used for optimal design of the cell geometry of honeycombs were constructed. The short fibre reinforcement in the cell walls is found to have significantly increased the load carrying capacity of the fibre reinforced honeycomb cores.

Acknowledgment The authors would like to thank the BioPolymer Network (BPN) and the Foundation for Research Science and Technology, New Zealand for their funding of this research.

References

1. Gibson LJ and Ashby MF (1997) Cellular Solids: Structure and Properties. Cambridge Univ. Press, Cambridge.
2. Wierzbicki T and Abramowicz W (1983) On the crushing mechanics of thin-walled structures. J. Appl. Mech. ASME, 50(4A):727–734.

3. McFarland RK (1963) Hexagonal cell structures under post-buckling axial load. AIAA J 1(6):1380–1385.
4. Zhang J and Ashby MF (1992) The out-of-plane properties of honeycombs. Int. J. Mech. Sci 34(6):475–489.
5. Banerjee S, Battley M and Bhattacharyya D (2008), Mech. Adv. Mat. Str., in press.
6. Reddy JN (2004) Mechanics of Laminated Composite Plates and Shells Theory and Analysis. CRC Press, Boca Raton.
7. Timoshenko SP and Gere JM (1961) Theory of Elastic Stability. McGraw-Hill Book Company, New York.

Strain-Space Solution for the Elasto-plastic Analysis of Adhesively Bonded Single Lap Joint

B.M. Dawari and P.C. Pandey

Abstract The inelastic phenomenon in multi-functional materials would be the topic of investigation in years to come. The two schools of plasticity based on conventional stress-space formulation and the evolving strain-space formulation would be the candidate concepts to be extrapolated to multi-functional materials. This work presents strain-space formulation and analysis of adhesively bonded single lap joint to predicts elasto-plastic behaviour. In strain-space formulation, yield criteria are functions of strain invariants whereas loading-unloading criteria is based on I'lyushin's postulate. Even though, it is equivalent to later one for work-hardening materials; in case of elastic-perfectly plastic and work-softening materials it has been proved computationally quite effective. The Drucker-Prager yield criterion in strain-space is adopted for adhesive to reflect the pressure sensitive nature of the material. Example calculations show that the predicted adhesive shear and peel profiles are well in agreement with previously published stress-space solutions.

Keywords Adhesive · Bonded joints · Drucker-Prager yield function · Nonlinear finite element analysis · Strain-space

1 Introduction

Even though, stress space formulation is commonly accepted in engineering applications, it has certain disadvantages. Naghdi and Trapp [1] pointed out two significant shortcomings of the stress space formulation that restricts the theory's usefulness: (1) A plasticity theory based on stress space cannot be directly reduced to represent the case of an elastic-perfectly plastic material, so a separate treatment is required. (2) There is some confusion in the formulation of the loading-unloading criteria in stress space for unstable or work-softening materials.

A strain space formulation, on the other hand, can avoid such difficulties and can therefore accommodate work hardening, work-softening, and elastic-perfectly

B.M. Dawari and P.C. Pandey (✉)
Department of Civil Engineering, Indian Institute of Science, Bengaluru, India
e-mail: dawari@iisc.civil.ernet.in; pcpandey@civil.iisc.ernet.in

B. Dattaguru et al. (eds.), *IUTAM Symposium on Multi-Functional Material Structures and Systems*, IUTAM Bookseries 19, DOI 10.1007/978-90-481-3771-8_23,

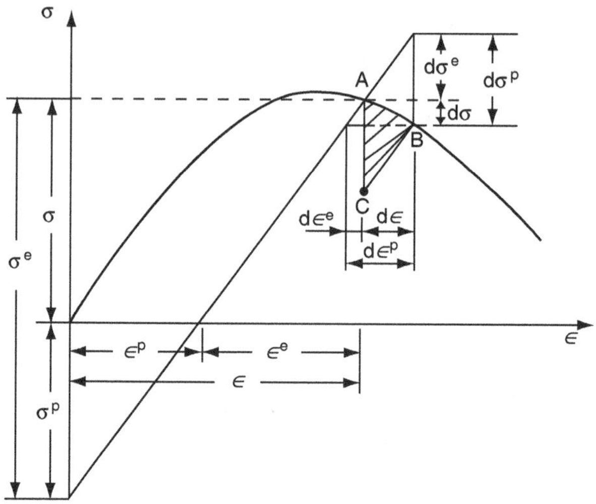

Fig. 1 Schematic description of plasticity formulation based on Il'yushin's postulate

plastic situations in one set of constitutive equations. The strain-space formulation has been discussed in the literature (Naghdi and Trapp 1975; Yoder and Iwan 1981; Casey and Naghdi 1981, 1983; Han and Chen 1986). The following formulation is based upon Il'yushin's [5] postulate (Fig. 1). This postulate considers a weak stability criterion that relaxes the requirements of Drucker's stability postulate and allows for an unstable behaviour.

From the viewpoint of implementation, the strain space formulation also has several advantages in certain situations. First, it avoids calculating the stress components to determine whether the material is in elastic or elastic-plastic state, which is an inevitable step in stress space theory regardless of the practical significance of these stresses. Second, it provides a more "natural" way to calculate the strain controlled loading path, just as stress space based plasticity models are natural to stress controlled loading path because of the input quantities in each cases are directly linked to the yield surface, loading/unloading criteria, and flow rule.

In the following discussion, the strain space plasticity is implemented to study the elasto-plastic behaviour of adhesive material in the single lap joint. It is assumed to follow Drucker-Prager plasticity for monotonic loading.

2 Strain Space Formulation of Plasticity

In a strain based plasticity model, the stress-strain relations, yield surface, loading criteria, flow rule, and hardening rule are formulated in terms of strains.

The additive decomposition law for the strain increment tensor is assumed

$$d\varepsilon_{ij} = d\varepsilon_{ij}^{e} + d\varepsilon_{ij}^{p} \tag{1}$$

Incompressibility for plastic flow is assumed,

$$d\varepsilon_{ii}{}^{p} = 0 \tag{2}$$

with

$$d\sigma_{ij} = C_{ijkl} d\varepsilon_{kl} - d\sigma_{ij}^{p} \tag{3}$$

where C_{ijkl} is the elastic constants tensor of material. The term $d\sigma_{ij}^{p} = C_{ijkl} d\varepsilon_{kl}^{p}$ is called the plastic relaxation stress.

Yield Criterion in Strain – Space The Drucker-Prager type failure surface f in stress – space can be written as

$$f = \alpha I_1 + \sqrt{J_2} - K = 0 \tag{4}$$

$$\text{Since} \quad P = \kappa\left(\varepsilon_{ii} - \varepsilon_{ii}^{p}\right) \tag{5}$$

$$\text{and} \quad S_{ij} = 2\mu\left(e_{ij} - e_{ij}^{p}\right) \tag{6}$$

where κ, μ, e_{ij} and $e_{ij}{}^{p}$ are the bulk modulus, the shear modulus, the deviatoric strain tensor and plastic deviatoric strain tensor respectively.

The stress invariants I_1 and J_2 are written as

$$I_1 = 3P = 3K\left(\varepsilon_{ii} - \varepsilon_{ii}^{p}\right) \tag{7}$$

$$J_2 = \frac{1}{2} S_{ij} S_{ij} = 2\mu^2\left(e_{ij} - e_{ij}^{p}\right)\left(e_{ij} - e_{ij}^{p}\right) \tag{8}$$

The strain invariants $\overline{I}_1$ and $\overline{J}_2$ are introduced as

$$\overline{I_1} = \left(\varepsilon_{ii} - \varepsilon_{ii}^{p}\right) \tag{9}$$

$$\overline{J_2} = \frac{1}{2}\left(e_{ij} - e_{ij}^{p}\right)\left(e_{ij} - e_{ij}^{p}\right) \tag{10}$$

The stress invariants I_1 and J_2 are rewritten as

$$I_1 = A\overline{I_1}; \quad \text{where } A = 3K \tag{11}$$

$$J_2 = B\overline{J_2} \quad \text{where } B = 4\mu^2. \tag{12}$$

Substituting Eqs. (11) and (12) into Eq. (7), the following Drucker-Prager type of yield surface φ in strain-space can be obtained.

$$\varphi = \alpha A\overline{I_1} + \sqrt{B\left(\overline{J_2}\right)} - \kappa = 0 \tag{13}$$

Loading criteria and the flow rule Casey, Naghdi and co-workers established the loading criteria in strain space (1975, 1981, and 1983). These criteria hold for work softening as well as for work hardening materials. The loading criteria are:

$\varphi < 0$ ...elastic deformation, $\varphi = 0, \frac{\partial \varphi}{\partial \varepsilon'_{ij}} d\varepsilon'_{ij} < 0$.........elastic unloading,

$$\varphi = 0, \frac{\partial \varphi}{\partial \varepsilon'_{ij}} d\varepsilon'_{ij} = 0 \text{ neutral loading, } \varphi = 0, \frac{\partial \varphi}{\partial \varepsilon'_{ij}} d\varepsilon'_{ij} > 0 \text{ plastic loading.} \quad (14)$$

Based on the work postulate in strain space [5], the following important result can be derived [6]:

$$d\sigma^p_{ij} - dC_{ijkl}\varepsilon^e_{ij} = d\eta \frac{\partial \varphi}{\partial \varepsilon_{ij}} \quad (15)$$

where $d\eta$ is a proportionality factor. For no elastic-plastic coupling,

$$d\sigma^p_{ij} = d\eta \frac{\partial \varphi}{\partial \varepsilon_{ij}} \quad \text{or} \quad d\varepsilon^p_{ij} = \frac{d\eta}{2G} \frac{\partial \varphi}{\partial \varepsilon_{ij}} \quad (16)$$

$$\text{We may assume } \frac{d\eta}{2G} = \frac{1}{h} \frac{\partial \varphi}{\partial \varepsilon'_{ij}} \quad (17)$$

$$\text{In which } d\varphi = \varphi\left(\varepsilon_{ij} + d\varepsilon_{ij}, \varepsilon^p_{ij}, \kappa\right) - \varphi\left(\varepsilon_{ij}, \varepsilon^p_{ij}, \kappa\right) = \frac{\partial \varphi}{\partial \varepsilon_{kl}} d\varepsilon_{kl} \quad (18)$$

$$\text{The stress relaxation increment is } d\sigma^p_{ij} = \frac{1}{h} \frac{\partial \varphi}{\partial \varepsilon_{ij}} \frac{\partial \varphi}{\partial \varepsilon_{kl}} d\varepsilon_{kl} \quad (19)$$

Incremental constitutive relations The incremental stress-strain relations can be obtained [7] as

$$d\sigma^p_{ij} = \left(C_{ijkl} - \frac{1}{h} \frac{\partial \varphi}{\partial \varepsilon_{ij}} \frac{\partial \varphi}{\partial \varepsilon_{kl}}\right) d\varepsilon_{kl} \quad (20)$$

$$\text{the tangent stiffness tensor } C^{ep}_{ijkl} = C_{ijkl} - \frac{1}{h} \frac{\partial \varphi}{\partial \varepsilon_{ij}} \frac{\partial \varphi}{\partial \varepsilon_{kl}} \quad (21)$$

Plastic compliance modulus The plastic strain increment is determined by the strain space normality flow i.e. (10). The scalar function $d\eta$ is evaluated with the consistency condition. From the loading criteria,

$$\partial \varphi = \frac{\partial \varphi}{\partial \varepsilon'} d\varepsilon_{ij} + \frac{\partial \varphi}{\partial \varepsilon^p_{ij}} d\varepsilon^p_{ij} + \frac{\partial \varphi}{\partial \kappa} \frac{\partial \varphi}{\partial \varepsilon^p_{ij}} d\varepsilon^p_{ij} = 0 \quad (22)$$

Substituting Eq. (10) back into Eq. (16) and recalling Eq. (11)

$$\frac{d\eta}{2G} = \frac{1}{h} \frac{\partial \varphi}{\partial \varepsilon'_{kl}} d\varepsilon_{kl} \quad (23)$$

$$\text{in which } h = -\frac{\partial \varphi}{\partial \varepsilon^p_{mn}} D_{mnpq} \frac{\partial \varphi}{\partial \varepsilon_{pq}} - \frac{\partial \varphi}{\partial \kappa} \frac{\partial \varphi}{\partial \varepsilon^p_{mn}} D_{mnpq} \frac{\partial \varphi}{\partial \varepsilon_{pq}} \quad (24)$$

where D_{mnpq} represents elastic compliance tensor

Table 1 Material properties of adhesive and adherends

Parameters	Adherend (steel)	Adhesive (FM-73)
E	211,364.0 MPa	2,210.0 MPa
ν	0.33	0.43
σ_y	Very high	40.0 MPa
λ	–	1.4

Material model for adhesive In the present analysis, the Ramberg-Osgood equation has been employed to model the stress-strain curve of adhesive FM73. The unknown parameters σ_{ch} and r of the Ramberg-Osgood curve were obtained as $r = 9.999$ and $\sigma_{ch} = 43.957$ in the equation

$$\frac{\varepsilon}{\varepsilon_{ch}} = \frac{\sigma}{\sigma_{ch}}\left[1 + \left[\frac{\sigma}{\sigma_{ch}}\right]^{r-1}\right]$$

Yield Criterion for adhesive In the present formulation, Drucker-Prager type yield criterion has been adopted for adhesive material $f = \alpha I_1 + \beta\sqrt{J_2} - \kappa = 0$ in which $\alpha = \frac{(\lambda-1)}{2\lambda}; \beta = \frac{\sqrt{3}(1+\lambda)}{2\lambda}$, I_1 = the first invariant of the general stress tensor, J_2 = the second invariant of the deviatoric stress tensor (S_{ij}) and λ = the ratio between the compressive and tensile yield stresses (Table 1).

3 Finite Element Implementation

A typical single lap joint and the finite element mesh are shown in Fig. 2. The two adherends are of the same size, but having the same or different thickness. The thickness of adherends is taken as 1.6 mm and is kept constant; while the thickness of the adhesive is considered 0.3 mm. A total of 176 eight-noded isoparametric elements with 678 nodal points have been taken to generate the mesh. The adhesive is described using 96 elements in two layers across the thickness. Finer mesh has been used to capture the stress concentration near the edges. The size of the element near the edge is of the order of 0.25 mm. The joints are considered wide enough in the transverse direction. Hence, even though the lap joint problem is three dimensional in nature; it has been treated as plane strain problem in the present study.

In the non-linear Finite Element analysis, infinitesimal dispacements are considered. The effect of material nonlinearity is only accounted. The formulation is done for Drucker-Prager plasticity. For testing the convergence, force convergence criteria is used and modified Newton-Raphson solution technique is used for terminating the iterations [9].

The results obtained show good agreement with stress-space results [10–13].

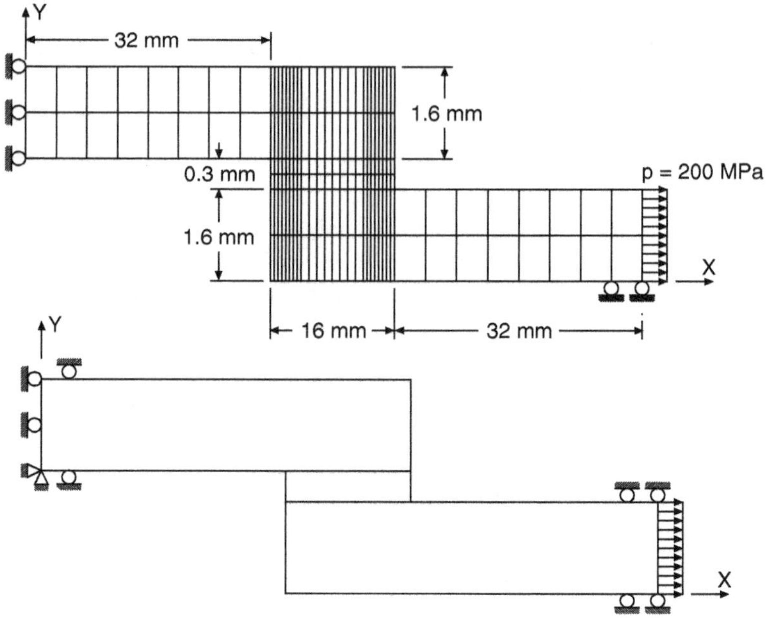

Fig. 2 A typical single lap joint with finite element mesh

4 Results and Discussions

The elasto-plastic analysis has been carried out with 12 load increments until the pressure 200 MPa is fully applied on the free edge as shown in Fig. 2. The following points are observed:

1. Strain-space formulation showed better convergence compared to its relevant stress-space counterpart.
2. The displacement and nodal stresses computed are on higher side for strain-space calculations. It shows enhanced ductility in this formulation.
3. The variation of peel stress and shear stress along the lap length at the center of adhesive layer is shown in Figs. 3 and 4. The strain-space solution is compared with the steady state elasto-viscoplastic solution reported in [10].

5 Conclusions

In this paper, the strain-space formulation is implemented for Drucker-Prager plasticity for adhesive (Fig. 5). Using the incremental theory of plasticity the equations are implemented in a finite element program to predict response of adhesively

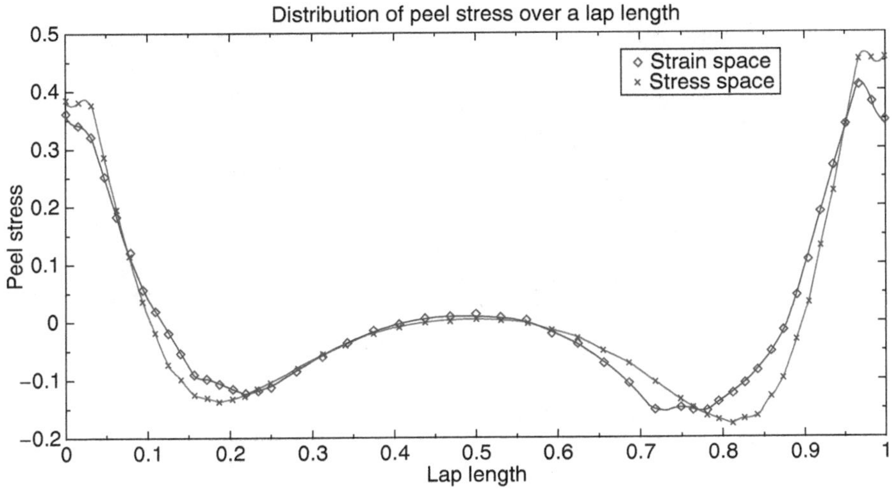

Fig. 3 Distribution of peel stress over a lap length

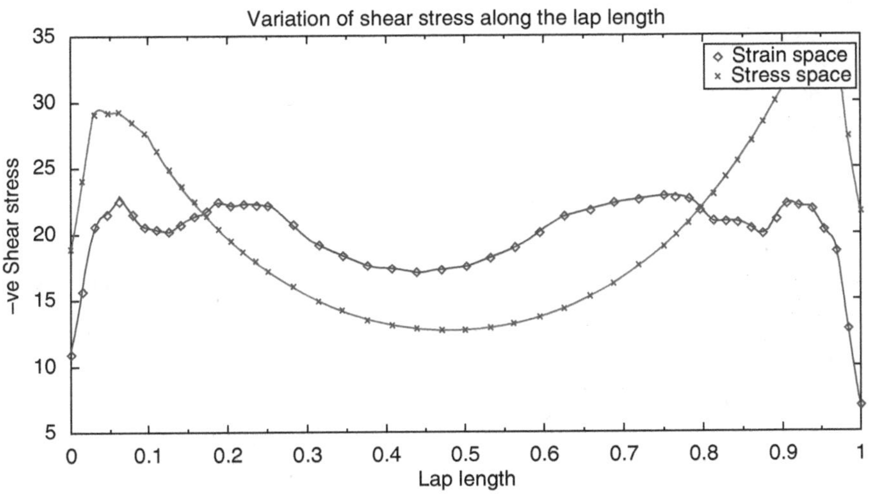

Fig. 4 Distribution of negative shear stress over a lap length

bonded single lap joint. The displacements are computed for small deformations. It has been found that strain-space formulation showed better performance during the convergence of solution and it showed enhanced ductility. Further study modeling 3-D and using coupled damage analysis are required to be carried out to have a better insight of failure of adhesive layer.

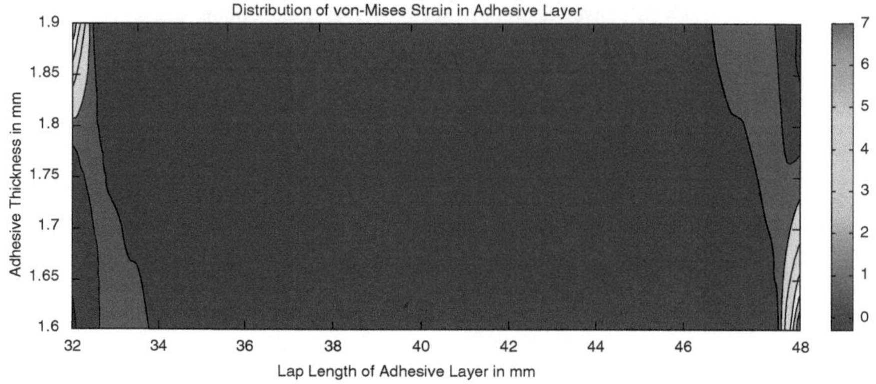

Fig. 5 von-Mises strain distribution over the adhesive layer

References

1. Naghdi PM, Trapp JA (1975) The significance for formulating plasticity theory with reference to loading surfaces in strain space. Int J Eng Sci 13:785–797
2. Yoder RJ, Iwan WD (1981) On the formulation of strain-space plasticity with multiple loading surface. ASME J Appl Mech 48:773–778
3. Casey J, Naghdi PM (1981) On the characterization of strain-hardening in plasticity. ASME J Appl Mech 48:285–296
4. Casey J, Naghdi PM (1983) A remark on the definition of hardening, softening and perfectly plastic behaviour. Acta Mech 48:91–94
5. Il'yushin AA (1961) On the postulate of plasticity. PMM 25:503–507
6. Khan AS, Huang S (1995) Continuum theory of plasticity. John Wiley, New York
7. Chen WF, Han DJ (1988) Plasticity for structural engineers. Springer-Verlag
8. Han DJ, Chen WF (1986) Strain-space plasticity formulation for hardening-softening materials with elastoplastic coupling. Int J solid struct 22:935–950
9. Owen DRJ, Hinton H (1980) Finite elements in plasticity: theory and practice. Pineridge Press, Swansea
10. Pandey PC et al. (1997) Nonlinear analysis of adhesively bonded lap joints considering viscoplasticity in adhesives. Comp & Struct 70:387–413
11. Pandey PC, Narasimhan S (2001) Three-dimensional nonlinear analysis of adhesively bonded lap joint considering viscoplasticity in adhesive. Comp & Struct 79:769–783
12. Andruet RH et al. (2001) Two- and three-dimensional geometrical nonlinear finite elements for analysis of adhesive joints. International Journal of adhesion & adhesives 21:17–34
13. Goncalves JPM et al. (2002) A three-dimensional finite element model for stress analysis of adhesive joints. International Journal of adhesion & adhesives 22:357–365

Design and Development of a Smart Composite T-Tail for Transport Aircraft

Samikkannu Raja

Abstract The experience gained in developing a smart T-Tail concept for low speed transport aircraft is presented. Multi-layered PZT stack actuators are employed to demonstrate the active vibration control of three aeroelastically critical modes. The limitations of stack actuation in terms of its output parameters, namely induced force and deflection are overcome by adopting a mechanical amplification system. Flexural hinges are employed to protect the stacks from shear failures. Two stack actuating mechanisms are optimally placed in the horizontal stabilizer and two are integrated onto the vertical stabilizer. A Linear Quadratic Gaussian control scheme is implemented in dSPACE 1104® to conduct the hardware in loop experiments. Independent modal control and multi-modes control concepts are experimentally demonstrated under an aircraft power simulated environment.

Keywords Active vibration control · Linear Quadratic Gaussian control (LQG) Piezoelectricity · Smart structure

1 Introduction

The aircraft design is conventionally driven by strength and stiffness requirements, keeping the weight as a major design constraint. Therefore, the light-weight and directionally efficient composite materials are used in the structural construction [1]. Low speed aircrafts need a clean airflow over the tail surfaces to have better pitch control; thus a *T-Tail* configuration is preferred due to its geometric location. Many flying aircrafts have T-Tail structures; for example Boeing 727, ATR-72, Q-400, CRJ700 and SARAS of NAL, India. Nevertheless, aeroelastic problems such as *flutter* and *gust* are the great concern for designers because the structurally heavy vertical stabilizer needs to carry the lift producing horizontal tail; which makes T-Tail a *structure of concern* in the low speed aircraft [2]. Therefore, the vertical

S. Raja (✉)
Scientist, Structural Technologies Division, National Aerospace Laboratories,
Benguluru – 560017, India
e-mail: raja@nal.res.in

B. Dattaguru et al. (eds.), *IUTAM Symposium on Multi-Functional Material Structures and Systems*, IUTAM Bookseries 19, DOI 10.1007/978-90-481-3771-8_24,

stabilizer (VT) has to be made considerably stronger and stiffer in order to support the elastic reactions developed by the horizontal stabilizer (HT) under aerodynamic loadings. Indeed, T-Tail may also bring aerodynamic and control related issues like deep stall condition, the complex control runs to the elevator, respectively.

Hence, it is necessary to enhance the structural performances of the T-Tail through advanced technologies without adding much weight penalty. Consequently, a smart T-Tail concept has been developed to improve the conventional design to have enhanced vibration and aeroelastic efficiencies. A cost effective *comfortable structure concept* is proposed using the PZT stack actuators and patch sensors in a closed loop feedback environment to reduce the vibration pain in the structure (dynamic strain). Smart structure is the emerging and enabling technology to build future aerospace structural systems with self-adaptable characteristics. A substantial amount of research effort has been put forth already in this multi-disciplinary field over the past 2 decades [3–6]. It is becoming evident that many engineering disciplines, especially aerospace industry shall potentially benefit from this technology in the coming years [7]. Vibration and aeroelastic controls are the prime applications of this technology for aircraft structures [8–12]. *Active Vibration Control* (AVC) and *Active Aeroelastic Control* (AAC) integrate hardware (actuators/sensors/DSP/FPGA) and control algorithms (feedback/feedforward) to offer a unique solution for aircraft structural discomfort.

In the present work, an all composite T-Tail model (20% scaled) is constructed using the spar-rib-skin arrangement. The skeletons of vertical tail (VT) and horizontal tail (HT) are built with three and two spars, respectively. The number of ribs is selected to provide the skin (CFRP), a sufficient buckling strength besides maintaining the required aerodynamic shape. The spars (CFRP) are made active by integrating four PZT stack actuating mechanisms and the ribs are fabricated using GFRP. The PZT patches are mounted on the skins and spars, near the actuating mechanisms to act as feedback sensors. Three steel blocks are designed to have slots for accommodating the composite spars of VT and these blocks are mounted on two steel plates to get fixed on a vibration block (see Fig. 1). The flexible components are modeled and meshed in HYPERMESH® to build a numerical T-Tail model, and then a normal mode analysis is carried out in ABAQUS® to establish its dynamics.

The plant modeling is done using elastic, piezoelectric coupling matrices, computed from ABAQUS. These second order system matrices are further reduced to first order system equations in state variable form, targeting three critical modes for vibration suppression studies. A Linear Quadratic Gaussian (LQG) controller is designed and implemented in modal domain using the updated plant model, after conducting the ground vibration testing.

2 PZT Amplification Mechanism

A Parallel pre-stressed stack actuator (CEDRAT®) consists of PZT active materials that deform in axial direction electro-mechanically with a limited amount of stroke length (d_{33} mode actuation). The deformation can be elastically constrained to generate induced force, which can be used in AVC/AAC applications. The blocked force (fully constrained state) or induced force (elastically constrained state) and free deflection (unconstrained state) are identified as design parameters to characterize the piezoelectric stack actuator. Stack actuator with a blocked force of 800 N and a free deflection of 40 μm has been considered to develop the amplification mechanism for the present application (see Fig. 1). The stack actuator is modeled in ABAQUS® using C3D8E elements, where the voltages are imposed as enforced electrical boundary conditions.

The voltages are applied at the nodes of top and bottom surfaces of each element with opposite potentials. This helps to introduce the opposite polarity across the elements to simulate a parallel connection. The free deflection of the stack is estimated with clamped-free boundary condition for the maximum operating voltage. It is observed that the present model predicts very closely the deflection with CEDRAT® data (see Table 1). The blocked force is then computed by constraining both the ends of actuator under maximum voltage condition. By introducing an appropriate and protective amplification mechanism, a self-actuating active structural

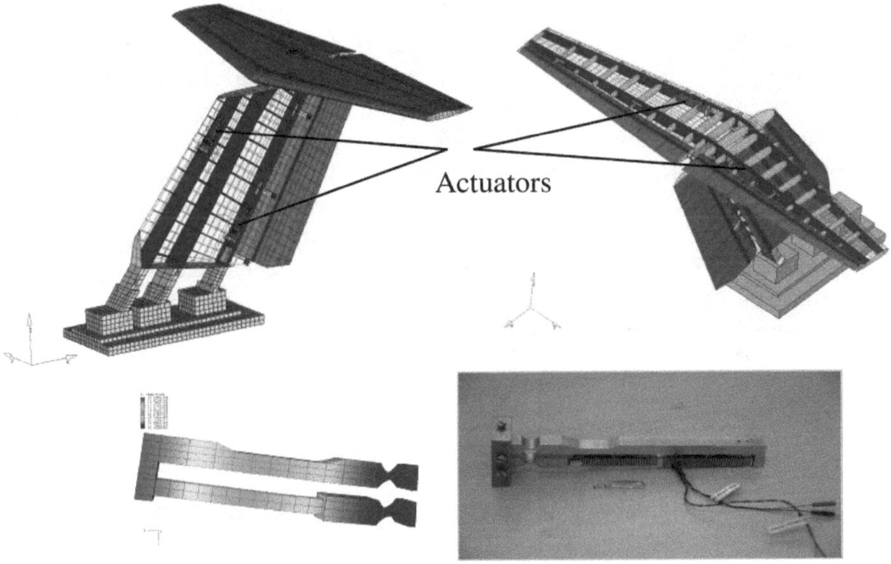

Fig. 1 Smart T-Tail with actuating systems

Table 1 Comparison of analysis results

Analysis	Displacement (μm)	Blocked force (N)
PPA40M CEDRAT®	40	800
Analytical	39.99	799.9
FEA-ABAQUS®	40	794.1

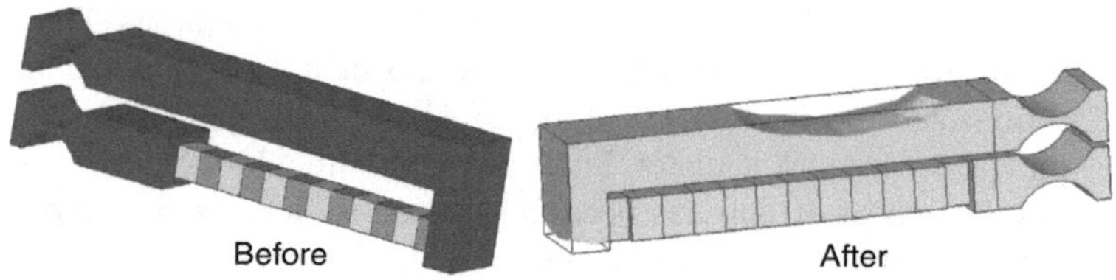

Fig. 2 Topology optimization on the mechanism

element (Smart spar) may be realized. Design of such optimized piezo-mechanical amplification system involves challenging tasks as follows:

- Concept idealization
- Material selection
- Optimized topology

After the completion of conceptual design and preliminary analysis, optimization needs to be performed, considering the weight reduction, besides strength, size and shape, depending upon the location of the amplification system and the space availability on the T-Tail structure. The proposed mechanism consists of an arm with a flexural hinge and a strut with a flexural hinge to which the rear end of the stack is fixed. The induced force of the stack is stored in the form of strain energy by the arm, which makes the actuating system to bend about the collocated flexural hinges. The flexural hinges deform and allow the mechanism to vibrate transversely. The designed hinge profile has a hinge thickness to length ratio as 1:10. Topology optimization (OPTISTRUCT®) is performed to remove the materials, wherever it is not necessary in mechanical parts of the amplification mechanism (see Fig. 2). The optimized geometry of the actuator mechanism is modeled in ABAQUS and further a static piezoelectric analysis is done to compute the induced deflection (215 μm) in the transverse direction and induced force (63 N). The analysis shows five folds amplification for the actuator deflection.

3 Smart T-Tail Analysis

Four numbers of developed mechanisms are integrated onto the T-Tail Skeleton, two each in HT and VT, respectively (see Fig. 1).

These locations are identified based on the following aspects:

- Space availability for integration
- Dynamic strain
- Location of elastic axis

The optimal locations for the actuating systems in the spars are chosen based on the static piezoelectric analysis; an iterative procedure is followed, taking more number of initial locations to arrive at four places, where these mechanisms may induce the required three elastic modes. Figure 3 gives a glimpse of the induced deflection patterns.

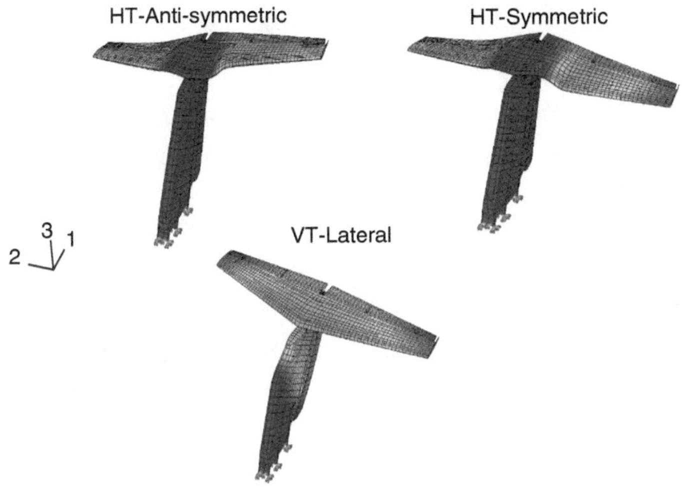

Fig. 3 Simulated mode shapes under piezoelectric loading (applied voltage 150 V)

The LQG control scheme is developed in MATLAB/SIMULINK environment and the same has been implemented using dSPACE DSP-1104 and RTW/Control Desk 2.1 modules. The LQR design assumes that the actuator power is fully available to perform a specified control action; Kalman filter (estimator) is used to estimate the unmeasured states from measured ones. The design also takes into account the process disturbance (β) and measurement noise (χ). The plant matrices obtained from finite element analysis have been employed to build the required state space model of the smart T-Tail. The measured experimental damping of the respective elastic mode from ground vibration testing is then introduced in the mathematical model (refer to Section 4). The control model is built in modal domain with appropriate actuator to sensor loop connections. In the first phase, the MATLAB code is developed to perform the necessary time (impulse) and frequency analysis, following the Linear Quadratic Gaussian control scheme. The simulation study is further extended by building the real time SIMULINK® models. Before these working models are downloaded in DSP platform for conducting hardware in loop experiments, simulation results are obtained in MATLAB/SIMULINK environment to make sure the workability of the modal controller.

4 Open and Closed Loop Experiments

In the first phase, the workability of the developed mechanism has been tested by integrating the actuating mechanism onto a composite spar. Figure 4(a), shows the picture of the test setup, where an accelerometer is mounted on the tip of the spar for response measurement. After examining the performance of the designed and developed actuating systems, they (four numbers) are integrated onto

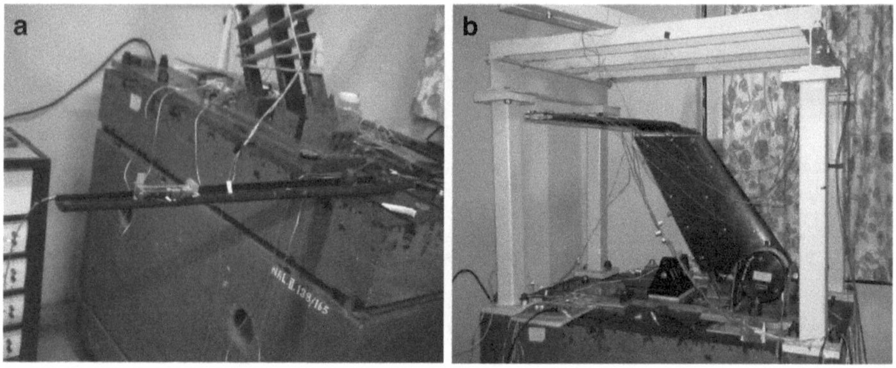

Fig. 4 Smart T-Tail testing

the T-Tail spars; wiring and insulation/isolation have been properly done to ensure the functionality of these smart devices. In order to measure the structural responses of the feedback control, the PZT sensor patches are surface bonded on the spar webs. Ground vibration testing is performed on the T-Tail to extract its frequencies, mode shapes, and damping (see Fig. 4(b)). A burst random test is conducted with multi-input (two shakers) and multi-output (accelerometers) configuration. Table 2 presents the dynamic characteristics of the T-Tail. Further the finite element model is revised in order to build a realizable plant, which could represent mathematically the actual T-Tail with integrated actuators and sensors.

4.1 Active Vibration Control Studies

After the completion of ground vibration studies, the mounted actuators and sensors are tested for their functionalities. Integration of in-flight electronics and the smart T-Tail in closed loop configuration is a challenging task, right from insulation/isolation of electronics to the operation of high powered stack actuators under the aircraft simulated power environment [13]. This is in fact due to sustainability of the power supply to actuators (−30 to 150 V) from 28VDC power source, besides any transients or spikes surging in the bus. In the closed loop configuration, four actuator electronic modules and two sensing units are used. Each actuator module drives a single actuator; however each sensing unit can receive signals from four patches. The SIMULINK model is then implemented with the help of dSPACE-DSP 1104 board, which is a multi-channel digital real time computing environment with ADC's and DAC's to collect sensor signals to output control signals, respectively. The smart T-Tail model is disturbed in two axes, namely along X and Y. The shaker that is positioned along X axis has excited the HT anti-symmetric bending and HT symmetric bending modes; on the other hand, the shaker along Y targets the VT lateral bending excitation (Refer to Fig. 4(b)). The global sensors (accelerometers) are mounted on HT tips (left and right; to measure the vibration along Z) and as well as VT tip (to measure the vibration along Y).

The final plant matrix of smart T-Tail is updated with the experimentally measured frequencies and damping for the modes of interest (Refer to Table 2).The open and closed loop experiments are conducted using RTW/Control Desk software and other necessary hardware. Two cases are handled, namely case (i): resonant mode control and case, (ii): combination resonance control.

The gains used in the resonant control have also been implemented for combination resonance study. In the combination resonance case, two modes are excited simultaneously, namely VT lateral bending and HT symmetric bending. The idea here is to understand the modal coupling between controlled and uncontrolled modes. As there is no filter is designed, the sensors' responses of this combination resonance contain both frequencies. Therefore, the control signal also will have both frequencies. Figure 5 shows that the targeted mode is properly controlled (only actuators in HT activated) and the other mode is stable and not affected. It indicates

Table 2 Dynamic characteristics of Smart T-Tail

Mode order	Mode shape	Frequency (Hz)	Damping (%)
1	HT Anti-symmetric bending	26.6	1.91
2	VT Lateral bending	62.6	1.35
3	HT Symmetric bending	86.6	1.01

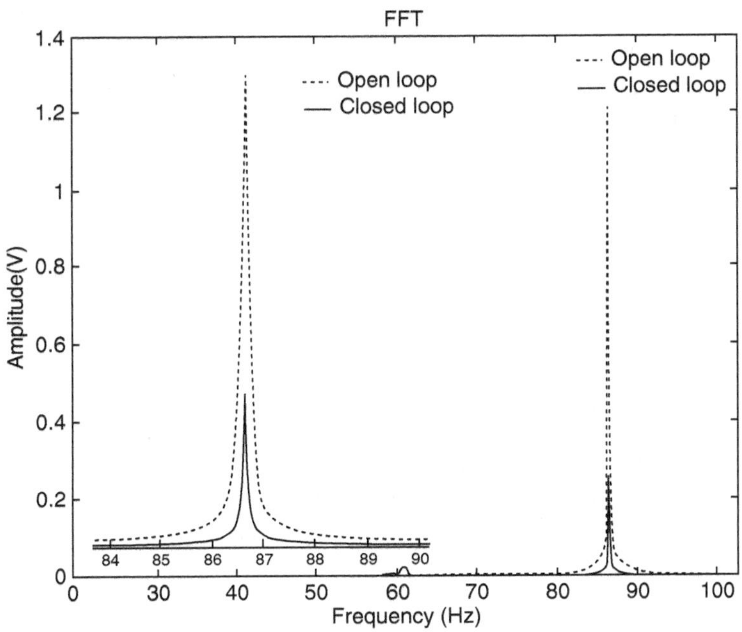

Fig. 5 Single mode target

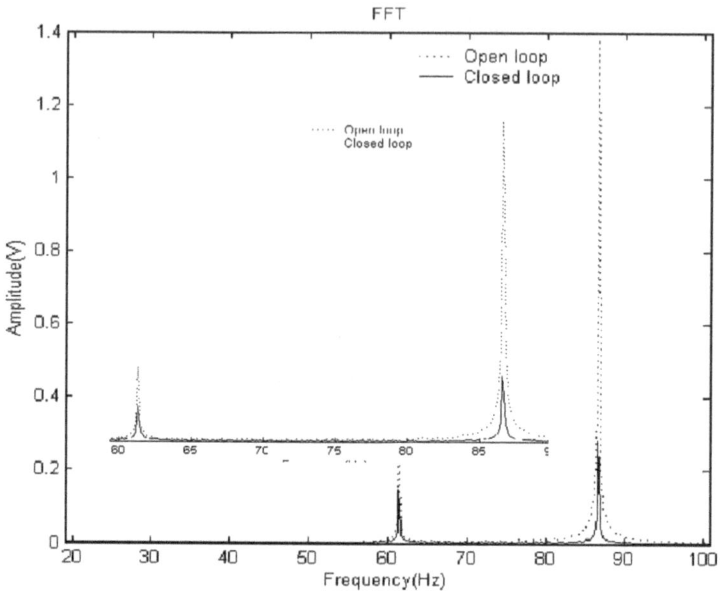

Fig. 6 Multi mode target

that if the target mode frequency is quiet away from the uncontrolled mode, the interference effect is insignificant. Next both the modes are targeted at the same time. Since the signal of the feedback sensors contain both frequencies, the control signals supplied to all the four actuators will have energies of both modes. It is noticed that both the modes are significantly controlled (see Fig. 6). Since these two modes are well separated, we could conduct the closed loop experiments without a filter. Resonant control may be preferred, when the target mode frequency is quiet far from the other modes; otherwise combination resonance is appeared to be efficient.

5 Conclusions

A smart composite T-Tail model is designed and fabricated with self-actuating and sensing capabilities. Parallel pre-stressed PZT stack actuators are employed to develop the actuating systems; whereas isotropic PZT patches are used as feedback sensors. The hardware in loop experiments are conducted under aircraft simulated power environment. Three aeroelastically important modes are identified and active vibration suppression has been demonstrated using LQG control scheme.

Acknowledgments I would like to thank Dr. A. R. Upadhya, Director, NAL, for permitting a part of the outcome of DISMAS sponsored project to be presented in IUTAM-MMSS08. The author greatly acknowledges the funding received from DISMAS, ADA, Benguluru, to carry out this work; in particular Mr. T. M. Naidu, the project coordinator and Dr. K. Vijayaraju, the coordinator,

DISMAS programme. I profoundly remember Mrs. Shashikala Rajappa, Mr. V. Shankar, Mr. S. Janardhanam, Mr. R. Sreedeep and H. M. Raghavendra, for their major contributions in the project. The author wants to place his appreciation to Mr.Ramachandra and his team members, who had fabricated the 20% smart T-Tail.

References

1. Jim Renton, et al., 2004, "Future of Flight Vehicle Structures," J. of Aircraft, Vol.41, No.5, pp. 986–998.
2. Raja, S., et al., April 2007, "Aeroelastic Testing of SARAS T-Tail," Project Document ST 07011, National Aerospace Laboratories, Benguluru, India.
3. Chopra, I., 2002, "Review of State of Art of Smart Structures and Integrated Systems," AIAA J., Vol. 40(11), pp. 2145–2187.
4. Gern, F., D. Inman, and R. Kapania, December 2005, "Computation of actuation power requirements for smart wings with morphing airfoils," AIAA J., Vol. 43, pp. 2481–2486.
5. Hurlebaus, S., Gaul, L., 2006, "Review Smart Structure Dynamics," Mechanical Systems and Signal Processing, Vol. 20, pp. 255–281.
6. Barrett, R., January 2006, "Adaptive Aero-structures: Curriculum Changes Needed to close the Aerospace Knowledge Gap," AIAA 2006-284, 44th AIAA Aerospace Sciences Meeting and Exhibit.
7. Barrett, R., September 2007, "Improvements to Commercial and General Aviation via Adaptive Aero-structures," AIAA 2007-7873, 7th AIAA Aviation Technology, Integration and Operations Conference, Belfast, Northern Ireland.
8. Changho Nam, Aditi Chattopadhyay, and Youndan Kim, November 2000, "Aeroelastic control of Composite Plate with Delaminations," J. of intelligent Material Systems and Structures, Vol. 11, pp. 868–876.
9. Suleman, A., Costa, A.P., "A RPV Adaptive Aeroelastic Demonstrator," Proc. SPIE European Workshop on Smart Structures in Engineering and Technology, Vol. 4763, 113 (2003).
10. Matsuzaki, Y., Torii, H., "Flutter Boundary Prediction of an Adaptive Smart Wing during the Process of Adaptation," Proc. SPIE - Smart Structures and Materials 2005: Smart Structures and Integrated Systems, San Diego, CA, USA, Vol. 5764, 203 (2005).
11. Raja, S., Upadhya, A.R, 2007, "Active Control of Wing Flutter Using Piezo-Actuated Surface," J. of Aircraft, Vol.44 (1), pp. 71–80.
12. Ed Pendleton, et al., April 2007, "The X-53: A Summary of the Active Aeroelastic Wing Flight Research Program," AIAA 2007-1855, 48th AIAA/ASME/ASCE/AHS/ASC Structures, Structural Dynamics, and Materials Conference, Honolulu, Hawaii.
13. Shankar, V., S. Raja, Shashikala Rajappa, and Janardhanam, September 2008, "Development of a System for In-flight Actuation and Sensing of Piezoelectric Based Smart Aircraft Structures," Project Document ST 08019, National Aerospace Laboratories, Benguluru, India.

Manufacturing of Multi-functional Composites

Prasad Potluri and Robert J. Young

Abstract This paper investigates the manufacturing routes for multi-functional materials. Multi-functionality includes functions such as sensing, actuation, heating, cooling, thermal conductivity, camouflage and stealth, in addition to load bearing. Composites have a distinct advantage in incorporating multi-functionality at the material scale. Dry fibre preforming has been identified as a key enabling technology for incorporating multifunctional elements into a fibre assembly before the resin is injected. This paper also investigates non-contact measuring techniques such as wireless sensing and Raman spectroscopy.

Keywords 3D Weaving · Non-contact sensing · Raman Spectroscopy · Sensor integration

1 Introduction

There is a large volume of literature on theoretical framework for multi-functional materials. However, there is limited published work on practical manufacturing and integration issues related to multi-functional materials. Multi-functional materials are not a simple collection of devices each fulfilling a discrete function. A truly multi-functional material should be a single material system with an intelligent incorporation of several functions at the manufacturing stage. Fibre reinforced composites are ideal candidates for multi-functionality since different materials can be combined to form a single structure. In fact, composites in the form of biological tissue are the nature's answer to multi-functionality.

Multi-functionality in composites is incorporated for a variety of reasons:

1. Stiffness and strength tailoring: stiffness and strength of composites can be effectively tailored by fibre placement in preferred orientations (in addition to

P. Potluri (✉) and R.J. Young
School of Materials, University of Manchester, Manchester M60 1QD, United Kingdom
e-mail: Prasad.Potluri@manchester.ac.uk, Robert.Young@manchester.ac.uk

B. Dattaguru et al. (eds.), *IUTAM Symposium on Multi-Functional Material Structures and Systems*, IUTAM Bookseries 19, DOI 10.1007/978-90-481-3771-8_25,

geometrical tailoring in isotropic materials). For example, bend-twist coupling in composite laminates has been exploited in aeroelastic structures.
2. Tailoring of thermal or electrical conductivity in different parts of the composite structure can be tailored by incorporating different fibres in different locations/layers. For example, incorporation of pitch based carbon fibres in carbon composites enhances thermal conductivity [1]. Similarly, thermal barrier, thermal camouflage or radar stealth can be achieved by incorporating appropriate fibres in the surface layers.
3. Sensors and actuators can be embedding in composite structures during manufacturing. Strain, temperature etc. can be sensed at a specific location or at a large number of distributed positions. Actuating materials such as SMAs, PZTs can be incorporated for active vibration damping or structural morphing.

2 Composites Manufacturing Techniques

Fibre reinforced composites in the aerospace and other high-end applications are traditionally manufactured using prepreg materials and cured with autoclave technology. However, in recent years, out-of-autoclave technologies are becoming popular in order to reduce manufacturing costs and improve productivity. From the point of view of multi-functionality, prepreg based systems are restrictive in terms of combining a variety of material systems. The authors believe that out-of-autoclave techniques involving dry fibre preforms offer significant flexibility to incorporate a variety of multi-functional hybrid material systems. A variety of sensors, actuators and other functional materials can be incorporated into a dry preform before resin infusion process. Consequently, this paper reviews the dry fibre preforming technology.

2.1 Dry Fibre Preform Technology

Dry fibres are supplied in the form of a yarn or roving with specified filament count (3k, 6k, 12k etc.). These yarn or roving are assembled together into two dimensional sheets or three dimensional multi-layer structures. A review of preforming techniques has been presented by Verpoest [2]. Textile manufacturing techniques such as weaving, braiding and stitching are popular for manufacturing dry textiles. Figure 1 shows typical 2D textile structures. Woven fabrics are produced by interlacing two sets of fibre tows orthogonal to each other. Braided structures are produced by interlacing two or three sets of tows in a non-orthogonal fashion. Stitch-bonded fabrics consist of biaxial or multi-axial tows stitched together by fine filaments.

A number of 2D plies are cut into required shapes and assembled together into a preform. Smart materials such as sensors and actuators may be placed between individual plies. However, placing dissimilar materials may lead to delaminations. Schaaf et al. [3] investigated the influence of embedded sensing elements

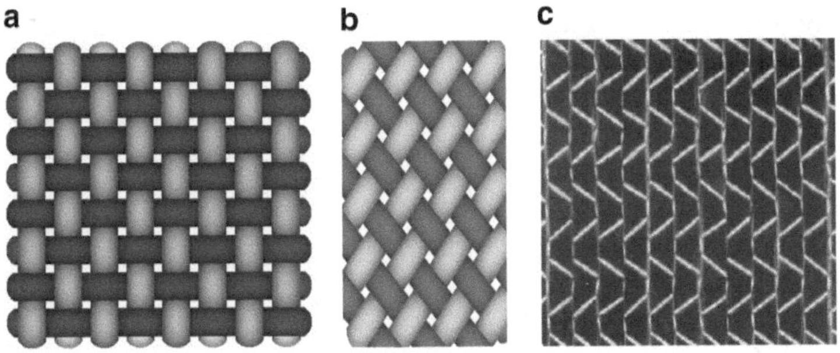

Fig. 1 (**a**) 2D woven (**b**) Braided (**c**) Stitch bonded NCF

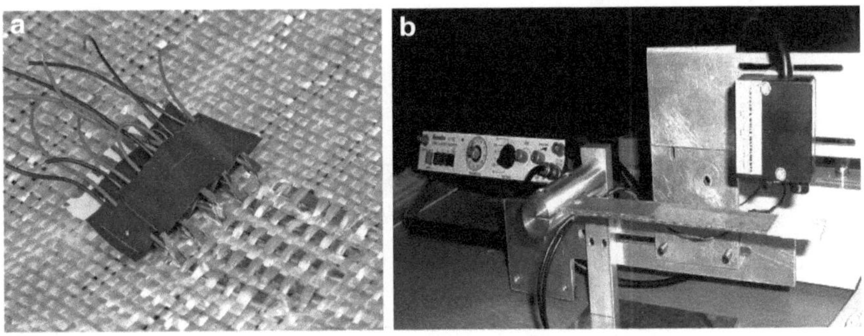

Fig. 2 (**a**) Integration of PVDF sensors into 2D fabrics, (**b**) cantilever beam with built-in vibration sensor [5]

on inter-laminar shear strength and fatigue life of composite laminates. They co-cluded that PVDF film sensors, simply placed between layers, reduces the short beam shear strength and significantly shortens the fatigue life. Precision punched cut-outs in prepregs tend to improve both the properties considerably.

Handling and accurate placement of sensors (such as fibre optic sensors) is another issue. Glisic and Inaudi [4] developed a sensing tape by sandwiching optical fibres between two thermoplastic films. This sensor tape may be placed between the plies in any preferred orientation and position within a ply stack. The sensor film assembly is about 200 μm thick, and hence may adversely affect the inter-laminar strength of the composite laminate. An interlacement technique has been developed (at the University of Manchester) to incorporate sensing elements directly into a woven fabric in order to improve inter-laminar strength. PVDF strips were co-woven on a loom (Fig. 2(a)). Since the sensing strips are locked in position and covered by the reinforcing fibres, interlaminar strength has improved significantly over the traditional method of placing the sensing elements between the layers. Figure 2(a) shows the connecting wires attached to the sensing elements subsequent to co-weaving. In future work, micro wires will be co-woven along with the PVDF film.

One layer of the glass woven preform with the bonded PVDF was stacked with several plies of plain woven glass fabric after which resin infusion was carried

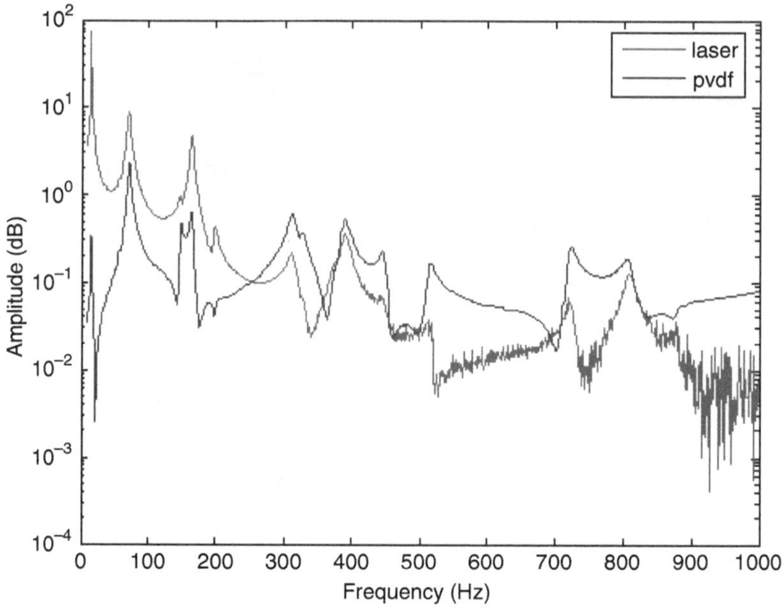

Fig. 3 Comparison of natural frequencies measured by integrated PVDF sensor and an external laser sensor [5]

out and the structure consolidated. The cantilevered beam with woven PVDF was excited with a shaker and the frequency response function of the beam was measured via the embedded PVDF films as well as a laser probe focussed at the free end of the cantilever (Fig. 2(b)). The results obtained are shown in Fig. 3. Natural frequencies measured by the embedded PVDF sensors and an external laser sensor are comparable.

2.1.1 3D Weaving

3D weaving has shown a lot of promise in creating multi-layer near-net preforms [6]. Through thickness reinforcement offered by the binding tows in 3D weaves improve the damage tolerance. Figure 4 shows typical 3D woven structures. Smart material can be incorporated between the layers without compromising the interlaminar strength, due to the presence of through thickness binders. Bogdanovich et al. [7] developed 3D woven preforms with built-in fibre optic sensors. Figure 5 shows a 3D woven laminate with optical sensors, developed at Manchester University. The optical fibres are protected by a fine braided sleeve prior to weaving.

3D weaving has additional advantages in terms of tailoring the thickness of the composite laminate by dropping some tows or by manipulating the cross-section of the weft tows (Fig. 6).

Figure 7 shows various stiffening sections (T, PI) produced using 3D weaving processes. Strain sensors can be incorporated in critical stress region such as fillets

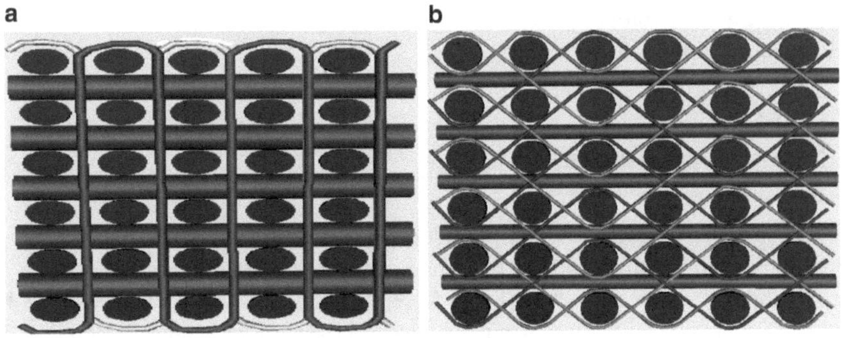

Fig. 4 3D weaves: (**a**) orthogonal, (**b**) layer-to-layer interlocked

Fig. 5 3D woven composite with fibre optic sensors

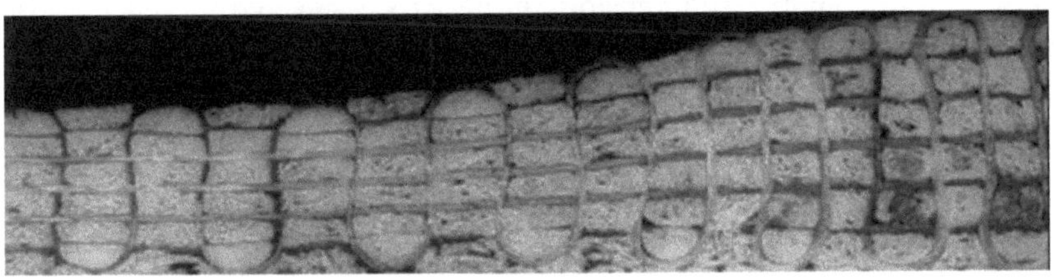

Fig. 6 Geometrical tailoring with 3D weaving [6]

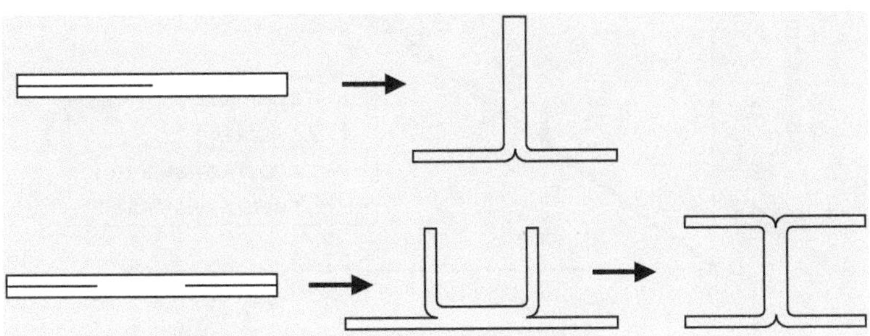

Fig. 7 3D weaving of stiffeners

in the stiffeners, without compromising the strength between skin and stiffeners (in a stiffened composite structure).

3 Non-contact Sensing

One of the main problems with embedded sensors is the connectors; the connecting wires are often bulky and get damaged in service. Passive wireless sensors are getting popular in recent years [8, 9]. Direct–write thermal spray technique has become a reality for creating strain gauges, thermocouples and RF elements for wireless sensing [10]. Young et al [11] pioneered the use of Raman spectroscopy in measuring strains generated in reinforcing fibres such as Kevlar and PBO. This is a non-contact optical technique; a 2 μm laser beam is focussed on a fibre and inelastic scattering is captured using a CCD array and analysed for Raman band shift. Raman band shift has a linear relationship with fibre strain. Figure 8 shows the comparison between Raman strain and the strain measured using a conventional strain gauge. In this case, a 12 μm Kevlar fibre is pasted on the surface of a carbon composite laminate. Raman spectroscopy is a laboratory based technique, and is generally difficult to use it on real composite structures. However, recently, Kezirian et al [12] reported the use of Raman spectroscopy on large Kevlar wound spherical pressure vessels for space applications. They used a remote Raman probe mounted at the end of a three-axes gantry to autofocus and scan the surface of a pressure vessel. Raman spectroscopy is the only non-contact technique that will enable measurement of complete strain history during processing and subsequent loading of a component/structure.

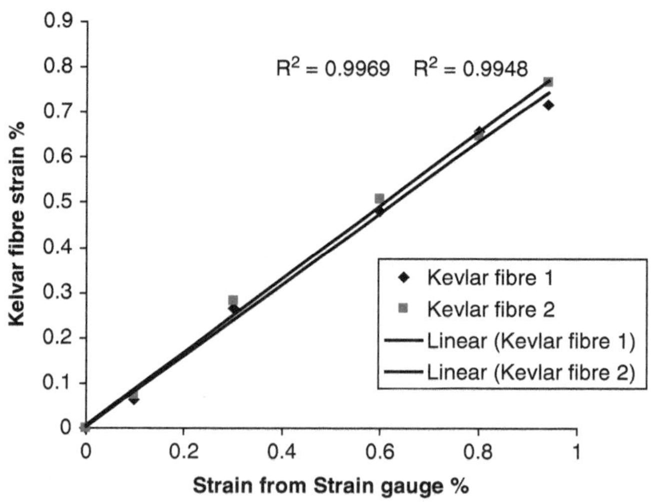

Fig. 8 Comparison of Raman strains with strain gauge strains

4 Conclusion

Various manufacturing strategies for incorporating multi-functional elements into composite structures have been investigated. Dry fibre preforming, especially, 3D weaving is a promising technique for incorporating smart elements into a fibre assembly prior to resin infusion. Wireless/non-contact measurement techniques are promising with the recent development of RF arrays for transmitting the measured signal. Direct-write techniques can produce sensors and RF arrays directly on a laminate surface, and these are less likely to effect inter-laminar strength in comparison to discrete sensors. Raman spectroscopy is another promising lab-based technique that needs further development for field applications.

References

1. Sharp K, Bogdanovich A, Tang W, Heider D, Advani S and Glowiana M (2008) High through-thickness thermal conductivity composites based on three-dimensional woven fiber architectures AIAA Journal 46 (11):2944–2954
2. Verpoest I (2000) Composite preforming techniques, Comprehensive Composite Materials. 2:623–669
3. Schaaf K, Rye P, Nemat-Nasser S (2007) Optimization of sensor introduction into laminated composites. Proceedings of the 2007 SEM Annual Conference and Exposition on Experimental and Applied Mechanics, 3–6 June 2007, Springfield, Mass, USA
4. Glisic B, Inaudi D (2003) Integration of long-gage fiber-reinforced composite sensing tape. SPIE International Symposium on Smart Structures and Materials, 2–6 March 2003, San Diego, USA
5. Dauda B (2007) Delamination of multilayer textile reinforced structural composites. PhD Thesis, University of Manchester
6. Jetavat DS, Potluri P (2008). Extension of 3D weaving concepts for near-net preforming, 49th AIAA SDM conference, 7–10 April 2008, Schaumburg, IL, USA
7. Bogdanovich AE, Wigent DE, Whitney TJ (2003) Fabrication of 3-D Woven Preforms and Composites with Integrated Fiber Optic Sensors. SAMPE Journal. 39 (4): 6–15
8. Wong ZJ, Kim CG (2008) Smart composite structure based on integrated passive wireless strain Sensors. Proc of SPIE. 6932
9. Matsuzaki R, Todoroki A (2006) Passive wireless strain monitoring of actual tire using capacitance-resistance change and multiple spectral features. Sensors and Actuators, A.126: 277–286
10. Longtin J, Sampath S, Tankiewicz S, Gambino RJ, Greenlaw, R (2004) Sensors for direct-write thermal spray. IEEE Sensors Journal. 4(1):118–121
11. Lei SY, Young RJ (2001) Deformation of PBO/epoxy plain weave fabric laminae followed using Raman spectroscopy. Composites A. 32:499–509
12. Kezirian MT, Phoenix SL, Eldridge JI (2009) Use of Raman spectroscopy and delta volume growth from void collapse to access overwrap stress gradients compromising the reliability of large Kevlar/epoxy COPVs. 50th AIAA SDM conference, Palm Springs, 4–7 May 2009 (AIAA 2009–2566)

Section VI
Computational Methods- I

Iso-Spectral Rotating and Non-Rotating Beams

Ananth Kumar and Ranjan Ganguli

Abstract It is shown that centrifugal stiffening effects in rotating beams can be countered by spanwise stiffness variation for a given mode. Smart materials such as magnetorheological elastomer can be used to vary the stiffness in a manner such that a given mode is destiffened. The flexural stiffness functions are demonstrated using a p-version finite element rotating beam code and represent a multifunctional rotating structure concept which can be used for active stiffness control.

Keywords Finite element method · Helicopter · Modal analysis · Rotating beam · Wind turbine

1 Introduction

Rotating beams are useful mathematical models for a number of structure such as helicopter rotor blades, wind turbine blades, gas turbine blade etc [1]. In these structures, rotation of the wing is used to create a velocity field which can lead to thrust or power creation. However, rotation causes increased centrifugal effects such as stiffening of the blades, which leads to changes in the vibration characteristics. In this paper, we propose an approach to destiffen rotating blades to alleviate the structural effects of centrifugal stiffening. The advent of smart materials has made it possible to vary stiffness [2]. For example, shape memory alloys have temperature dependant stiffness [3]. Shape memory polymers [4] and magnetoactive elastomer [5, 6], can also be used for varying the stiffness.

A. Kumar and R. Ganguli (✉)
Department of Aerospace Engineering, Indian Institute of Science, Benguluru – 560012, India
e-mail: kooolant@gmail.com; ganguli@aero.iisc.ernet.in

B. Dattaguru et al. (eds.), *IUTAM Symposium on Multi-Functional Material Structures and Systems*, IUTAM Bookseries 19, DOI 10.1007/978-90-481-3771-8_26,

2 Formulation

Rotating beams such as helicopter beams can be modeled as shown in Fig. 1(a) where the vibrations perpendicular to the plane of rotation and along the axis of rotation are of primary interest. The dynamics of the out-of-the-plane vibrations of this beam is governed by

$$(EI(x)w'')'' + m(x)\,\ddot{w} - (T(x)w')' = 0 \tag{1}$$

where $T(x) = \int_x^L m(x)\,\Omega^2\,(R + x)\,dx + F$ constitutes the centrifugal stiffening term. Due to centrifugal stiffening, the natural frequencies of rotating beams increases with speed of rotation. The non-dimensional form of Eq. (1) for normal modes of vibration can be written as

$$(\Phi(\xi)\overline{w}'')'' - \frac{m_0\Omega^2L^4}{EI_0}(\Gamma(\xi)\overline{w}')' - \frac{m_0\omega^2L^4}{EI_0}\Psi(\xi)\overline{w} = 0 \tag{2}$$

where $'$ now denotes derivatives w.r.t $\xi = \frac{x}{L}$. Also, $\Phi(\xi) = \frac{EI(\xi L)}{EI_0} = \frac{EI(x)}{EI_0}$, $\Gamma(\xi) = \frac{T(x)}{m_0\Omega^2L^2}$, $\Psi(\xi) = \frac{m(x)}{m_0}$ and $\overline{w} = \frac{w}{L}$. Here Eq. (1) has been non-dimensionalised with respect to reference stiffness and mass values equal to EI_0 and m_0 respectively. Non-dimensional quantities $\lambda = \frac{m_0\Omega^2L^4}{EI_0}$ and $\mu = \frac{m_0\omega^2L^4}{EI_0}$ represent the non-dimensional rotating speed and the natural frequencies of the beam. For non-rotating beams ($\Omega = 0$), Eqs. (1) and (2) reduce to

$$(EI(x)w'')'' + m(x)\ddot{w} = 0 \tag{3}$$

$$(\Phi(\xi)\overline{w}'')'' - \mu\Psi(\xi)\overline{w} = 0 \tag{4}$$

where $\Phi(\xi)$ and $\Psi(\xi)$ now correspond to the non-rotating beam.

A rotating beam will be called equivalent, in the *jth* mode, to a corresponding non-rotating beam of same length and mass distribution if and only if the

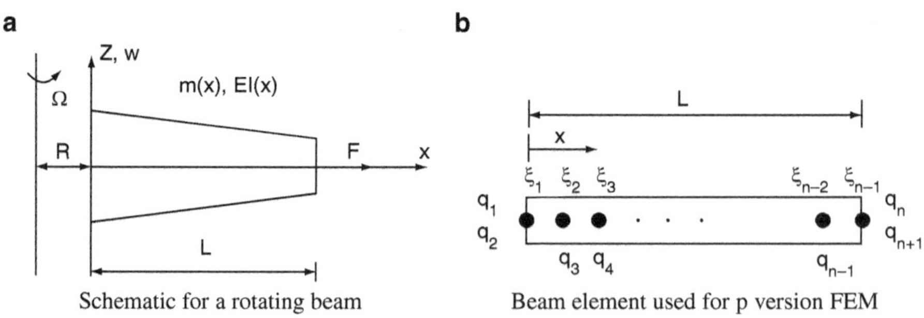

Fig. 1 Finite element modeling of rotating beam

jth mode-shapes and the *jth* frequencies of the two beams are exactly the same. In such rotating beams, the effect of rotation in increasing the frequency of the *jth* mode is countered by a characteristic variation of flexural stiffness along their length to give natural frequencies corresponding to a non-rotating beam. By this definition, note that two equivalent beams have different $\Phi(\xi)$, but same $\Psi(\xi)$. If ϕ_j and μ_j are the *jth* mode-shape and the *jth* non-dimensional frequency, respectively, of a non-rotating beam and an equivalent rotating beam in the *jth* mode, then

$$(\Phi_R(\xi)\phi_j'')'' - \lambda(\Gamma(\xi)\phi_j')' - \mu_j\Psi(\xi)\phi_j = 0 \tag{5}$$

$$(\Phi_{NR}(\xi)\phi_j'')'' - \mu_j\Psi(\xi)\phi_j = 0 \tag{6}$$

For the sake of simplicity, we shall only consider uniform non-rotating beams (uniform mass as well as flexural distribution). Let $EI_0 = EI_{NR}$ and $m_0 = m_{NR}$. Then, $\Phi_{NR} = \Psi(\xi) = 1$, $\Gamma(\xi) = 1 - \xi^2$ and we obtain

$$(\Phi_R(\xi))_j = \frac{\int\left(\int\left(\phi_j^{(4)} + \lambda((1-\xi^2)\phi_j')'\right)dx\right)dx}{\phi_j''} \tag{7}$$

$(\Phi_R(\xi))_j$ is the flexural stiffness function (FSF) of a rotating beam equivalent in the *jth* mode to its corresponding non-rotating beam. Note that since ϕ_j is different for different modes, the FSFs are different for different modes. An FSF for the *jth* mode has $j-1$ internal singularities and one tip singularity, where its value blows up to infinity. If the nodal coordinates of the *jth* mode of the beams, where $\phi_j = 0$, are ξ_1, $\xi_2, \ldots, \xi_{j-1}$, then it can be shown that $\phi_j'' = 0$ at $\xi = 1-\xi_1, 1-\xi_2, \ldots, 1-\xi_{j-1}$. Hence the FSFs must vanish at each of these points and also at $\xi = 1$, where the free end boundary condition forces ϕ_j'' to vanish. The advantage of uniform non-rotating beams is that their mode-shapes and frequencies can be obtained until any degree of accuracy. Hence rotating beams equivalent to such beams can be developed easily from Eq. (7).

The generic FSFs for uniform non-rotating beams are given below for the first four modes.

$$(\Phi_R(\xi))_j = \frac{N_i}{D_j} \tag{8}$$

where

$$\begin{aligned} N_1 = {} & (-3.5160 + 0.78441\lambda)\cos(z_1) - 3.5160\cosh(z_1) + 2.5811\sinh(z_1) \\ & + (2.5811 - 0.57583\lambda)\sin(z_1) + 0.53330\lambda\xi\sin(z_1) - 0.07094\lambda\xi e^{z_1} \\ & + 0.46240\lambda\xi e^{-z_1} + 0.06647\lambda\xi^2 e^{z_1} + 0.43352\lambda\xi^2 e^{-z_1} \\ & - 0.02866\lambda e^{z_1}\xi^2\sin(z_1) - 0.5\lambda\xi^2\cos(z_1) - 0.18692\lambda e^{-z_1} + c_1 x + c_2 \end{aligned} \tag{9}$$

$$D_1 = -3.5160\cos(z_1) - 3.5160\cosh(z_1) + 2.5811\sin(z_1) + 2.5811\sinh(z_1) \tag{10}$$

$$
\begin{aligned}
N_2 = {} & (-22.034 + 0.54538\lambda)\cos(z_2) - 22.034\cosh(z_2) + 22.441\sinh(z_2) \\
& + (22.441 - 0.55546\lambda)\sin(z_2) + 0.21697\lambda\xi\cos(z_2) + 0.00420\lambda e^{z_2} \\
& + 0.21303\lambda\xi\sin(z_2) + 0.50923\lambda\xi^2\sin(z_2) - 0.5\lambda\xi^2\cos(z_2) \\
& - 0.45881\lambda e^{-z_2} + 0.00197\lambda\xi e^{z_2} + 0.21500\lambda\xi e^{-z_2} + 0.00462\lambda\xi^2 e^{z_2} \\
& + c_1 x + c_2 + 0.50462\lambda\xi^2 e^{-z_2}
\end{aligned} \tag{11}
$$

$$
D_2 = -22.034\cos(z_2) - 22.034\cosh(z_2) + 22.441\sin(z_2) + 22.441\sinh(z_2) \tag{12}
$$

$$
\begin{aligned}
N_3 = {} & (-61.697 + 0.51621\lambda)\cos(z_3) - 61.697\cosh(z_3) + 61.649\sinh(z_3) \\
& + (61.649 - 0.51581\lambda)\sin(z_3) + 0.12731\lambda\xi\sin(z_3) - 0.48360\lambda e^{-z_3} \\
& + 0.12721\lambda\xi\cos(z_3) - 0.00018\lambda e^{z_3} + 0.49961\lambda\xi^2\sin(z_3) \\
& - 0.5\lambda\xi^2\cos(z_3) - 0.49365\times 10^{-4}\lambda\xi e^{z_3} + 0.12726\lambda\xi e^{-z_3} \\
& + 0.49981\lambda\xi^2 e^{-z_3} + 0.19388\times 10^{-3}\lambda\xi^2 e^{z_3} + c_1 x + c_2
\end{aligned} \tag{13}
$$

$$
D_3 = -61.697\cos(z_3) - 61.697\cosh(z_3) + 61.649\sin(z_3) + 61.649\sinh(z_3) \tag{14}
$$

$$
\begin{aligned}
N_4 = {} & (-120.90 + 0.50827\lambda)\cos(z_4) - 120.90\cosh(z_4) + 120.90\sinh(z_4) \\
& + (120.90 - 0.50829\lambda)\sin(z_4) + 0.09095\lambda\xi\sin(z_4) + 0.09095\lambda\xi\cos(z_4) \\
& - 0.49174\lambda e^{z_4} + 0.50001\lambda\xi^2\sin(z_4) - 0.5\lambda\xi^2\cos(z_4) - 0.49174\lambda e^{-z_4} \\
& - 0.15258\times 10^{-5}\lambda\xi e^{z_4} + 0.09095\lambda\xi e^{-z_4} + 0.50002\lambda\xi^2 e^{-z_4} \\
& + 0.83883\times 10^{-5}\lambda\xi^2 e^{z_4} + c_1 x + c_2
\end{aligned} \tag{15}
$$

$$
D_4 = -120.90\cos(z_4) - 120.90\cosh(z_4) + 120.90\sin(z_4) + 120.90\sinh(z_4) \tag{16}
$$

Here $z_1 = 1.875\xi$, $z_2 = 4.694\xi$, $z_3 = 7.855\xi$, $z_4 = 10.996\xi$ and c_1 and c_2 are arbitrary constants resulting from integration. It can be observed from these equations that as $\lambda \to 0$, $\Phi_R(\xi) \to 1$.

The FSFs for the first four modes, with $c_1 = c_2 = 0$ and $\lambda = 1$ are shown in Fig. 2.

3 Finite Element Solution

These FSFs derived above are of interest from a fundamental science perspective and also yield test functions which could be used to validate approximate methods for rotating beams. By inserting the FSFs into these codes, we must obtain frequencies equal the known values of a uniform rotating beam ($\mu_{NR} = 3.5160, 22.0345, 61.6972, 120.902\ldots$).

In this section, we validate a particular p-version finite element code for rotating beams. In this model, a single element is used for the entire beam. A set of four boundary conditions (two displacement conditions and two slope conditions) at the

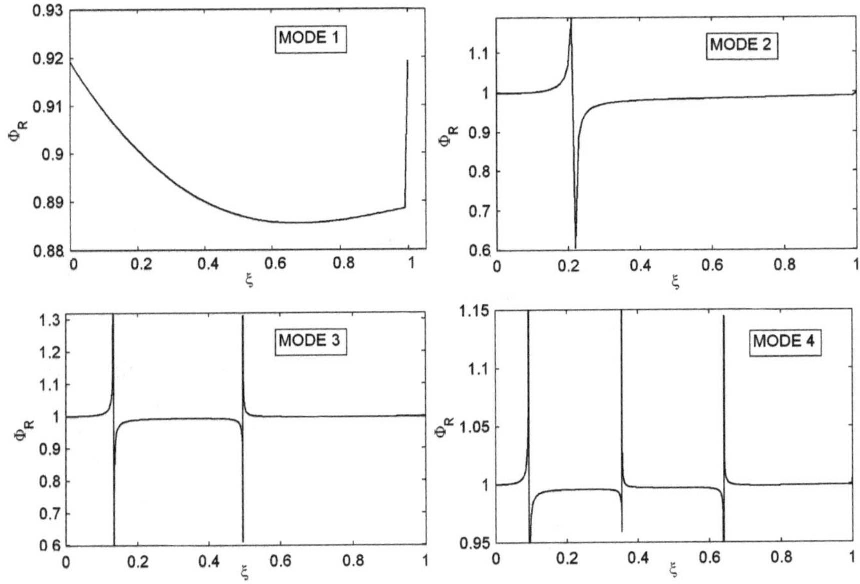

Fig. 2 FSFs, Φ_R, for first the four modes of a rotating beam with $\lambda = 1$

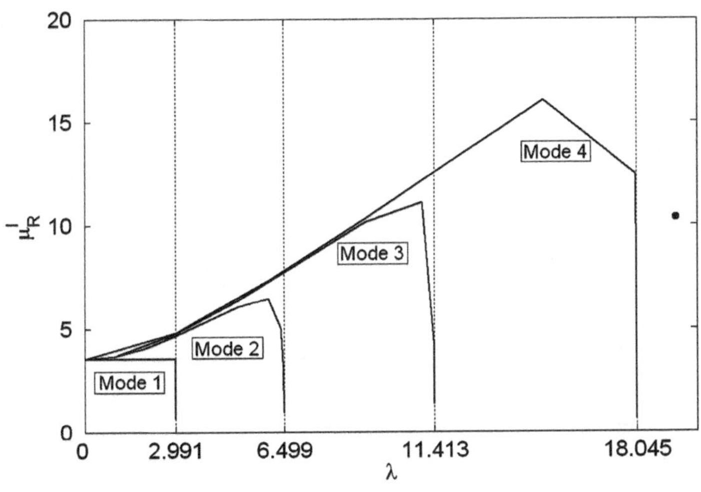

Fig. 3 Non-dimensional frequency with respect to rotation speed

extreme nodes and displacement conditions at the interior nodes are used to generate shape functions which are polynomials of degree $n \geq 3$. A general beam element is shown in Fig. 3 with equally spaced interior nodes. The interpolation polynomials are of the form $\{\psi_i(\xi) = \sum_{j=0}^{n} p_{ij}\xi^j\}_{i=1,2,\ldots,n+1}$. These polynomials satisfy the conditions $\beta_1\psi_i(\xi_j) + \beta_2\psi_i'(\xi_j) = \delta_{ij}\beta_k$ where δ is the Kronecker delta function.

$(\beta_1, \beta_2) = (1, 0)$ for $j = 1, 2, \ldots, n-1$ and for each i, two additional conditions are enforced with $(\beta_1, \beta_2) = (0, 1)$ for $j = 1$ and $n-1$. $k = 2$ for $i = 1, n-1$ and $k = 1$ otherwise.

The global mass and stiffness matrices (non-dimensional) are then given by

$$\mathbf{M}_{ij} = \int_0^1 \psi_i \psi_j \, dx \tag{17}$$

$$\mathbf{K}_{ij} = \int_0^1 \left(\Phi_R(\xi)\psi_i''\psi_j'' + \Gamma(\xi)\psi_i'\psi_j'\right) dx = \int_0^1 \left(\Phi_R(\xi)\psi_i''\psi_j'' + (1-\xi^2)\psi_i'\psi_j'\right) dx \tag{18}$$

The first term of the integral in Eq. (10) cannot be explicitly calculated, owing to the complicated nature of the function $\Phi_R(\xi)$ obtained from the FSFs. Hence we use Gaussian quadrature for numerical estimation of the integral. While doing so, we must ensure that the Guass-Legendre points are sufficiently distant from the singularity points. Such an approach is typically used to address singularities using numerical methods. With 30 points in the domain, this condition is ensured for the first four modes.

The influence of the rotational speed, λ, is not so much in altering the convergence rate as in defining the limit of λ for convergence. Convergence is uniform over the range of λ that the code converges and occurs at polynomial degree 5, 8, 10, 12 respectively for the first four modes. We observe that for each mode, there exists an upper limit on λ when the code ceases to converge to μ_{NR}. For the first four modes, these limits are found to be $\lambda_{lim} = 2.991, 6.499, 11.413$ and 18.043 respectively. At these values of λ, the fundamental frequency of the equivalent beam was found to decrease rapidly. A plot of μ_R^I (superscript I refers to the fundamental mode) versus λ is given in Fig. 4. At these limits it can be observed that the fundamental frequencies of the equivalent rotating beams approach 0. Until these given limits, the p-version FEM can be validated.

4 Application to Multifunctional Structures

The FSFs derived here can be useful in the purview of multifunctional structures which are systems intended to perform more than one function as a structure. Smart materials may be used to modify the stiffness variation of the beam to one of the FSFs, so that the beam performs the role of rotation and load support while alleviating the effects of centrifugal stiffening. Mathematically, all the centrifugal effects will then have been nullified by appropriate tailoring of the flexural stiffness variation.

As we have noted earlier, real beams with an FSF variation cannot exist due to singularities. But the FSFs give a basis for construction of beams that are nearly equivalent to non-rotating beams. For instance, the FSF for the first mode has only an end-point singularity, that does not hinder construction of such real beams. For

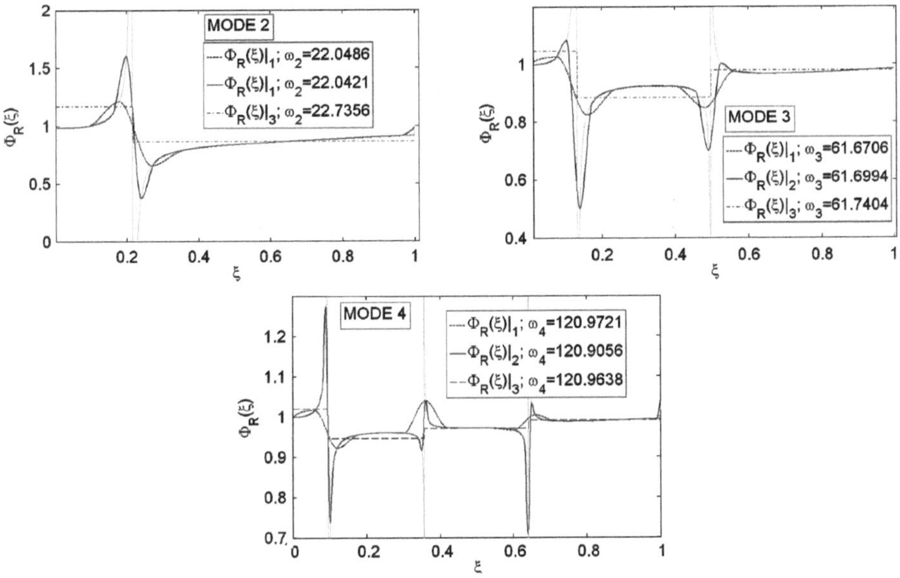

Fig. 4 FSFs, Φ_R, for first the four modes of a rotating beam with $\lambda = 1$

higher modes, approximations to FSFs can be built such that they are practically useful. We have approximated the higher mode FSFs, $\Phi_R(\xi)$ by $\widetilde{\Phi}_R(\xi)$ such that $\widetilde{\Phi}_R(\xi_j) = \Phi_R(\xi_j)$ for a discrete number of points in the domain and intermediate values are obtained by cubic Hermite interpolation. Another type of approximation was obtained by taking the average of the FSFs between nodal points.

The approximations for the second, third and the fourth modes, for a rotating speed of $\Omega = 3$ are given in Fig. 4 with their corresponding frequencies.The departure of these frequencies from that of a non-rotating beam is not large, the percentage deviations being less than 0.01% for each approximation.

It has been shown that sandwich beams with conductive skins and a soft core of magnetorheological elastomer (MRE) can be made where magnetic fields can be used to vary the field-dependant shear modulus [6]. It was found that the resonant frequencies can be changed by upto 30 percent by application of the magnetic field. Therefore, it is possible to alleviate the effects of centrifugal forces by actuating the rotating beam using the FSFs derived in this paper. Also, different modes can be targetted by varying the magnetic field. Shear modulus of MS elastomers is given by [7, 8]

$$\mathrm{G} = G_{\circ}(1 + 2.5\phi + 14.1\phi^2) + 0.601\phi\mu_{\circ}\mathrm{M}^2$$

where $G_{\circ}$ is the shear modulus of the elastomers, ϕ is the volume fraction, $\mu_{\circ}$ is the permeability in vacuum and M is the magnetization.

We should point out that the attempt to tailor any one frequency would also change the other frequencies. However, for rotating structures, the key problem is to move a critical frequency away from multiples of the rotating speed, which can be done using the proposed FSF's.

5 Conclusions

The centrifugal stiffening effect of rotating beam can be overcome by designing the beam with special flexural stiffness variations which are obtained in this paper. Smart material such as magnetosensitve elastomers allow such stiffness variation to be given in an active manner. The results in the paper show that active suppression of centrifugal stiffening effects are possible by designing rotating beams as multi-functional structures.

References

1. Gunda, J.B., and Ganguli, R., "Stiff String Basis Functions for Vibration Analysis of High Speed Rotating Beams", *ASME Journal of Applied Mechanics*, Vol. 75, No. 2, Jan 2008.
2. Krylov, V., and Sorokin, S.V., "Dynamic of elastic beams with controlled distributed stiffness", *Smart Materials and Structures*, Vol. 6, No. 5, 1997, pp. 573–582.
3. Christoph, C., Bernd, H and Masoud, M., "RC beam with variable stiffness and strength", *Construction and Bulding Materials*, Vol. 20, No. 9, 2006, pp. 824–833.
4. Farhan, G., and Sang-Guk, K., "Beams with controllable flexural stiffness", *Smart Materials and Structures*, Vol. 16, No. 4, 2007, pp. 1179–1184.
5. Farshad, M., and Andre, B., "Magnetoactive elastomer composites", *Polymer Testing*, Vol. 23, No. 3, 2004, pp. 347–353.
6. Zhou, G.Y., and Wang, Q., "Use of magnetorheological elastomer in an adpative sandwich beam with conductive skins. Part II: Dynamic properties", *Solids and Structures*, Vol. 43, No. 17, 2006, pp. 5403–5420.
7. Davis, L.C., "Model of magnetorheological elastomer", *Journal of Applied Physics*, Vol. 85, No. 6, 1999, pp. 3348–3351.
8. Patil, K.D., and Venkatraman, K., "Vibrations Characteristics of Thin-Walled Magneto-Sensitive Elastomer Beams", *INCAST* 2008.

Innovative Energy Absorbing Composite Tubes Incorporating Extension-Torsion Coupling, Stitch Ripping, and Foam Crushing

Chandrashekhar Tiwari, Edward C. Smith, and Charles E. Bakis

Abstract A lightweight energy absorbing device consisting of two extension-twist-coupled, concentric, cylindrical composite tubes with opposite angle of fibers is presented. The tubes are stitched to each other by threads and a lightweight foam filling is placed inside the inner tube. At certain values of axial force, the stitches rip and the foam crushes dissipating energy. Wide dimensionality of the design parameters ensures that the device can obtain attractive force-displacement characteristics and energy absorbing behavior.

Keywords Crashworthy · Crush tube · Energy absorption · Extension-twist · Load limiting · Stitch ripping

1 Introduction

1.1 Background and Motivation

During hard or crash landings of rotorcraft, large inertial forces are exerted by heavy payloads on the mounts connecting them to the structural frame, leading to failure of the mount and potential loss of life of the crew. A payload mount with tailored force-displacement characteristics can limit the load transferred to the supporting structure and absorb the kinetic energy of cargo during hard landings. Therefore, new innovations in energy absorbing payload mounts are needed.

C. Tiwari and E.C. Smith (✉)
Department of Aerospace Engineering, The Pennsylvania State University, University Park, PA, USA
e-mail: cut2@psu.edu; ecs5@psu.edu

C.E. Bakis
Department of Engineering Science and Mechanics, The Pennsylvania State University, University Park, PA, USA
e-mail: cbakis@psu.edu

B. Dattaguru et al. (eds.), *IUTAM Symposium on Multi-Functional Material Structures and Systems*, IUTAM Bookseries 19, DOI 10.1007/978-90-481-3771-8_27,

Over the past few decades, composites have been proven to be efficient replacements of metals in various structural applications. They have higher strength-to-weight ratio than metals and have much more design flexibility. By exploiting their anisotropic properties, elastic coupling effects can easily be incorporated in the composites. One such important effect is extension-twist coupling [1,2], which could be designed to cause the twisting of composite tubes on the application of axial force. Due to this twist, tubes can apply tangential forces on attachments holding them circumferentially, thus offering the potential for a compact energy dissipating device for axial tensile loads. The rate of twist can be variable depending upon material selection and fiber angle.

One way to apply circumferential constraints on composite tubes is by stitching them with threads. Hagon et al. [3] have shown that energy can be absorbed by the mechanism of stitch slip and rupture. The specific energy absorption (SEA) of such so-called stitch ripping devices (SRDs) can be substantially higher than that of currently used energy absorbers based on metal plasticity. Further gains could be achieved by tailoring the tubes to exhibit high Poisson's contraction to crush an inelastic core as tension is applied. These two concepts – stitch ripping and core crushing, form the basis of the innovative crashworthy payload mounting system proposed in this investigation.

1.2 Objectives

The overall objective of this research is to analytically explore the potential of a novel type of energy absorbing device – the tension-torsion stitch-ripping device with crushable core – for improving the crashworthiness of payload mounts for heavy-lift rotorcraft. The emphasis is on maximization of the SEA of the device and parametric study of the device for variable available stroke and limit loads.

2 Governing Concepts

The device comprises two concentric cylindrical extension-twist coupled composite tubes having angle of fibers opposite to each other, stitched with high specific strength thread, and filled with a lightweight core material such as foam (Figs. 1 and 2). Upon being loaded axially in tension, these tubes twist in opposite

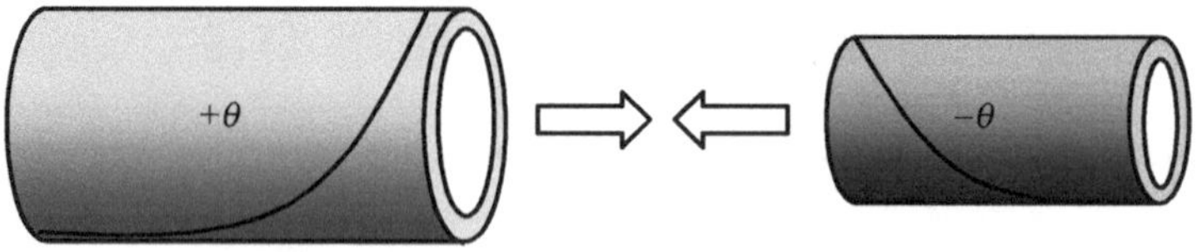

Fig. 1 Extension twist coupled concentric tubes

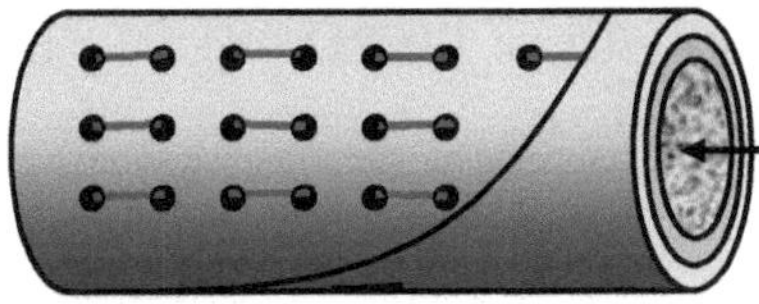

Fig. 2 Sample stitch pattern of the TTSRD-crush device

directions and load the threads in tension. At a certain peak load, stitch rupture initiates and the elastic energy stored in the threads is dissipated. In this phase of actuation, the device is denoted a tension-torsion stitch-ripping device (TTSRD). The load corresponding to stitch rupture is called the peak load for TTSRD and the end-to-end elongation of the device is called the TTSRD-stroke of the device. The TTSRD phase concludes when all the stitches have ruptured. The area under the force-displacement curve is considered to be the energy absorbed in this phase of the device. Since the tubes contract transversely due to Poisson's effect, further application of tensile force results in fiber reorientation, a change in stiffness of the tubes, and radial compression of the foam. At a certain load, the foam begins to be inelastically crushed, dissipating additional energy. This phase of actuation is denoted as the crush phase of the device and the load at which the crushing commences is denoted as the crush load. The crush phase of the device concludes when the foam is fully compacted and negligible further decrease in diameter of the tubes can be achieved with additional axial force. At the end of the crush phase, the end-to-end elongation of the device is called the over-all stroke of the device and the corresponding total area under the force-displacement curve is defined as the total energy absorbed by the device. Hence, energy absorption is achieved via dual mechanisms: the ripping of stitches and the crushing of foam.

3 Modeling and Results

The mechanical model of the device involves many material and geometric parameters, and has the ability to simulate a wide range of device designs:

Tube Materials (properties can be found in [4–6]): carbon/elastomeric matrix (flexible matrix composite, FMC), E-glass/epoxy

Material of the threads used: V-138 polyester

Tube parameters: fiber angle, θ, inner tube thickness, t, radius, R, length, L, number of stitches in a cross section, n_c, and along the length, n_l, number of threads in a stitch, n_t,

Thread parameters [3]: breaking strength, σ_{ult}, diameter of thread, d

Foam parameters (brittle polyurethane, density 48 kg/m^3): crush stress, P_c, Young's modulus, E_f, Poisson's ratio, $\nu_f \sim 0$

The device can be modeled in two stages: TTSRD, and crush tube. These stages are described in the following sections.

3.1 Tension-torsion Stitch Ripping Device (TTSRD)

As discussed in [1], since cylindrical tubes have axial symmetry, a differential element of the inner tube can be analyzed as a lamina having fibers at an angle θ to the global axial direction of the tube, as shown in Fig. 3.

In general, the plane stress constitutive equations for an off-axis lamina are

$$\begin{Bmatrix} N_{xx} \\ N_{ss} \\ N_{xs} \end{Bmatrix} = \begin{bmatrix} A_{11} & A_{12} & A_{16} \\ A_{12} & A_{22} & A_{26} \\ A_{16} & A_{26} & A_{66} \end{bmatrix} \begin{Bmatrix} \varepsilon_{xx} \\ \varepsilon_{ss} \\ \gamma_{xs} \end{Bmatrix} \tag{1}$$

where N are plate force resultants, A is the in-plane stiffness matrix of the lamina, and the ε and γ are normal and shear strains, respectively. The loadings on a differential element of the stitched region of the inner tube are shown in Fig. 4.

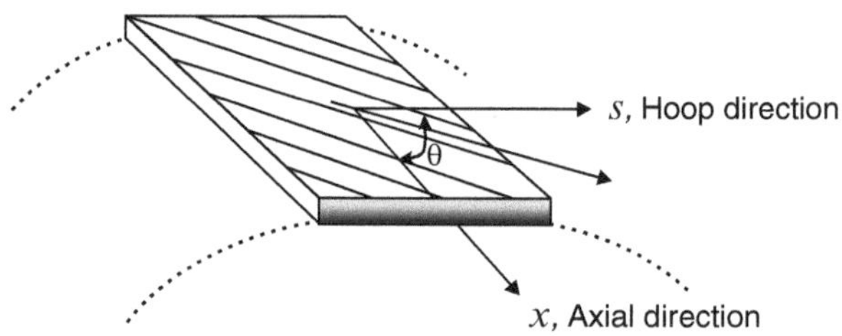

Fig. 3 Differential element of the composite tube, laminate coordinates

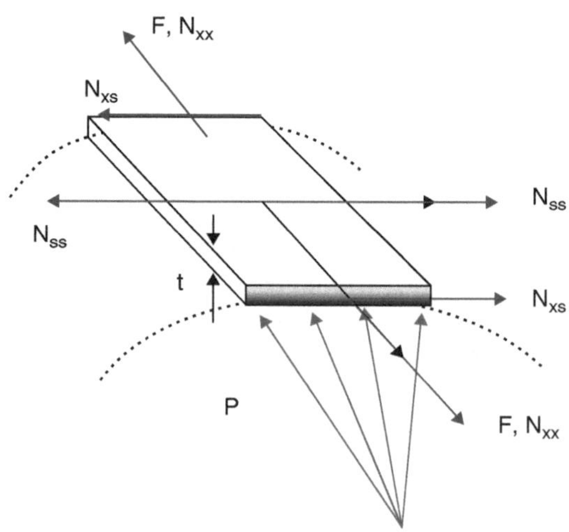

Fig. 4 Loadings on the stitched plate element of the inner tube

For the region of the tubes stitched together,

$$
\begin{Bmatrix} N_{xx} \\ N_{ss} \\ N_{xs} \end{Bmatrix} = \begin{bmatrix} A_{11} & A_{12} & A_{16} \\ A_{12} & A_{22} & A_{26} \\ A_{16} & A_{26} & A_{66} \end{bmatrix} \begin{Bmatrix} \varepsilon_{xx} \\ \varepsilon_{ss} \\ 0 \end{Bmatrix}
$$

$$
\begin{aligned}
N_{xx} &= F/(2\pi R) \\
N_{ss} = PR &= \frac{E_f(\varepsilon_{ss} + \nu_f \varepsilon_{xx})}{2\nu_f^2 + \nu_f - 1} \\
N_{xxb} &= 2n_t n_c \sigma_{ult}(\pi d^2/4)/(2\pi R) \\
\gamma_{xs} &= 0
\end{aligned} \tag{2}
$$

where,

σ_{ult} = Breaking strength of the thread.
N_{xsb} = Shear force resultant when a stitch breaks.
F = Axial force on the inner tube.
P = Internal pressure on the inner tube by core foam.

Equation (2) has been derived assuming equal uniform axial strain in the tube and the foam core and radial displacement continuity between the inner tube and foam core. For the unstitched part of the tube, the constitutive equations are:

$$
\begin{Bmatrix} N_{xx} \\ N_{ss} \\ 0 \end{Bmatrix} = \begin{bmatrix} A_{11} & A_{12} & A_{16} \\ A_{12} & A_{22} & A_{26} \\ A_{16} & A_{26} & A_{66} \end{bmatrix} \begin{Bmatrix} \varepsilon'_{xx} \\ \varepsilon'_{ss} \\ \gamma_{xs} \end{Bmatrix} \tag{3}
$$

Axial strains for the stitched and unstitched parts of the inner tube can be determined using Eqs. (2) and (3), respectively. The stroke and force of the TTSRD phase are then calculated as:

$$
\begin{aligned}
D &= \varepsilon_{xx}(n_l - n)(L/n_l) + \varepsilon'_{xx}(n)(L/n_l) \\
F_x &= 2F
\end{aligned} \tag{4}
$$

where, F_x is the total axial force acting on the device. The assumptions taken for the TTSRD phase include:

1. Shear strain $\gamma_{xs} = 0$ in the stitched part of the tube.
2. Shear force resultant $N_{xs} = 0$ in the un-stitched part of the tube.
3. The foam core deforms elastically until the TTSRD phase is concluded.

Variation of the parameters mentioned at the beginning of section 3 results in a wide range of calculated SEAs for the device along with a wide range of strokes. Figure 5 shows three examples. Every peak on the force-displacement curve corresponds to a stitch rupture. In the stitch ripping part of the analysis, fiber re-orientation is not taken into account. The decreasing stiffness behavior with increasing axial load is caused by the reduction in cross-sectional area of the tube with applied tension, which is accounted for in the model. A more detailed analysis of the TTSRD phase of the device can be found in [7].

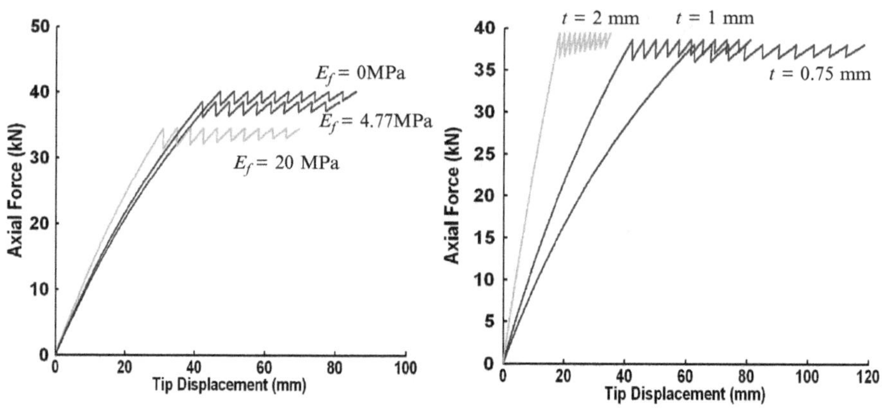

Fig. 5 Predicted variation of force versus displacement curve for stitch ripping part of device with crush foam modulus, E_f and inner tube thickness, t (Initial radius = 1 cm, fiber angle = 45 degrees, number of stitches across cross section = 50, length = 10 cm, thread strength = 985 MPa, thread diameter = 0.36 mm)

3.1.1 Optimization of TTSRD Phase

Due to the presence of multi-dimensional parametric space for these devices, it becomes useful to optimize the devices for maximum SEA. A simulated annealing algorithm [8] was used to find the optimal combination of previously mentioned parameters for TTSRD phase, under the constraints of ply failure stresses. The problem is stated as:

Objective Function: maximization of specific energy absorption (SEA)
Constraints: ply failure stresses, maximum stress ply failure criterion
Design parameters varied: t: (0.5–4 mm), R: (1– 4 cm), θ: (5–60 deg), n_c: (5–200), n_l: (5–80)
Design parameters held constant: L: 10 cm, Type of thread: V-138 Polyester ($d = 0.36$ mm, $\sigma_{ult} = 985$ MPa, $E_f = 4.77$ MPa, $P_c = 0.294$ MPa, $\nu_f = 0.076$)

The algorithm was executed for different material systems and results corresponding to two of them are presented in Fig. 6. It is seen that different force-displacement responses can be obtained as functions of the material system chosen in accordance with the design requirements of available stroke, peak load, etc.

3.2 *Crush Tube Part of Device*

The tube parameters are chosen in such a way that the Poisson's ratio of the tube is very high in order to obtain a large decrease in diameter. The crush tube concept is illustrated in Fig. 7.

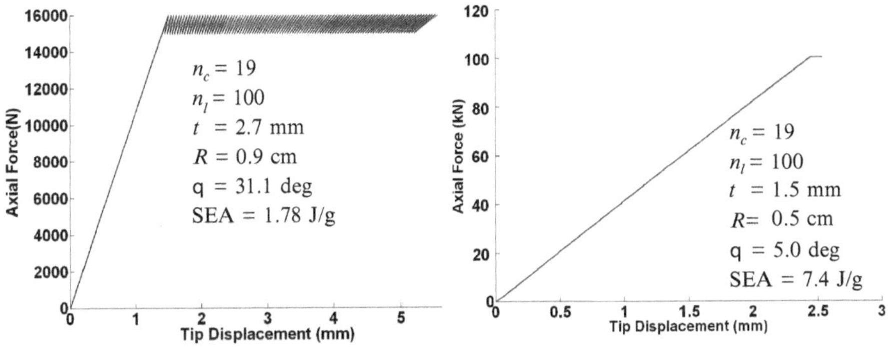

Fig. 6 Optimized force-displacement curves for FMC (*left*) and E-glass/epoxy (*right*) TTSRD

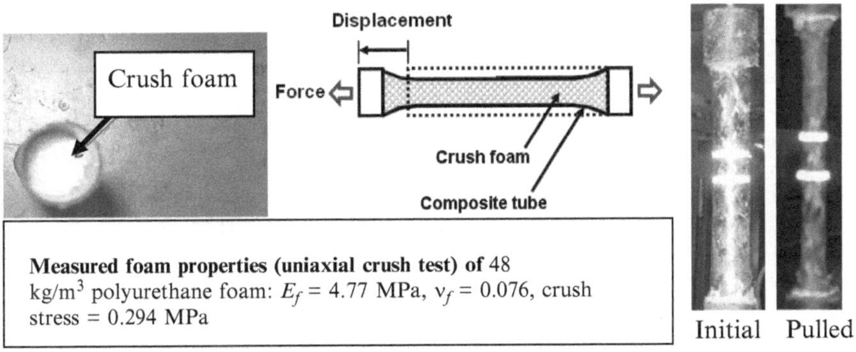

Fig. 7 Composite crush tube with foam filling (*left*), schematic shape of crushed tube (*center*) and photographs of a ±30 FMC tube curing a tensile test (*right*)

In the analysis it is assumed that the tube experiences a re-orientation of fibers and change in stiffness after the matrix fails subcritically. This causes additional rapid decreases in diameter of the tube and adds to the crushing of the core material. An analytical model was developed to study the behavior of such devices. The crush tube model features various design parameters which have been discussed before. In the paper by Kydoniefs [9], change in the angle of the fibers is formulated in terms of the stretch ratios along the longitudinal and transverse directions of the tube.

Experimental results for a balanced ±30 deg FMC crush tube are shown in Fig. 8. It can be seen that this crush tube device undergoes multiple phases during the course of loading, beginning with an elastic regime and followed by matrix cracking based upon the maximum stress failure criterion. After matrix cracking occurs, fiber reorientation and diameter reduction proceed rapidly. At roughly 4 kN, foam crushing commences and continues until the foam is fully compacted inside the tube or there is fiber breakage in the tube wall.

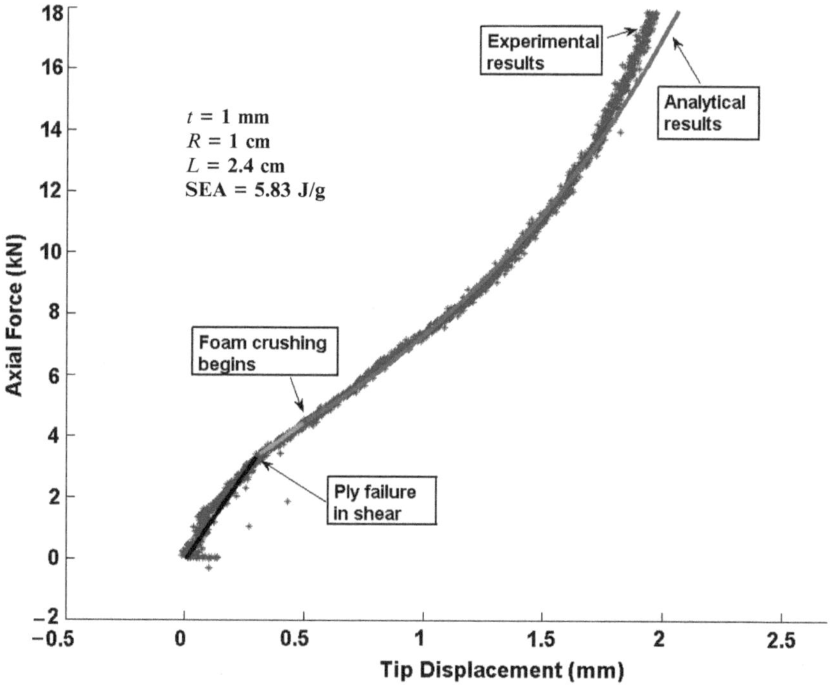

Fig. 8 Force-displacement curves for tensile-loaded ±30 deg FMC crush tube

3.3 TTSRD with Crush Foam Filling

The devices discussed above are now combined together as shown in Fig. 2, to obtain higher volumetric energy absorption and superior SEA. The modeling of this combined TTSRD-crush device is done in such a way that the ply failure does not occur before the TTSRD phase has concluded. This helps in maximizing the energy absorption in the TTSRD phase. Figure 9 shows behavior of such a device with FMC, exhibiting TTSRD and crush phases.

3.3.1 Optimized Results for TTSRD with Crush Foam

The optimization of TTSRD-crush device was carried out a using similar simulated annealing algorithm, with constraints of ply failure for TTSRD phase of the device. The parameters obtained from TTSRD optimization were used to develop the force-displacement characteristics of the crush phase.

As seen from Fig. 9, the TTSRD-crush device serves the purpose of load limiter over a certain range of displacement as well as large energy absorber. The SEA of this device is far better (four to five times) than a typical energy absorber based on metal plasticity, such as a wire bending device.

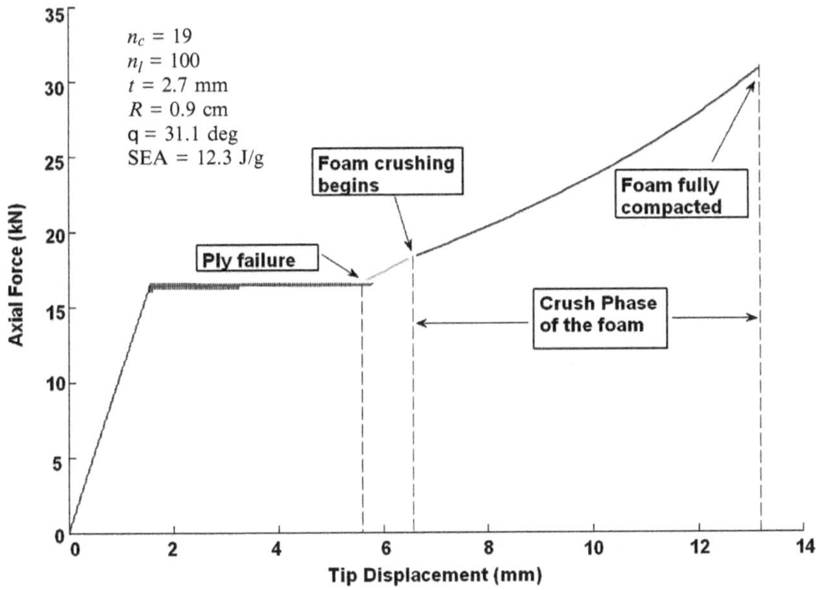

Fig. 9 Optimized force-displacement curve for FMC TTSRD-crush tube

4 Conclusions and Future Work

A study of novel load limiter devices was conducted. It was found that TTSRD-crush devices can be useful in many applications because of their high adaptability to meet different design requirements. They provide an efficient means of energy absorption and are effective load limiting devices. For the specified material systems and constraints considered in the present study, mounting devices exhibited optimized SEAs ranging from 1.8 to 12.3 J/g. Certain parameters such as diameter of the thread were held constant but can also be varied according to the design constraints and requirements. Ongoing work involves improved optimization routines, fabrication and testing of the TTSRD part of the device, and comparing experimental results of TTSRDs with the analysis.

Acknowledgements This research was partially funded by the Office of Naval Research under the Award No. N00014-08-C-0420. The U.S. Government is authorized to reproduce and distribute reprints not withstanding any copyright notation thereon. Any opinions, findings and conclusions or recommendations expressed in this publication are those of the authors and do not necessarily reflect the views of the Office of Naval Research. The authors thank Prof. Ashok Belegundu of Penn State for his help in parametric optimization routines.

References

1. Rehfield LW, Atilgan AR, Hodges DH (1990) Nonclassical behavior of thin walled composite beams with closed cross sections, J Am Helicopter Soc 35:42–50
2. Nampy SNN (2005) Structural behavior and design of flexible matrix composite box beams with extension twist coupling. MS thesis. Penn State University, University Park, PA
3. Hagon MJ, Kong W, Bakis CE, Yukish MA, Smith EC (2008) Energy-absorbing textile devices for heavy cargo restraints. In: Proc 64th Annu Forum. Am Helicopter Soc, Alexandria, VA
4. Herakovich CT (1998) Mechanics of fibrous composites. John Wiley & Sons, New York
5. Daniel IM, Ishai O (2006) Engineering mechanics of composite materials, 2nd edn. Oxford University Press
6. Shan Y (2006) Flexible matrix composites: dynamic characterization, modeling and potential for driveshaft application. PhD thesis. Penn State University, University Park, PA
7. Tiwari C, Smith EC, Bakis CE, Yukish MA (2009) Energy absorption and load limiting via extension-torsion coupled stitch ripping composite tubes. In: Proc 50th AIAA/ASME/ASCE/AHS/ASC Struct, Struct Dynamics, and Mater Conf. Am Inst Aeronautics Astronautics, Reston, VA
8. Belegundu AD, Chandrupatla TR (1999) Optimization concepts and applications in engineering. Prentice Hall, Upper Saddle River, NJ
9. Kydoniefs AD (1970) Finite axisymmetric deformations of an initially cylindrical membrane reinforced with inextensible cords. Q J Mechanics Appl Math 23:481–488

A Pseudo-dynamical Systems Approach to Inverse Problems

B. Banerjee and D. Roy

Abstract A pseudo-dynamical approach for inverse problems involving static model equations and statically acquired measurements is proposed. The normal equations are converted to a system of linearized ordinary differential equations (ODE-s) whose steady-state solutions yield the desired reconstruction. Thus we establish a deterministic reconstruction strategy without an explicit use of regularization. We follow this up with a stochastic strategy that exploits the forward model of the inverse problem within the measurement equations and uses an ensemble Kalman filter to obtain the desired reconstruction.

Keywords Ensemble Kalman filters · Inverse problems · Linearization · Pseudo-dynamical systems · Quasi-Newton methods

1 Introduction

We consider a class of inverse problems that involve determining unknown coefficient (parameter) functions of imperfectly known mathematical models, given partial and statically obtained noisy measurements of the system response. Available numerical algorithms for such problems are often challenged by the inherent non-uniqueness and ill-posedness. For nonlinear inverse problems, the deterministic inversion algorithm is usually evolved using a (variant of the) Gauss-Newton iterative approach. The ill-posed nature of the operator is usually handled through regularization, e.g. Tikhonov regularization [4]. Unlike a deterministic approach, a stochastic filtering scheme, traditionally applicable within a dynamical systems setting, provides a more rigorous probabilistic framework for obtaining estimates as well as error characteristics. While the Kalman filter (KF) provides the optimal

B. Banerjee and D. Roy (✉)
Structures Lab, Department of Civil Engineering, Indian Institute of Science, Benguluru – 560012, India
e-mail: bishu@civil.iisc.ernet.in; royd@civil.iisc.ernet.in

B. Dattaguru et al. (eds.), *IUTAM Symposium on Multi-Functional Material Structures and Systems*, IUTAM Bookseries 19, DOI 10.1007/978-90-481-3771-8_28,

solution to a linear inverse problem, nonlinear problems may be tackled either via variants of the optimal particle filter [3] or via several sub-optimal strategies including extended or ensemble Kalman filters [5, 6]. An advantage of such filters, especially the ones using Monte Carlo simulations through an ensemble of trajectories, is that they do not require explicit regularization.

We presently explore a novel family of pseudo-dynamical systems whose numerical integration in time provides an asymptotic solution to the inverse problem at hand. In particular, given the Gauss-Newton equation for updates, we convert it into a pseudo-dynamical form, propose appropriate linearizations for the resulting dynamical system (for nonlinear inverse problems) and arrive at some strategies for integration. We evaluate the performance of this approach (vis-à-vis a few existing approaches) in the context of an exponentially ill-posed linear inverse problem. Following this, we exploit another form of a pseudo-dynamical strategy within a sub-optimal ensemble Kalman filter to solve the inverse elasticity problem. A common feature in these methods is the avoidance of an explicit regularization while integrating the dynamical system.

2 A Deterministic Pseudo-dynamical Approach

Consider the problem of reconstructing the parameter vector $\boldsymbol{\mu} \in V_\mu \subset \hbar^p$, which denotes the finite-dimensional discretization of the associated function $\mu(\boldsymbol{x})$ via finite element (FE) shape functions with $\boldsymbol{x} \in \hbar^q$ denoting the spatial variable. A similar discretization of the response function $u(\mu(\boldsymbol{x}))$ (typically governed by a partial differential equation (PDE)) yields the nodal vector $\mathbf{U} \in V_U \subset \hbar^d$. Assuming that the elements of the (noisy) measurement vector $\boldsymbol{u}^m \in \hbar^m$ are at nodal locations, we define $\mathbf{u} \in V_u \subset \hbar^m$ to be the subset of $\mathbf{U}$ corresponding to $\boldsymbol{u}^m$; $m \leq d$. The usual route to reconstruct $\boldsymbol{\mu}$ is by minimizing the squared error norm between the measured data and response computed through a mathematical model of the system. The resulting operator cannot be inverted without regularization. A common (Tikhonov) regularization technique is to penalize the functional with some extra terms leading to the following optimization problem

$$\min_{\boldsymbol{\mu} \in \hbar^p} \frac{1}{2} \|\mathbf{u}(\boldsymbol{\mu}) - \mathbf{u}^m\|^2 + \frac{\lambda}{2} \|\mathbf{L}(\boldsymbol{\mu})\|^2 \tag{1}$$

where $\mathbf{u}(\boldsymbol{\mu})$ is computed using $\mathbf{K}(\boldsymbol{\mu})\mathbf{U} = \mathbf{f}$ (1a,b).

$\boldsymbol{L}$ denotes the penalization operator on $\boldsymbol{\mu}$ ($\| \; \|$ is the L_2 norm) and λ the regularization parameter, $\boldsymbol{K}$ the stiffness matrix and $\boldsymbol{f}$ the discretized forcing vector. The normal equation is:

$$\boldsymbol{g}(\boldsymbol{\mu}) = \boldsymbol{J}(\boldsymbol{\mu})^T \left(\mathbf{u}(\boldsymbol{\mu}) - \boldsymbol{u}^m\right) + \lambda \boldsymbol{L}'(\boldsymbol{\mu}) = 0 \tag{2}$$

$\boldsymbol{J}(\boldsymbol{\mu}) = \partial \mathbf{u}(\boldsymbol{\mu})/\partial \boldsymbol{\mu}$ is the Jacobian matrix and $\boldsymbol{L}'(\boldsymbol{\mu}) = \partial \boldsymbol{L}(\boldsymbol{\mu})/\partial \boldsymbol{\mu}$. The above nonlinear equation is generally solved using successivelinearization with the

Hessian approximated as $\boldsymbol{J}^T\boldsymbol{J}$ (Gauss-Newton approximation). This yields the following linear equation in $\Delta\boldsymbol{\mu}$:

$$\left(\boldsymbol{J}^T(\boldsymbol{\mu})\boldsymbol{J}(\boldsymbol{\mu}) + \lambda\boldsymbol{L}''(\boldsymbol{\mu})\right)\Delta\boldsymbol{\mu} + \boldsymbol{J}^T(\boldsymbol{\mu})\left(\mathbf{u}(\boldsymbol{\mu}) - \boldsymbol{u}^m\right) + \lambda\boldsymbol{L}'(\boldsymbol{\mu}) = 0 \quad (3)$$

with $\boldsymbol{\mu}$ being the linearization point. Instead of explicitly computing $\boldsymbol{J}^T(\boldsymbol{\mu})\boldsymbol{J}(\boldsymbol{\mu})$, the inverse of the Hessian may be directly approximated only through information on the gradient $\boldsymbol{g}(\boldsymbol{\mu})$ over successive iterations, leading to the so-called quasi-Newton method (QNM). The acceptability of a regularized solution is greatly affected by the parameter λ.

We obtain a recursive solution to the inverse problem by introducing an artificial time derivative term and solving the following ODE in $(t, t + \Delta t]$:

$$\dot{\boldsymbol{\mu}}(t) + \left[\boldsymbol{J}^T(\boldsymbol{\mu}^*)\boldsymbol{J}(\boldsymbol{\mu}^*)\right]\left(\boldsymbol{\mu}(t) - \boldsymbol{\mu}^*\right) + \left[\boldsymbol{J}^T(\boldsymbol{\mu}^*)\right]\left(\mathbf{u}\left(\boldsymbol{\mu}^*\right) - \boldsymbol{u}^m\right) = 0 \quad (4)$$

where $\boldsymbol{J} \in \hbar^{m\times p}$. $\boldsymbol{\mu}^*$ the linearization point in the p-dimensional vector space V_μ. The ODE contains $\boldsymbol{u}^m$, which is noisy. The solution of the linearized ODE (4) may be written as:

$$\boldsymbol{\mu}(t + \Delta t) = \exp\left(\left[-\boldsymbol{J}^T(\boldsymbol{\mu}^*)\boldsymbol{J}(\boldsymbol{\mu}^*)\right](\Delta t)\right)\boldsymbol{\mu}(t) + \int_t^{t+\Delta t} \exp\left(\left[-\boldsymbol{J}^T(\boldsymbol{\mu}^*)\boldsymbol{J}(\boldsymbol{\mu}^*)\right] \times (t + \Delta t - s)\right) f ds \quad (5)$$

where $\boldsymbol{f} = [\boldsymbol{J}^T(\boldsymbol{\mu}^*)\boldsymbol{J}(\boldsymbol{\mu}^*)]\boldsymbol{\mu}^* - [\boldsymbol{J}^T(\boldsymbol{\mu}^*)](\boldsymbol{u}(\boldsymbol{\mu}^*) - \boldsymbol{u}^m)$. Following [11], the linearization point $\boldsymbol{\mu}^*$ could be chosen anywhere in $[t, t + \Delta t]$ without affecting the formal error order. While choosing $\boldsymbol{\mu}^* = \boldsymbol{\mu}(t)$ yields the explicit phase space linearization (PSL), $\boldsymbol{\mu}^* = \boldsymbol{\mu}(t + \Delta t)$ results in the implicit locally transversal linearization (LTL) [12]. Denoting $h = t_{k+1} - t_k$ to be the time step and $\boldsymbol{\mu}_k := \boldsymbol{\mu}(t_k)$, the explicit PSL map is:

$$\boldsymbol{\mu}_{k+1} = \exp\left(\left[-\boldsymbol{J}^T(\boldsymbol{\mu}_k)\boldsymbol{J}(\boldsymbol{\mu}_k)\right](t_{k+1} - t_k)\right)\boldsymbol{\mu}_k + \int_{t_k}^{t_{k+1}} \exp\left(\left[-\boldsymbol{J}^T(\boldsymbol{\mu}_k)\boldsymbol{J}(\boldsymbol{\mu}_k)\right] \times (t_{k+1} - s)\right) f_k ds \quad \text{(6a)}$$

with

$$\boldsymbol{f}_k = \left[\boldsymbol{J}^T(\boldsymbol{\mu}_k)\boldsymbol{J}(\boldsymbol{\mu}_k)\right]\boldsymbol{\mu}_k - \left[\boldsymbol{J}^T(\boldsymbol{\mu}_k)\right]\left(\mathbf{u}(\boldsymbol{\mu}_k) - \boldsymbol{u}^m\right) \quad \text{(6b)}$$

We now consider some numerical schemes to integrate Eq. (4), which is rewritten as:

$$\dot{\boldsymbol{\mu}} + \boldsymbol{M}(\boldsymbol{\mu}^*)(\boldsymbol{\mu} - \boldsymbol{\mu}^*) + \boldsymbol{V}(\mu^*) = 0 \quad (7)$$

where $\boldsymbol{M} = [\boldsymbol{J}^T(\mu^*)\boldsymbol{J}(\mu^*)]$ **and** $\boldsymbol{V} = [\boldsymbol{J}^T(\mu^*)](\mathbf{u}(\mu^*) - \boldsymbol{u}^m)$. The forward (explicit) Euler discretization of Eq. (7) leads to the well-known damped Landweber iteration:

$$\mu_{k+1} = \mu_k - h\boldsymbol{V}(\mu_k) \tag{8}$$

A difficulty with an explicit method is that the choice of the time step is restricted by the spectral radius of the locally linearized operator. Thus we also explore a backward Euler scheme given by the following equation over $(t_k, t_{k+1}]$:

$$\mu_{k+1} - h\boldsymbol{V}(\mu_{k+1}) = \mu_k \tag{9}$$

Of current interest is the implicit Euler scheme derivable through a backward Taylor expansion of the linearized vector field of ODE (7) over $(t_k, t_{k+1}]$ with $\mu^* = \mu_k$. This leads to the implicit map:

$$\begin{aligned} \mu_k &= \mu_{k+1} - h\left(\boldsymbol{M}(\mu_k)(\mu_{k+1} - \mu_k) + \boldsymbol{V}(\mu_k)\right) \\ \Rightarrow \mu_{k+1} &= \mu_k + h\left[\boldsymbol{I} - h\boldsymbol{M}(\mu_k)\right]^{-1}\boldsymbol{V}(\mu_k) \end{aligned} \tag{10, 11}$$

Curiously, this expression may be shown to be similar to the Gauss-Newton update Eq. (3) if we choose $\lambda = (1/h)$ and $\boldsymbol{L} = \boldsymbol{I}$. We may thus look upon h as a regularization parameter. A somewhat analogous form of artificial dynamics is also proposed in [1]. For a linear inverse problem, we have the pseudo-dynamical system:

$$\dot{\boldsymbol{\mu}} + \boldsymbol{M}_L\boldsymbol{\mu} + \boldsymbol{V}_L = \mathbf{0} \tag{12}$$

The subscript 'L' represents $\boldsymbol{\mu}$-independence. The implicit Euler scheme for this case is given by:

$$\boldsymbol{\mu}_{k+1} = [\boldsymbol{I} + h\boldsymbol{M}_L]^{-1}(\boldsymbol{\mu}_k + \boldsymbol{V}_L) \tag{13}$$

2.1 Numerical Experiment on a Linear Inverse Problem

We consider an exponentially ill-posed, linear inverse problem with the forward integral equation:

$$g(x) = \int_0^1 k\left(x - x'\right)f(x')dx' \quad \textit{with}\ 0 \le x \le 1 \tag{14}$$

$k(x) = C\exp\left(-x^2/2\gamma^2\right)$ is the kernel function (point spread function) with C, γ being positive parameters (see [14]), f the light source intensity and g the image. A discrete form of the above equation is $\boldsymbol{Kf} = \boldsymbol{d}$, where $K_{ij} = hC\exp\left(-((i-j)h)^2/2\gamma^2\right); 1 \le i, j \le n$. Given the noisy measurement, the inverse

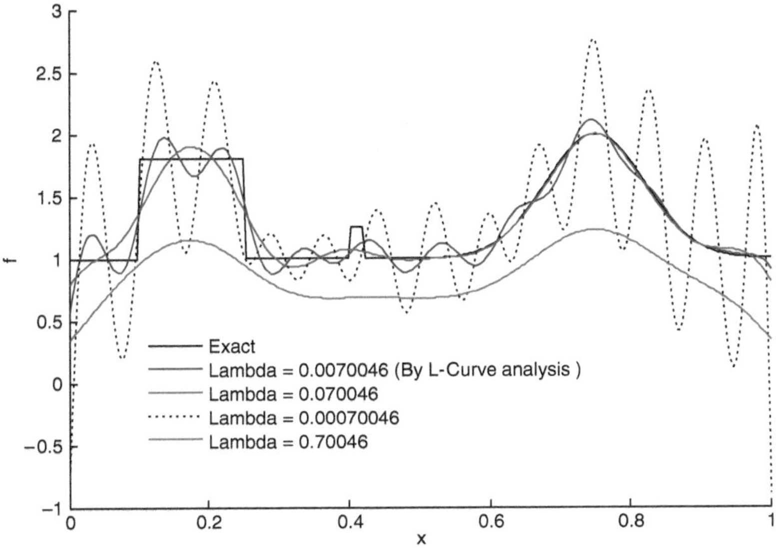

Fig. 1 Sensitivity of reconstructions to Tikhonov regularization parameter

problem is to find out the source profile. For numerical work, we take $\gamma = 0.05$, $C = 1/(\gamma\sqrt{2\pi})$ and $n = 257$. The QNM-based reconstructed profiles of $\boldsymbol{f}$, using different Tikhonov parameters, are shown in Fig. 1 and high sensitivity of solutions to variations in this parameter is self evident. Similarly, we consider the effect time step-size on pseudo-dynamic iterations. Figure 2 reveals that solutions are not significantly influenced by the time step. The pseudo-dynamical approach is thus seen to possess better stability vis-à-vis the regularization-based method, which exhibits more oscillations.

3 A Pseudo-dynamical Ensemble Kalman Filter (EnKF) for Elastography

'Elastography' [10], a non-invasive imaging modality, aims at mapping localized changes in the stiffness of a tissue structure to spatial variations of elasticity parameters, e.g. the shear modulus. It is a potentially useful tool for early detection of breast cancer. In the absence of body forces, the governing (Navier's) differential equations for the static forward elasticity problem are given by:

$$\nabla\hbar\left(\mu\nu/\left(1-2\nu\right)tr\left(\left(\nabla\mathbf{u}+\nabla\mathbf{u}^T\right)\right)\mathbf{I}+\mu\left(\nabla\mathbf{u}+\nabla\mathbf{u}^T\right)\right)=0 \text{ in } \Omega$$
$$\left(\mu\nu/\left(1-2\nu\right)tr\left(\left(\nabla\mathbf{u}+\nabla\mathbf{u}^T\right)\right)\mathbf{I}+\mu\left(\nabla\mathbf{u}+\nabla\mathbf{u}^T\right)\right)\mathbf{n}=\mathbf{t} \text{ in } \Gamma_{\mathrm{t}}$$
$$\text{and} \quad \mathbf{u}=\mathbf{r} \text{ in } \Gamma_{\mathrm{u}} \qquad (15\text{a, b, c})$$

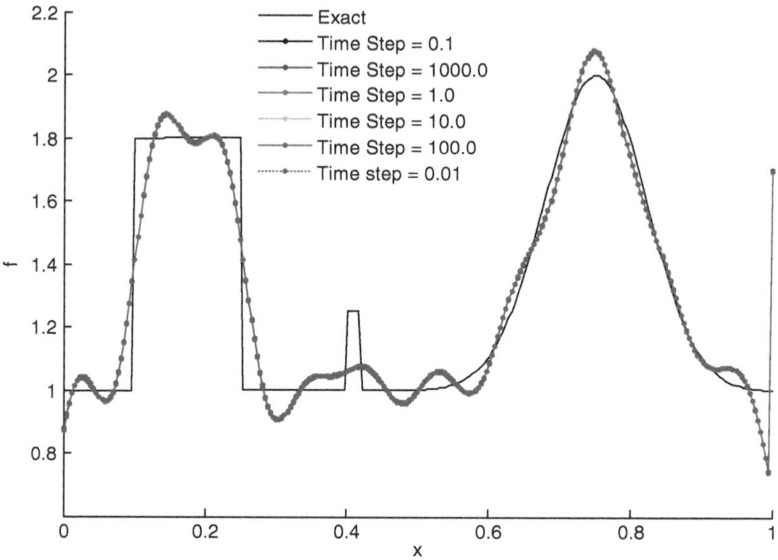

Fig. 2 Reconstructions via the exponential time stepping with different time steps

u is the displacement vector field over Ω, μ the shear modulus, ν the Poisson's ratio, **t** the applied boundary traction, **r** the boundary displacement field and **n** the outward normal. Γ_t and Γ_u respectively denote the traction (natural) and displacement (essential) parts of the boundary $\Gamma := \partial\Omega = \Gamma_t \cup \Gamma_u$. We assume the material to be isotropic, linear elastic solid undergoing small isochoric deformation. In order to counter numerical difficulties with nearly-incompressible tissue material (as $\nu \to 0.5$), we use linearized Cosserat point (four-noded) elements to discretize Eq. (28.15). Cosserat point formulation bypasses the conventional weak form and the associated numerical integration (whilst remaining within the usual FE discretization) and is known to arrest volumetric locking/hour-glassing [8]. For a detailed exposition of the four-noded Cosserat point element, adapted for the small-deformation case, we refer to [2]. This finally yields the following discretized system of algebraic equations:

$$\mathbf{K}(\boldsymbol{\mu})\boldsymbol{\delta} = \mathbf{f} \tag{16}$$

$\boldsymbol{\delta} \in \hbar^n$ represents the nodal displacement vector, $\mathbf{K} \in \hbar^{n\times n}$ the global stiffness matrix, $\boldsymbol{\mu} \in \hbar^p$ the discretized (nodal) parameter vector of the shear modulus field μ and $\mathbf{f} \in \hbar^n$ the discretized force vector.

The aim of a stochastic filter is to obtain the probability distribution $\Pi_{k|1:k} := \Pi(\boldsymbol{\mu}_k|\mathbf{z}_{1:k} := \{\mathbf{z}_1^{\mathrm{T}}, \ldots, \mathbf{z}_k^{\mathrm{T}}\}^T)$ (or the associated density $p_{k|1:k}$) of the states (in this case, $\boldsymbol{\mu}(t)$) conditioned on the measurements $\mathbf{z}_{1:k}$ available till the current time instant t_k. The estimate of $\boldsymbol{\mu}$ at t_k may then be given by $\hat{\boldsymbol{\mu}}_k = E_{\Pi_{k|1:k}}(\boldsymbol{\mu}_k)$, where E denotes the expectation operator. The main components of a filter are the process SDE-s (that govern the evolution of the states) and measurement equations that relate the observed variables to a subset of the states (modulo noise). For non-trivial updates of the parameters to be reconstructed, the σ-algebra (filtration) generated by the measurement processes must increase with time. This is however not consistent with static elastography, wherein one typically has only one set of (statically recorded) measurements of (a subset of) the displacement vector. We address this difficulty by adding a fictitious, low-intensity Wiener noise to the measurement equation. $\mu(t)$ is declared as states that fictitiously evolve as Wiener martingales leading to a filtering problem wherein the process SDE and measurement equation are given by:

$$d\mu(t) = d\xi(t);\; \mathbf{z}(t) = \mathbf{D}\delta(\mu(t)) + \Delta\eta(t);\; \Delta\eta(t) := \eta(t) - \eta_{k-1} \tag{17a, b}$$

$\mathbf{z} \in \hbar^p$ is the measured displacement components (only axial components of $\boldsymbol{\delta}$) and $\mathbf{D} \in \hbar^{p\times n}$ is a coefficient matrix (with entries 0 or 1). $\boldsymbol{\xi}(t)$ and $\eta(t)$ are vectors of independently evolving Wiener processes. Equation (17b) is nonlinear, as $\boldsymbol{\delta}(\mu)$ is nonlinear in μ. The aim is to obtain the estimate $\hat{\mu}_t := E(\mu_t|\mathbf{z}(s); 0 < s \le t)$. Measurements are typically available in discrete time and it is thus convenient to write the discretized forms of process and measurement equations as:

$$\mu(t_k) = \mu_k = \mu_{k-1} + \boldsymbol{\xi}_k \tag{18}$$

$$\mathbf{z}(t_k) = \mathbf{z}_k = \mathbf{D}\delta(\mu_k) + \Delta\eta_k \tag{19a}$$

$t_k = kh$, h is the (uniform) time step and $\Delta\eta_k := \eta_k - \eta_{k-1}$. $\{\boldsymbol{\xi}_k\}$, $\{\eta_k\}$ are discretized noise sequences with covariance matrices $\mathbf{Q}_k$ and $\mathbf{R}_k$ respectively. For using the KF update, Eq. (19a) is linearized and brought to the form:

$$\mathbf{z}_k = \mathbf{H}_k\mu_k + \Delta\eta_k \tag{19b}$$

where $\mathbf{H}_k = \mathbf{D}\,[\partial\delta(\mu, t)/\partial\mu]_{\mu=\hat{\mu}_{k-1}}$. Let $\hat{\mu}_k := \hat{\mu}_{t_k}$, $\hat{\mu}_{k|k-1} := E(\mu_k|\mathbf{z}_{1:k-1})$, $\mathbf{P}_{t|t} := E\left[(\mu_t - \hat{\mu}_{t|t})(\mu_t - \hat{\mu}_{t|t})^T\right]$ and $\mathbf{P}_{k|k-1} := E[(\mu_k - \hat{\mu}_{k|k-1})(\mu_k - \hat{\mu}_{k|k-1})^T]$. Since the pseudo-dynamic EnKF (PD-EnKF) requires Monte Carlo simulations, we denote by N the ensemble size. Now, the error covariance ($\mathbf{P}$) is approximated as:

$$\mathbf{P} = (1/N - 1)\sum_{i=1}^{N} (\Sigma_i - \bar{\mu})^T (\Sigma_i - \bar{\mu}) = \mathbf{A}^{\mathrm{T}}\mathbf{A} \tag{20}$$

Here $\Sigma = [\mu(1) \quad \mu(2) \quad \ldots\ldots. \quad \mu(N)] \in \hbar^{m\times N}$ is the matrix of realizations $\mu(j) \in \hbar^p$, $j \in [1, N]$ and $\bar{\mu} = (1/N)\sum_{i=1}^{N}\mu(i)$ is the ensemble mean. $\mathbf{A}(i) = \mu(i) - \bar{\mu}$ denote the anomalies. We now describe PD-EnKF algorithm.

Step (1): Using last analyzed ensemble $\Sigma_{k-1|k-1}$, we obtain the forecast ensemble as:

$$\Sigma_{k|k-1} = \Sigma_{k-1|k-1} + \boldsymbol{\Theta}_k \tag{21}$$

The forecast mean ($\bar{\mu}_{k+1|k}$) is computed as the ensemble mean as noted above. The j-th column of the matrix $\mathbf{\Theta}_k$ contains the j-th realization of Wiener vector increment $\Delta\xi_k(j) = \boldsymbol{\xi}_k(j) - \boldsymbol{\xi}_{k-1}(j)$. Now we obtain forecast anomalies as:

$$\mathbf{A}_{k|k-1} = \Sigma_{k|k-1}(\mathbf{I} - \mathbf{1}_N) \tag{22}$$

$\mathbf{I} \in \hbar^{N\times N}$ is the identity matrix and $\mathbf{1}_N \in \hbar^{N\times N}$ is a matrix with each entry being $1/N$.
Step (2): Kalman gain is computed as:

$$\mathbf{G}_k = \mathbf{P}_{k|k-1}\mathbf{H}_k^T \left[\mathbf{H}_k\mathbf{P}_{k|k-1}\mathbf{H}_k^T + \mathbf{R}_k\right]^{-1} \tag{23}$$

where $\mathbf{P}_{k|k-1} = (1/N-1)\mathbf{A}_{k|k-1}^T\mathbf{A}_{k|k-1}$ is the predicted error covariance.
Step (3): For obtaining the estimate $\hat{\mu}_k$ (mean of the filter density), we use the so-called deterministic EnKF strategy [13] and the fact that $\bar{\mu}_{k|k-1} = \hat{\mu}_{k-1}$.

$$\bar{\mu}_{k|k} := \hat{\mu}_k = \hat{\mu}_{k-1} + \mathbf{G}_k(\mathbf{z}_k - \hat{\mu}_{k-1}) \tag{24}$$

The filter anomalies, needed for the ensemble spread over pseudo time, are obtained as:

$$\mathbf{A}_{k|k} = \mathbf{A}_{k|k-1} - 0.5\mathbf{G}_k\mathbf{H}_k\mathbf{A}_{k|k-1} \tag{25}$$

The factor 0.5 in the second term of the RHS of Eq. (25) produces realizations whose error covariance is the same as that via a KF-based Riccati update modulo a linear approximation in h [13]. Then we obtain the realizations from the filter density as:

$$\Sigma_{k|k} = \mathbf{A}_{k|k} + [\hat{\mu}_k \quad \hat{\mu}_k \quad \ldots\ldots\ldots \quad \hat{\mu}_k] \tag{26}$$

Note that we could similarly use a pseudo-dynamic extended Kalman filter (PD-EKF) for reconstruction. However, the sensitivity of the EKF to process noise covariance could degrade the accuracy of reconstruction and this is demonstrated below through some numerical tests.

3.1 Numerical Experiments

We consider a planar object, 10×10 cm, with a single central inclusion. It is allowed to compress by 0.1 cm at the top and is fixed at the bottom against vertical movement [2]. We solve the forward problem with a finer mesh (100×100 elements) and thus generate the artificially noise-added 'measured' displacement field over a subset of uniformly spaced 21×21 grid points. The reference shear modulus profile is taken to be element-wise constant, i.e., 20 units for the background tissue and a maximum of 100 units (maximum contrast ratio is 5) for the inclusion representing cancer. We use 5% noise (defined below) to generate the 'measured' displacement vector. Figure 3

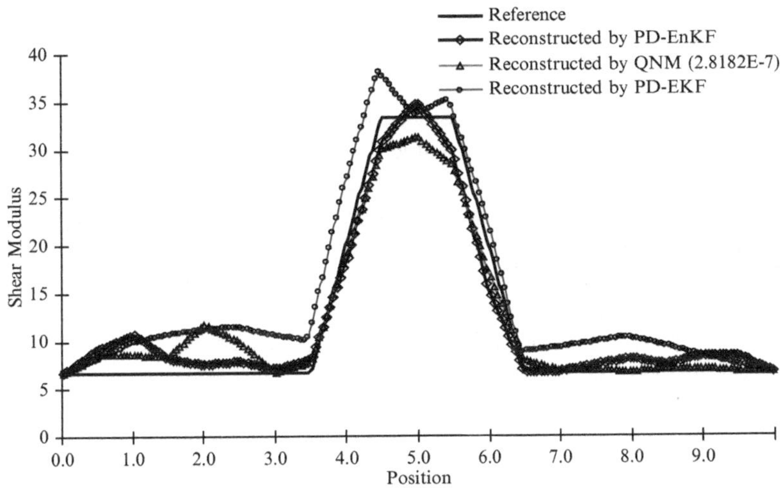

Fig. 3 PD-EnKF-reconstructed shear modulus profile

shows cross-sectional snapshots of the recovered μ (along A-A that is a horizontal line passing through the centre of the object). While PD-EnKF solutions remain unaffected by reasonable variations in process noise covariance, this is not the case with PD-EKF and we only show in Fig. 3 the PD-EKF estimate with the best tuned noise covariance matrix. We also compare the results via PD-EnKF and a Tikhonov-regularized QNM, a deterministic inversion that employs nonlinear optimization [9]. Specifically, we use a limited-memory quasi-Newton (L-BFGS) scheme for bound-constrained minimization [15]. We use the background value as the lower bound on the reconstructed μ and it appears to play a crucial role in the quantitative accuracy of QNM-based reconstruction. Even then, the sensitive dependence of QNM on the regularization parameter remains and an L-curve analysis reveals the optimal regularization parameter to be around 2.8182×10^{-7}. This may be compared with the PD-EnKF, which demands neither bound-constraints nor tuning.

The basic thrust of the current presentation has been on the effectiveness of the pseudo-dynamic inversion strategy that not only bypasses the explicit regularization procedure but also smoothes the numerical solution by projecting the inverse operator on a Lie manifold. Moreover the stochastic setup has enabled a more rational treatment of noises and a better quality of reconstruction through Monte Carlo simulations.

The general nature of the proposed scheme suggests wide range applicability to various inverse problems of engineering interest, especially those involving functionally graded materials. Topology optimization of multifunctional structural systems (e.g. multi-walled carbon nanotubes, design of protein cages, etc.) is one such area wherein the user does not have any prior idea on the material distribution. Multifunctional materials, as constituents of structural systems, help achieve multiple functionalities involving mechanical, electronic, optical, magnetic, acoustic, and/or biological aspects and are capable of exhibiting diverse physical behavior when subjected to various external conditions. The designs of such systems are practically directed via user requirements that effectively govern the stiffness and thus bulk material requirements. In order to devise such material systems with user specific requirements, one needs to solve the associated inverse problems [7] with practically no a-priori knowledge of the material profile. An economical design of this kind with little *a-priori* knowledge of the material profile and response often requires solving an ill-posed optimization problem that in turn needs regularization. Based on the insights available from the study conducted in this work, we anticipate that the proposed regularization-free pseudo-dynamic deterministic/stochastic setups should provide a more robust computational framework for the design of multifunctional material systems.

References

1. Ascher, U., Haber, E., and Huang, H. (2006) "On effective methods for implicit piecewise smooth surface recovery". SIAM J. Sci. Comput., 28:339–358.
2. Banerjee B, Roy D, and Vasu R. M. (2009) "A Pseudo-Dynamic Sub-Optimal Filter for Elastography under Static Loading and Measurements",*Phys. Med. Biol,* Vol. 54, pp. 285–305.
3. Doucet A., Godsill S. J., Andrieu, C. (2000), "On sequential Monte Carlo methods for Bayesian filtering", Statist. Comput. Vol. 10, pp. 197–208.
4. Engl W.H, Hanke M, Neubauer A (2000), "Regularization of the Inverse problem", Kluwer Academic Publishers.
5. Evensen G. (1994). "Sequential data assimilation with a nonlinear quasi-geostrophic model using Monte-Carlo methods to forecast error statistics", J. Geophys. Res. 99: 10143–10162.
6. Evensen, G. (2003). "The ensemble Kalman filter: theoretical formulation and practical implementation" Ocean Dyn. 53: 343–367.
7. Guest J K., Prévost J. H (2006), "Optimizing multifunctional materials: Design of microstructures for maximized stiffness and fluid permeability", Int J Solids Struct., Vol. 43, pp. 7028–7047.
8. Nadler B, Rubin MB (2003), "A new 3-D finite element for nonlinear elasticity using the theory of a Cosserat point", Int J Solids Struct., Vol. 40, pp. 4585–4614.
9. Oberai A.A., Gokhale N. H., Doyley M. M. and Bamber J. C. (2004), "Evaluation of the adjoint equation based algorithm for elasticity imaging", *Phys. Med. Biol.*, Vol. 49, pp. 2955–2974.
10. Ophir J., C'espedes I., Ponnekanti H., Yazdi Y. and Li X. (1991), "Elastography: A quantitative method for imaging the elasticity of biological tissues", Ultrasonic Imaging, Vol. 13, pp. 111–134.
11. Roy D. (2001) "A numeric-analytic technique for non-linear deterministic and stochastic dynamical systems", Proc. R. Soc. A 457, 539–566.
12. Roy D. (2004) "A family of lower and higher order transversal linearization techniques in nonlinear stochastic engineering dynamics", Int. J. Numer. Methods Eng. 61, 764–790.
13. Sakov P. and Oke P.R. (2008) "A deterministic formulation of the ensemble Kalman filter: an alternative to ensemble square root filters" Tellus, 60A, pp. 361–371
14. Vogel C. (2002) "Computational methods for inverse problem", SIAM, Philadelphia.
15. Zhu C, Byrd R H, Nocedal J (1997), "L-BFGS-B: algorithm 778: L-BFGS-B, FORTRAN routines for large scale bound constrained optimization", *ACM Trans. Math. Softwa.* Vol. 23, pp. 550–60.

Force Reconstruction for Wave Based Damage Detection

Agam Sharan and Mira Mitra

Abstract In this paper, a new methodology for damage detection is proposed based on the concept of force reconstruction of applied load from the measured wave response. The force reconstruction is performed using a wavelet spectral finite element (WSFE) model of the waveguide. The model is based on a transfer function approach relating the output (wave response) with the input (force). In this problem, the experimentally measured wave response of a damaged structure is given as input to the WSFE modeled of the undamaged structure to reconstruct the applied load. This reconstructed force is similar to the applied force except that there are few additional waves resulting from the damage. The position and also the extent of the damage can be predicted from the time of arrival and amplitude of these additional waves appearing in the reconstructed force history. The efficiency of the proposed technique is more prominent for certain realistic cases like presence of damage near the boundaries or damage in short waveguides, where, the additional waves due to damage get cluttered with the incident waves and/or waves reflected from the boundaries. In such cases, it is not possible to directly obtain the time of arrival and amplitude from the measured wave response. The problem is more prominent for composite structures, where due to elastic coupling several wave modes participate in the response. The above statements have been emphasized by performing numerical experiments, first for isotropic rod with degraded zone and next for asymmetric de-laminated composite beam. In these experiments, however, wave responses simulated using finite element (FE) are used as surrogate experimental results.

Keywords Damage prediction · Force Reconstruction · Wavelet based spectral finite element

---

A. Sharan
Aeronautical Development Agency, Benguluru – 560093
e-mail: agam@jetmail.ada.gov.in

M. Mitra (✉)
Indian Institute of Technology Bombay, Powai, Mumbai 400076
e-mail: mira@aero.iitb.ac.in

B. Dattaguru et al. (eds.), *IUTAM Symposium on Multi-Functional Material Structures and Systems*, IUTAM Bookseries 19, DOI 10.1007/978-90-481-3771-8_29,

1 Introduction

Recently, wave propagation is finding efficient applications in structural health monitoring (SHM) [1] where, diagnostic waves are used for damage detection. In this paper, a damage detection methodology has been proposed to predict the position of damage from measured wave response of both metallic and composite structures. Presence of damage in a structure results in additional waves. The position of the damage can be obtained from the time of arrival of these additional waves, knowing the wave speeds in the waveguide. In many realistic cases, however, the prediction is not so straightforward. For example, in short waveguides, multiple reflections from the boundaries occur within a short time window. In such cases, the additional wave due to the damage get cluttered with the waves reflected from the boundaries and it is not possible to determine the time of arrival of this wave directly from the response. Similar problems also occur when the damage is located near the boundaries. These problems are more prominent in anisotropic composite structures, where, multiple wave modes are present. Such composite structures are however widely used for aircraft construction in the recent years.

The motivation of the present work is to develop a force reconstruction based damage detection technique which will address the above issues. In this method, the experimentally measured wave response of damaged structure is given as input to the wavelet based spectral finite element (WSFE) [2] model of the undamaged structure. In absence of damage, the reconstructed force should be exactly similar to the applied force. In presence of damage, however, the reconstructed force exhibits certain additional waves that have resulted due to reflection from the damage. These waves are now uncluttered and the time of arrival can be calculated easily.

Identification of applied dynamic force from the experimentally measured response at some point is a problem of wide applicability. In many cases like high velocity impact of aircraft by bird hit, gust or tool drop, etc., it is difficult to measure the impacting force and the only way is to obtain it inversely from the measured response. Spectral element method is particularly efficient for handling such inverse problems of force reconstruction since the formulation is based on a transfer function based approach [3]. However, the conventional Fourier transform based spectral element formulation requires the complete trace of the wave response for accurate prediction of force. Experimentally recorded signals are bound to be truncated and this results in serious error in the reconstructed force. On the contrary, the WSFE technique used in this work, reconstructs the applied force accurately irrespective of the truncation point [2].

The paper is organized as follows. In section 2, the detail of the formulation of WSFE for the inverse problem of force reconstruction are presented. The following section explains the numerical experiments performed. First, the damage detection in an isotropic rod with a degraded zone is studied for cases where the damages are located near the boundaries and also in waveguides of smaller dimensions. Next, similar example of delaminated composite beam with asymmetric ply orientations are presented. The paper ends with important conclusions and scope for future work.

2 Mathematical Formulation

In this section, the formulation of WSFE for composite Timoshenko beam and its implementation for force reconstruction is described briefly. The detailed formulation is given in earlier work of the author [4]. The governing equations for a composite Timoshenko beam is given as,

$$I_0\frac{\partial^2 u}{\partial t^2} - A_{11}\frac{\partial^2 u}{\partial x^2} + B_{11}\frac{\partial^2 \phi}{\partial x^2} = 0 \quad (1)$$

$$I_0\frac{\partial w^2}{\partial t^2} - A_{ss} = \left(\frac{\partial^2 w}{\partial x^2} - \frac{\partial \phi}{\partial x}\right) = 0 \quad (2)$$

$$I_2\frac{\partial^2 \phi}{\partial t^2} - A_{ss}\left(\frac{\partial w}{\partial x} - \phi\right) + B_{11}\frac{\partial^2 u}{\partial x^2} - D_{11}\frac{\partial^2 \phi}{\partial x^2} = 0 \quad (3)$$

where, $u(x,t), w(x,t)$ and $\phi(x,t)$ are the axial, transverse and shear deformation respectively. The stiffness and inertial coefficients $[A_{ij} B_{ij} D_{ij}]$ and $[I_0 I_2]$ carry the conventional definition.

The first step in WSFE formulation is the approximation of the displacement variables using Daubechies scaling functions [5] as follows,

$$u(x,t) = u(x,\tau) = \sum_k u_k(x)\varphi(\tau - k) \quad (4)$$

where, $u_k(x)$ are the approximation coefficient at any spatial dimension x and $\varphi(t)$ is the Daubechies scaling function. $\tau = 0, 1, \ldots, n-1$ are the discrete time sampling points and are given as,

$$t = \Delta t\tau \quad \text{where, } \Delta t \text{ is the time sampling rate}$$

Substituting the approximation of $u(x,t)$ given by Eq. 4 and similarly for $w(x,t)$ and $\phi(x,t)$ in the governing differential equations given by Eqs. 1–3, one gets reduced ordinary differential equations (ODEs) of the form,

$$-I_o\gamma_j^2\hat{u}_j - A_{11}\frac{d^2 5\hat{u}_j}{dx^2} + B_{11}\frac{d^2\hat{\phi}_j}{dx^2} = 0_j = 0, 1, \ldots, n-1 \quad (5)$$

$$-I_o\gamma_j^2\hat{w}_j - A_{ss}\left(\frac{d^2\hat{w}_j}{dx^2} - \frac{d\hat{\phi}_j}{dx}\right) = 0 \quad (6)$$

$$-I_2\gamma_j^2\hat{\phi}_j - A_{ss}\left(\frac{d\hat{w}_j}{dx} - \hat{\phi}_j\right) + B_{11}\frac{d^2\hat{u}_j}{dx^2} - D_{11}\frac{d^2\hat{\phi}_j}{dx^2} = 0 \quad (7)$$

There are several intermediate steps in deriving the above ODEs from the original partial differential equations given by Eqs. 1–3 using Daubechies scaling function approximation given by Eq. 4. These steps use orthogonality properties of the scaling functions $\varphi(t)$ and eigenvalue analysis. The parameter γ_j is obtained during these steps and depends only on the properties of Daubechies scaling function used. The ODEs are then solved exactly for each j using polynomial eigenvalue problem (PEP) to obtain solution of the form,

$$\hat{u}_j = \sum_{l=1}^{8} C_j^i e^{-ik^i{}_j x} \tag{8}$$

where, C_j^i are unknown constants which are derived in terms of nodal displacements and nodal forces. This results in a finite element (FE) like solution for each j as,

$$[\hat{K}_j]\{\hat{U}_j\} = \{\bar{F}_j\} \tag{9}$$

where,

$$\hat{U}_j = \left\{\hat{u}_{1j}\hat{w}_{1j}\hat{\phi}_{1j}\hat{u}_{2j}\hat{w}_{2j}\hat{\phi}_{2j}\right\}$$

are the transformed nodal displacements at node 1 and 2 of the spectral element. Similarly, $\hat{F}_j$ are the transformed nodal force vector. Equation 9 can be solved to obtain $\hat{U}_j$ for given $\hat{F}_j$. This is termed as the forward problem and the time domain response is obtained by performing inverse wavelet transform. Similarly, for inverse problem of force reconstruction, the same equation given by Eq. 9 can be used to obtain the force $\hat{F}_j$ for known displacement (response) $\hat{U}_j$. This approach is used here to reconstruct applied force from measured wave response (displacement) and applied for damage detection. The use of force reconstruction for damage detection has been demonstrated through numerical examples in the next section.

3 Numerical Experiments

The numerical experiments are performed for two different types of damaged structures, i.e. isotropic rod with degraded zone and de-laminated composite beam. A schematic of rod or beam is shown in Fig. 1 representing the reflected waves due to damage and ends.

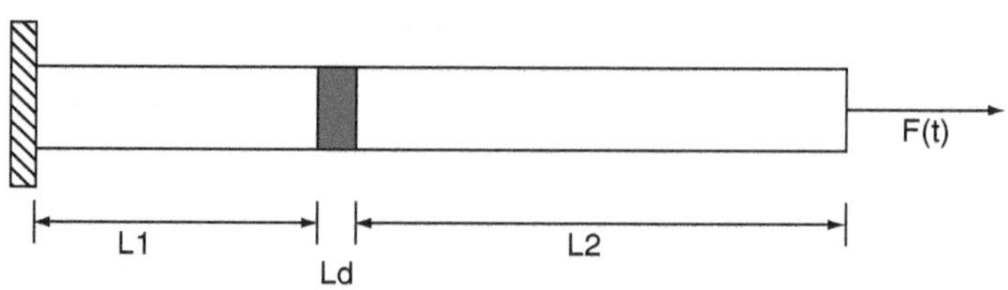

Fig. 1 Schematic of the damaged rod or beam

3.1 Isotropic Rod with Degraded Zone

The wave propagation analysis is done for an aluminum rod where the axial wave velocity is $C_g = 5091.3 m/s$. The first example is having $L_1 = 0.199$ m, $L_d = 0.001$m and $L_2 = 0.8$ m. The damaged zone has a area degradation of 70%. The axial velocity measured at the free end of the rod due to the modulated sinusoidal pulse at 50 kHz is plotted in Fig. 2(a) and is compared with the response of the undamaged rod. These responses are simulated using 2-D FE model. It can be observed from the Fig. 2(a), that the position of the damage cannot be predicted directly from this response since the wave resulting due to the damage is mingling with the waves reflected from the fixed boundaries. Next, this response is given as an input to the WSFE model for force reconstruction. The reconstructed force is plotted in Fig. 2(b) and also compared with the applied force. It can be seen that the reconstructed force is having two uncluttered waves. The time of arrival (t_a) of the first additional wave is 416 μs and thus the distance of the damage predicted is $L_2 = C_g \times t_a/2 = 0.804$ m. This is very near to the actual position of $L_2 = 0.8$ m. Hence, it can be said that the method can accurately predict the position of the damage.

Next, a short waveguide with $L_1 = 0.25$ m, $L_d = 0.001$ m and $L_2 = 0.25$ m. The axial wave response for this case is plotted in Fig. 3 and compared with the undamaged response. Even here, there is no observable difference between the damaged and undamaged response. Again the reconstructed force is plotted in Fig. 4, there are several additional waves which have resulted from the damage. It can be seen that wave resulting from first reflection from the damage is cluttered with the incident wave and hence its time of arrival cannot be obtained. Here, the time of arrival of the second set of waves is considered which is equal to 510μs and this gives a damage position of $L_2 = 0.253$ m which is a very good prediction of the damage position.

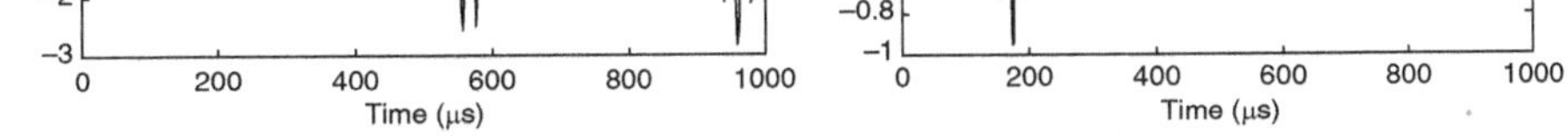

Fig. 2 (**a**) Axial velocity and, (**b**) Reconstructed force for a fixed-free rod with and without degraded zone

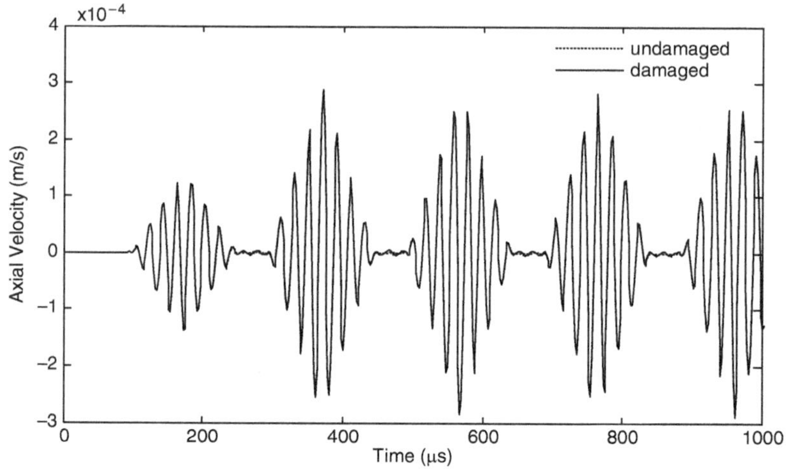

Fig. 3 Axial velocity for a fixed-free rod of shorter length with and without degraded zone

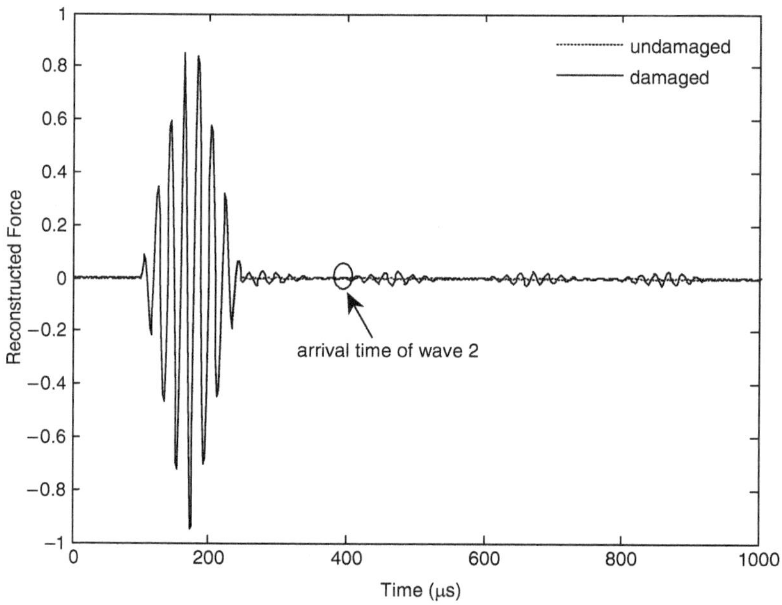

Fig. 4 Reconstructed force for a fixed-free rod of shorter length with and without degraded zone

3.2 De-laminated Composite Beam

Next, the technique is used to predict the position of de-lamination in an asymmetric graphite-epoxy composite beam under transverse wave propagation. The example considered is a beam with $[0_4/45_4]$ ply lay-up and cross-sectional dimension of

$0.01\times0.01\,m^2$. From the dispersion curve, the group velocity at 50 kHz for this beam is $C_g = 1956\,m/s$ for flexural mode. In the first example, $L_1 = 0.745\,m$, $L_d = 0.005\,m$, $L_2 = 0.25\,m$. The response or transverse velocity of beam is taken at free end due to applied load is shown in Fig. 5(a). It can be observed that the position of damage cannot be predicted from this response as the waves reflected from damage are cluttered with the wave resulting from different coupled modes. Next, this forward response is input into the WSFE force reconstruction model. The reconstructed force is plotted in Fig. 5(b) and also compared with applied force. It can be seen that reconstructed force has well separated reflected wave resulting from de-lamination. The arrival time (t_a) of this wave is 358 μs and thus the distance of damage is $L_2 = C_g \times t_a/2 = 0.2523\,m$ which is very near to the actual value of $L_2 = 0.25\,m$.

The second example is of short wave guide with shifted de-lamination between 3th and 4th ply, $L_1 = 0.242\,m$, $L_d = 0.008\,m$, $L_2 = 0.25\,m$. The transverse wave response for this case is plotted in Fig. 6(a) and compared with undamaged response. It can be observed that wave reflected from damage is getting cluttered with the mode resulting from coupling and its arrival time cannot be obtained. So,

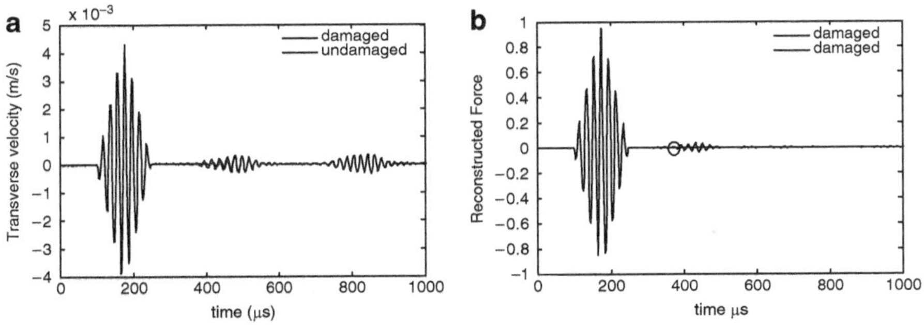

Fig. 5 (**a**) Flexural velocity and, (**b**) Reconstructed force for a fixed-free $[0_4/45_4]$ graphite-epoxy composite beam with and without de-lamination

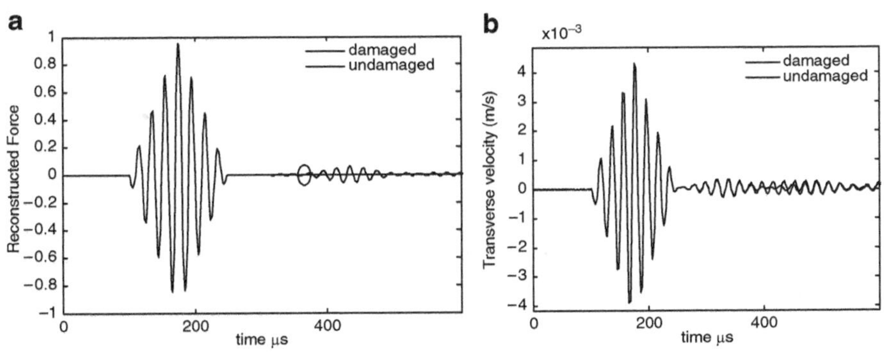

Fig. 6 (**a**) Flexural velocity and, (**b**) reconstructed force for a fixed-free $[0_4/45_4]$ graphite-epoxy composite beam with and without de-lamination

the force is reconstructed from the same response and is shown in Fig. 6(b). Now, the arrival time of wave reflected from de-lamination can be obtained clearly after elimination of coupling blob which is 362μs. Thus the distance of de-lamination is $L_2 = C_g \times t_a/2 = 0.2562$ m, which is very near to the actual value of $L_2 = 0.25$ m.

4 Conclusions

In this work, a novel wave based damage detection technique is proposed and validated through numerical experiments. The method is based on the principle of force reconstruction using spectral finite element method. Here, a wavelet based spectral finite element method is used for this inverse problem of force reconstruction as the method can handle finite length waveguide unlike its Fourier transform based counterpart. The efficiency of the proposed method is that it can predict the position of damage from wave responses where the waves resulting from the damage are cluttered with the incident waves or waves reflected from the boundaries. In these cases, the waves due to the damage are not visually distinguishable from responses.

Here, numerical experiments are performed on isotropic rods with degraded zone representing the damages and also for coupled composite beams with embedded delamination. In futures the method will be developed for Lamb waves in asymmetric 2-D composite laminates.

References

1. Farrar, C. R., Worden, K.: An introduction to structural health monitoring. Phil. Trans. Royal Soc. A 365, 303–315 (2007)
2. Mitra, M., Gopalakrishnan, S.: Spectrally formulated wavelet finite element for wave propagation and impact force identification in connected 1-D waveguides. Int. Jl. 3. Solids Struc. 42, 4695–4721 (2005)
3. Doyle, J. F.: Force identification from dynamic response of a bi-material beam. Exp. Mech. 33, 64–69 (1993)
4. Mitra, M., Gopalakrishnan, S.: Wavelet based spectral finite element modelling and detection of de-lamination in composite beams. Proc. Royal Soc. A 462, 1721–1740 (2006)
5. Daubechies, I.: Ten lectures on wavelets. CBMS-NSF Series in Applied Mathematics. Philadelphia: SIAM. (1992)

On Numerical Integration of Discontinuous Approximations in Partition of Unity Finite Elements

Sundararajan Natarajan, Stéphane P.A. Bordas, and D. Roy Mahapatra

Abstract This contribution presents two advances in the formulation of discontinuous approximations in finite elements. The first method relies on Schwarz-Christoffel mapping for integration on arbitrary polygonal domains [1]. When an element is split into two subdomains by a piecewise continuous discontinuity, each of these polygonal domains is mapped onto a unit disk on which cubature rules are utilized. This suppresses the need for the usual two-level isoparametric mapping. The second method relies on strain smoothing applied to discontinuous finite element approximations. By writing the strain field as a non-local weighted average of the compatible strain field, integration on the surface of the finite elements is transformed into boundary integration, so that the usual subdivision into integration cells is not required, an isoparametric mapping is not needed and the derivatives of the shape (enrichment) functions do not need to be computed. Results in fracture mechanics and composite materials are presented and both methods are compared in terms of accuracy and simplicity. The interested reader is referred to [1,6,13] for more details and should contact the authors to receive a version of the MATLAB codes used to obtain the results herein.

Keywords Composites · Discontinuous enrichment · GFEM · Material interfaces · Numerical integration · Open source MATLAB code · Schwarz-Christoffel conformal mapping · Singularity · Strain smoothing · XFEM

S. Natarajan
PhD Research Student, Theoretical, Applied and Computational Mechanics, Cardiff School of Engineering, U.K
e-mail: sundararajan.natarajan@gmail.com

S.P.A Bordas (✉)
Professor, Theoretical Applied and Computational Mechanics, Cardiff School of Engineering, Cardiff, U.K. Leverhulme/Royal Academy of Engineering Senior Research Fellow
e-mail: stephane.bordas@alumni.northwestern.edu

D.R. Mahapatra
Department of Aerospace Engineering, Indian Institute of Science, Benguluru 560012, India
e-mail: droymahapatra@aero.iisc.ernet.in

B. Dattaguru et al. (eds.), *IUTAM Symposium on Multi-Functional Material Structures and Systems*, IUTAM Bookseries 19, DOI 10.1007/978-90-481-3771-8_30,

1 Introduction

The eXtended Finite Element Method (XFEM) is a numerical method to model arbitrary discontinuities or internal boundaries (cracks, material interfaces, voids, domain boundary, etc.) in continuous bodies that does not require the mesh to conform to the discontinuities [2, 3, 10] and hence a single mesh suffices to capture the evolution of material interfaces and cracks in two and three dimensions. The finite element framework is retained, and a single-field variational principle is used. The final discretized algebraic system of equations has the following form:

$$\mathbf{K}^{\text{XFEM}}\mathbf{d} = \mathbf{f} \tag{1}$$

The basic idea of the XFEM is to enrich a classical finite element space with some additional functions. The general XFEM approximation on a domain Ω, reads

$$\forall \mathbf{x} \in \Omega \quad \mathrm{u}^h(\mathbf{x}) = \sum_{I \in \mathcal{N}^{fem}} N_I(\mathbf{x})\mathrm{q}_I + \sum_{J \in \mathcal{N}^{enr}} N_J(\mathbf{x})\mathcal{E}(\mathbf{x})\mathrm{a}_J \tag{2}$$

where $\mathcal{E}$ is an enrichment function. For cracks it is defined by

$$\mathcal{E}(\mathbf{x}) = \begin{cases} \mathrm{H}(\mathbf{x}) & \text{Heaviside function at split nodes} \\ \mathrm{B}(\mathbf{x}) & \text{Branch functions at tip nodes} \end{cases} \tag{3}$$

The choice of enrichment functions depends on the type of problem considered and the enrichment is able to take on a local form by only enriching those nodes whose support intersect a region of interest. Because of the presence of discontinuous functions, the numerical integration of Eq. (1) poses difficulty. This is overcome by sub-dividing the element that contains these functions into triangles and using Gauss quadrature on those triangles [4].

We present two new techniques to numerically integrate a discontinuous approximation that eliminates the need for two level isoparametric mapping. The first method relies on Schwarz-Christoffel conformal mapping (SCCM) [1, 13]. In this method, instead of sub-dividing into triangles, the polygonal domain is mapped onto a unit disk and cubature rules on this disk are used to obtain the integration points. Interested readers are referred to the recent papers [1, 13], where the method is described in detail.

The second method relies on strain smoothing. The smoothed finite element method [5, 15, 16] based on strain smoothing [14] is combined to the XFEM framework to obtain the Smoothed eXtended Finite Element Method (SmXFEM) [6]. In this method, the conventional strains obtained using displacement compatible equations in FEM are replaced by their smoothed counterpart. Due to strain smoothing, only shape functions are involved in the calculation (no isoparametric transformation is required and the derivatives of the shape or enrichment functions are not needed).

The paper is organized as follows: Section 2 briefly discusses the Schwarz-Christoffel conformal mapping as applied to numerical integration of discontinuous functions and Section 3 gives an overview of the Smoothed XFEM. Numerical results are presented in Section 4, followed by conclusions and scope for future work.

2 Schwarz-Christoffel Conformal Mapping

In this section, we describe the method to map an arbitrary *n-gon* onto a unit disk using Schwarz-Christoffel conformal mapping (SCCM). The basic Schwarz-Christoffel formula [7] is a recipe for a conformal map f from the complex upper half plane to the interior of a polygon. A conformal map f (angle preserving) of a region in the complex plane is an analytic (smooth) function whose derivative vanishes within the region [7]. The Schwarz-Christoffel map f is given by

$$\mathbf{f(z)} = \mathbf{A} \int_{\mathbf{z}} \Pi_{\mathbf{j}=1}^{\mathbf{n}} (\mathbf{z}' - \mathbf{z_j})^{-\beta_j} \, \mathbf{dz}' + \mathbf{B} \tag{4}$$

Figure 1 shows the conformal mapping of an arbitrary polygon onto a unit disk. The main idea presented in [1, 13] is to eliminate the two level isoparametric mapping usually required to integrate on polygonal domains, by mapping the polygonal domain onto a unit disk and to use cubature rule on a circle is used to obtain the Gauss points. Once the integration points or Gauss points are obtained, the numerical integration of Equation (1) is carried out as in conventional XFEM. More details are provided in [1, 13].

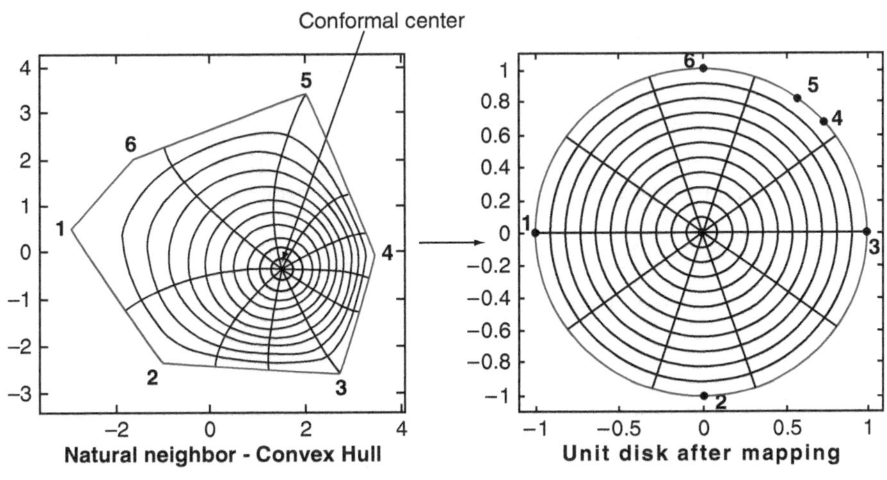

Fig. 1 Mapping physical domain onto a unit disk. This figure was produced with the MATLAB SC Toolbox [8]

3 Strain Smoothing in XFEM

Liu et al. [5] proposed a smoothed finite element method (SFEM) by combining the standard, displacement based FEM technology and the strain smoothing [14] technique which originated in meshfree methods [17]. In this method, elements are used as in the FEM, but the strain field is written as a spatial average of the compatible strain [9]. The smoothing is done over smoothing cells which form a partition of the elements. The final discretized algebraic system of equations has the following form:

$$\mathbf{K}^{\mathrm{SFEM}}\mathbf{d} = \mathbf{f} \tag{5}$$

where $\mathbf{K}_{\mathrm{ij}}^{\mathrm{SFEM}}$ is the smoothed stiffness matrix given by

$$\boldsymbol{K}_{ij}^{\mathbf{SFEM}} = \sum_{C=1}^{SC} \int_{\Omega_c} \tilde{\boldsymbol{B}}_{Ci}^{T} \boldsymbol{D} \tilde{\boldsymbol{B}}_{Cj} \, d\Omega = \sum_{C=1}^{SC} \tilde{\boldsymbol{B}}_{Ci}^{\mathrm{T}} \boldsymbol{D} \tilde{\boldsymbol{B}}_{Cj} A_C \tag{6}$$

where $\mathbf{A}_{\mathrm{C}} = \int_{\Omega_{\mathrm{C}}} \mathbf{d}\Omega$ is the area of the domain, SC is the number of subcells that divides the domain Ω such that $\Omega = \Omega_1 \cup \Omega_2 \cup \ldots \cup \Omega_{\mathrm{SC}}$ and $\Omega_1 \cap \Omega_2 \cap \ldots \cap \Omega_{\mathrm{SC}} = \phi$, and $\widetilde{\mathbf{B}}_{\mathrm{Ci}}$ is the smoothed strain matrix.

In this paper, we propose to combine the strain smoothing method proposed by Liu et al. [5] with the XFEM proposed by [2] to solve linear elastic fracture mechanics and multi-material problems.

The element stiffness matrix is given by Eq. (6) and all entries in the matrix $\widetilde{\mathbf{B}}_{\mathrm{C}}$ in Eq. (6) are constants over each subcell Ω_{C} – each of these entries are line integrals calculated along the boundaries of the subcells. The explicit expression for $\widetilde{\mathbf{B}}_{\mathrm{C}}$ is given in [6] and the interested reader is referred to the recent review paper on strain smoothing [6] for details, where the main results are summarized and applications to plates, shells, plasticity, three-dimensional problems and fracture are discussed.

As in standard XFEM, there are five types of elements, viz., Tip elements, Split elements, Tip-Blending elements, Split-Blending elements and standard elements. Since approximation of the displacement field differs from element to element, care has to be taken while numerically integrating Eq. (6). Different integration rules must be adopted to ensure proper integration of the non-polynomial functions.

4 Numerical Examples

In this section, we illustrate the effectiveness and accuracy of the proposed methods by solving a benchmark problem taken from linear elastic fracture mechanics (LEFM) and a multi-material problem. We first consider the case of an infinite plate with a crack under uniform far field tension. This problem has an analytic solution and the results are compared with the analytical solution. We then examine the case of a place with one and two circular inclusions.

4.1 Infinite Plate

Consider an infinite plate containing a straight crack of length a and loaded by a remote uniform stress field. Along ABCD the closed form displacements are imposed. The closed form near-tip displacement field is given by

$$u_x(r,\theta) = \frac{2(1+\nu)}{\sqrt{2\pi}}\frac{K_I}{E}\sqrt{r}\cos\frac{\theta}{2}\left(2-2\nu-\cos^2\frac{\theta}{2}\right) \tag{7a}$$

$$u_y(r,\theta) = \frac{2(1+\nu)}{\sqrt{2\pi}}\frac{K_I}{E}\sqrt{r}\sin\frac{\theta}{2}\left(2-2\nu-\cos^2\frac{\theta}{2}\right) \tag{7b}$$

In the two previous expressions $\mathbf{K_I} = \boldsymbol{\sigma}\sqrt{\boldsymbol{\pi}\mathbf{a}}$ denotes the stress intensity factor, $\boldsymbol{\nu}$ is Poisson's ratio and $\mathbf{E}$ is Young's modulus. All simulations are performed with $\mathrm{a} = 100\,\mathrm{mm}$ and $\boldsymbol{\sigma} = 10^4\,\mathrm{N/mm^2}$ on a square mesh with sides of length 10 mm. The geometry, loading and boundary conditions are shown in Fig. 2. The displacements of nodes on the boundary are prescribed by Eq. (7). A structured quadrilateral mesh was used for the comparative study. The convergence of relative error in displacement and normalized stress intensity factor is shown in Fig. 3. The relative error in displacement is given by

$$\mathbf{R_{ed}} = \sqrt{\frac{\sum_{i=1}^{\mathbf{ndof}}\left(\mathbf{u_i^h} - \mathbf{u_i^{analytical}}\right)^2}{\sum_{i=1}^{\mathbf{ndof}}\left(\mathbf{u_i^{analytical}}\right)^2}} \tag{8}$$

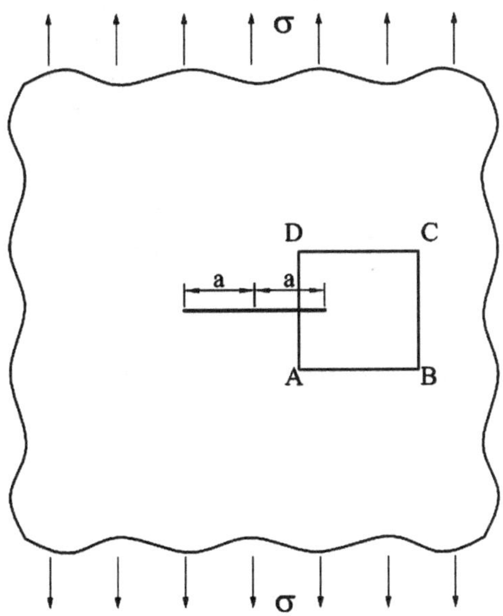

Fig. 2 Griffith Problem: geometry, loading and boundary conditions

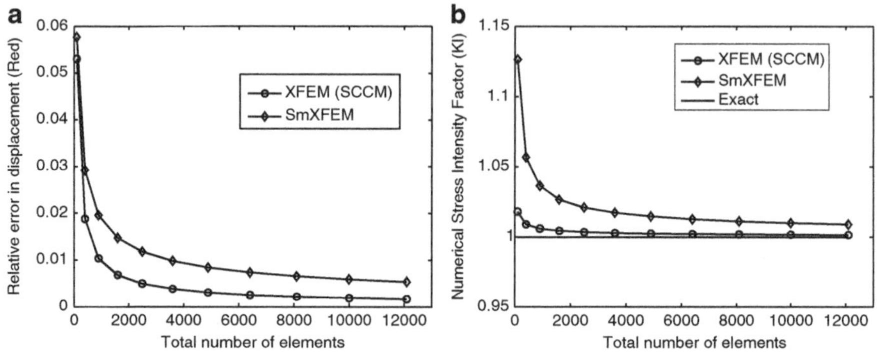

Fig. 3 Griffith Problem: (**a**) convergence in the displacement norm (L_2) and (**b**) the convergence of the numerical stress intensity factor to the analytical stress intensity factor

It is seen that the XFEM (SCCM) performs better than XFEM with strain smoothing, but with mesh refinement both converge to the exact solution. These convergence rates in the displacement and for the SIFs are suboptimal both for the standard XFEM and its strain smoothing version. This is due to the fact that only the tip element is enriched (topological enrichment).

4.2 *Plate with Circular Inclusion*

Inclusions are inhomogeneities in a matrix with different material properties. The modeling of inclusions requires the satisfaction of the Hadamard condition. Krongauz et al. [11] proposed a technique for incorporating discontinuities in derivatives for the meshless methods. In this paper, material discontinuity is defined by level set function. The XFEM displacement approximation for two dimensional domains with inclusions assumes the form:

$$\mathbf{u}^{\mathrm{h}}(\mathbf{x}) = \sum_{\mathrm{I} \in \mathrm{N}^{\mathrm{FEM}}} \mathrm{N}_{\mathrm{I}}(\mathbf{x})\mathbf{q}_{\mathrm{I}} + \sum_{\mathrm{J} \in \mathrm{N}^{\mathrm{enr}}} \mathrm{N}_{\mathrm{J}}(\mathbf{x})\psi(\mathbf{x})\mathbf{a}_{\mathrm{J}} \tag{9}$$

The level set function [12] given by Eq. (10) is used to describe the inclusions. For more than one inclusion, every distinct material pair interface is associated with a level set function, ϕ. The enrichment function $\psi(\phi(\mathbf{x})) = |\phi(\mathbf{x})|$ is used in the elements that intersect the material interface, and it is smoothed away from the interface.

$$\phi_{\mathrm{I}} = \min_{\mathbf{x}_{\mathrm{c}}^{\mathrm{i}} \in \Omega_{\mathrm{c}}^{\mathrm{i}}} \left\{ ||\,\mathbf{x}_{\mathrm{I}} - \mathbf{x}_{\mathrm{c}}^{\mathrm{i}}|| - \mathrm{r}_{\mathrm{c}}^{\mathrm{i}} \right\} \tag{10}$$

Numerical simulations are performed on a square mesh with sides of length 10 cm with a uniform far field tension, $\sigma = 1\,\mathrm{N/cm^2}$ and radius of circular inclusion $\mathrm{r} = 1$ cm for one inclusion and $\mathrm{r} = 0.5$ cm for two inclusions. Figures 4 and 5 show

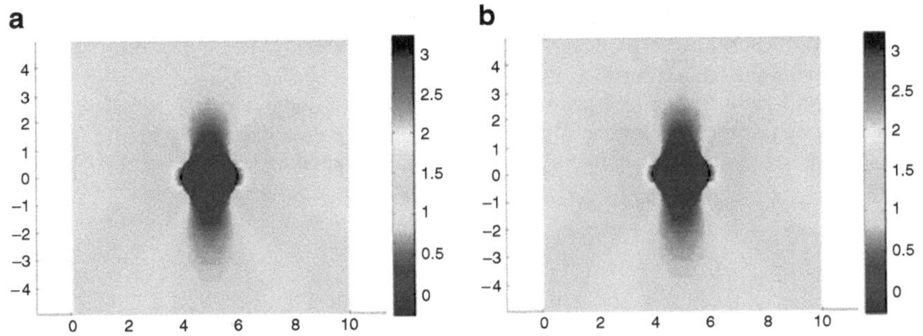

Fig. 4 Plate with one circular inclusion, stress in Y-direction: (**a**) XFEM (SCCM) and (**b**) SmXFEM

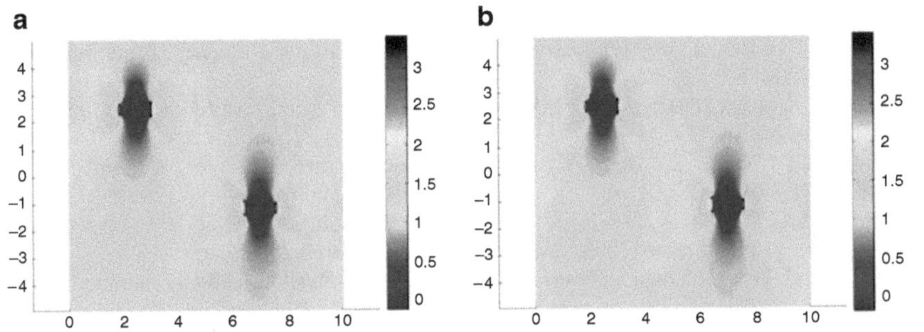

Fig. 5 Plate with two circular inclusions, stress in Y-direction: (**a**) XFEM (SCCM) and (**b**) SmXFEM

the stress distribution plot in y-direction for the plate with one and two circular inclusions, respectively. It is seen that both methods yield comparable results.

5 Conclusion

In this work, two new methods are proposed for numerical integration of discontinuous approximations. The first method relies on conformal mapping. The main advantage of this method is that it eliminates the need for two level isoparametric mapping, usually required to integrate over polygonal domains.

The second method relies on strain smoothing. The main advantage of this method is that it avoids the need to integrate the singular functions present in the XFEM stiffness matrix in linear elastic fracture mechanics problems. Another strong point of strain smoothing is that it should provide less sensitivity to mesh distortion and allow arbitrary polygonal elements to be used in meshing. Although this needs to be verified, it would provide a relatively simple way to perform quad(oct)-tree like adaptivity or/and relocation adaptivity.

Acknowledgements The first author acknowledges the financial support of (1) Overseas Research Students Awards Scheme; (2) the Faculty of Engineering, for period Jan. 2008 – Sept. 2009 and of (3) School of Engineering (Cardiff University) for the period Sept. 2009 onwards.

The second author gratefully acknowledge the financial support of the Royal Academy of Engineering and of the Leverhulme Trust Senior Research Fellowship (2009-2010) [Towards the Next Generation Surgical Simulators] http://www.raeng.org.uk/research/researcher/leverhulme/current.htm

References

1. S Natarajan, S Bordas, DR Mahapatra, International Journal of Numerical Methods in Engineering, **80**(1), 103 (2009), DOI:10.1002/nme.2589
2. T Belytschko, T Black, International Journal of Numerical Methods in Engineering, **45**, 601 (1999)
3. J Dolbow, N Moes, T Belytschko, Computer Methods in Applied Mechanics and Engineering, **19**, 6825 (2001)
4. N Moes, J Dolbow, T Belytschko, International Journal of Numerical Methods in Engineering, **46**, 131 (1999)
5. GR Liu, TT Nguyen, KY Dai, KY Lam, International Journal of Numerical Methods in Engineering, **71**(8), 902 (2007)
6. SPA Bordas, T Rabczuk, NX Hung, VP Nguyen, S Natarajan, T Bog, DM Quan, NV Hiep, Computers and Structures (2009), DOI: 10.1016/j.compstruc.2008.07.006
7. AT Driscoll, NL Trefethen, Cambridge Monographs on Applied and Computational Mathematics **8**(1) (2002)
8. AT Driscoll, NL Trefethen, ACM Transactions on Mathematical Software, **22** (2), 168 (1996)
9. J Chen, C Wu, S Yoon, Y You, International Journal of Numerical Methods in Engineering, **50**, 435 (2001)
10. S Bordas, V Nguyen, C Dunant, ND Hung, A Guidoum, International Journal of Numerical Methods in Engineering (2007), DOI:10.1002/nme.1966
11. Y Krongauz, T Belytschko, International Journal of Numerical Methods in Engineering, **41**(1), 1215 (1998)
12. N Sukumar, DL Chopp, N Moes, T Belytschko, Computer Methods in Applied Mechanics and Engineering, **190** (46–47), 6183 (2001), DOI: 10.1016/S0045-7825(01)00215-8
13. S Natarajan, DR Mahapatra, S Bordas, International Journal of Numerical Methods in Engineering (2009), DOI: 10.1002/nme.2798
14. JS Chen, HP Wang, Computational Mechanics, **25**, 137 (2000)
15. H Nguyen-Xuan, S Bordas, H Nguyen-Dang, International Journal of Numerical Methods in Engineering, **74**, 175 (2008). DOI: 10.1002/nme.2146
16. S Bordas, S Natarajan, International Journal of Numerical Methods in Engineering, (2009), DOI: 10.1002/nme.2713
17. V Nguyen Phu, T Rabczuk, S Bordas, M Duflot, Mechanics and Computer in Simulation, **79**, 763 (2008), DOI: 10.1016/j.matcom.2008.01.003

Section VII
Computational Methods II

Approximate Evaluations of the Modal Effective Electromechanical Coupling Coefficient

Ayech Benjeddou

Abstract This contribution reviews and derives, in a unified manner using energy concepts, available formulas in the smart structures literature for the approximate evaluation of the *modal effective electromechanical coupling coefficient* (EMCC). A *new approximate formula* is then proposed; it has the advantage to use only *short-circuit* and *open-circuit* elastic properties of the piezoelectric patches, thus allowing the use of *general-purpose* structural finite element (FE) codes for the modal effective EMCC approximate evaluation. Besides, it avoids several usual assumptions (similar mode shapes, mass-less patches, etc.). An *interpretation* of the modal effective EMCC is also given using *an analogy* with the *effective viscoelastic loss factor*, allowing also the use of classical FE codes for the modal effective EMCC approximate evaluation.

Keywords Electromechanical coupling · Free vibration · Model strain energy · Piezoelectric materials · Shunted damping · Smart structures

1 Introduction

The *electromechanical coupling coefficient* (EMCC) is a non dimensional *material* parameter useful for [1]: (i) the description of a particular piezoelectric material under a particular stress and electric field configuration, (ii) the conversion of stored energy to mechanical or electric work, and (iii) the comparison between different piezoelectric materials. The *effective* EMCC is a *structural* parameter that characterizes the piezoelectric devices – host structure interaction and measures their coupling effects. During the last decade, it became an important and useful parameter for designing smart structures applications. For example, it was shown that the *modal* effective EMCC is the most influencing parameter in the *piezoelectric*

A. Benjeddou (✉)
Supméca – Paris, Structures, 3 rue Fernand Hainaut, 93407 Saint Ouen CEDEX, France
e-mail: benjeddou@supmeca.fr

B. Dattaguru et al. (eds.), *IUTAM Symposium on Multi-Functional Material Structures and Systems*, IUTAM Bookseries 19, DOI 10.1007/978-90-481-3771-8_31,

shunted damping (PSD) concept [2,3]; it was also used as an active damage indicator [4], a free-vibration benchmarking comparator [5,6] and in added PSD optimisation (patches positioning and sizing) [7].

After a brief literature review of modal effective EMCC approximations, available approximate formulas are derived in a unified manner using modal strain energies (MSE) under *open-circuit* (OC) and *short-circuit* (SC) electric conditions. Then, a new approximate formula is proposed. Finally, an interpretation of the modal effective EMCC is given using an analogy with the modal effective viscoelastic loss factor of a damped structure.

2 Review of Modal Effective EMCC Approximations

The *dynamic* EMCC of a piezoelectric body is commonly evaluated from anti-resonant (a) and resonant (r) eigen frequencies, extracted from measured electrical impedance by an impedance analyser, using Mason's formula

$$k_d^2 = \frac{\omega_a^2 - \omega_r^2}{\omega_a^2} \tag{1}$$

The *modal* EMCC of a piezoelectric body can be also evaluated from OC and SC eigen frequencies, resulting from finite element (FE) modal analyses under OC and SC electric conditions, using the following *numerical* formula [8]

$$k_n^2 = \frac{\omega_{oc}^2 - \omega_{sc}^2}{\omega_{sc}^2} \tag{2}$$

This expression can be linked to the above one by this relation which has to be considered when comparing experimental and numerical EMCC values [8] (see its demonstration in the subsequent section)

$$k_d^2 = \frac{k_n^2}{k_n^2 + 1} \tag{3}$$

It is clear from this expression that dynamic and numerical EMCCs are identical for low numerical values i.e.: for $k_n^2 \ll 1$, $k_d^2 \approx k_n^2$. This approximation is valid for most practical piezoelectric smart structures.

Early in the 1990s, an approximate numerical formula has resulted from *single-mode* impedance modelling of an elastic beam with shunted piezoelectric patches. It evaluated the modal effective EMCC in terms of the material EMCC (k_{ij}) and the SC patch-to-total (base structure K_s + SC patch K_{jj}^{sc}) *modal stiffness* ratio [9]

$$K_{ij}^2 = \frac{k_{ij}^2}{1 - k_{ij}^2} \frac{K_{jj}^{sc}}{K_s + K_{jj}^{sc}} \tag{4}$$

With $ij = 31, 33, 15$ for transverse, longitudinal and shear mode responses. Approximate formula (4) was widely used in modelling transverse [9–15] longitudinal [16] and shear [2] mode response–based PSD. It is valid for *uniaxial loading* in the jth direction and assumes *similar modal shapes of the beam and the beam with SC patches.*

Denoting $K_T = K_s + K_{jj}^{sc}$, the total modal stiffness resulting from the modal analysis of the beam with the SC patch, the SC patch modal stiffness was evaluated from the difference between the total modal stiffness and that (K_s) of the beam, i.e., $K_{jj}^{sc} = K_T - K_s$. Hence, the modal effective EMCC is computed as [17]

$$K_{ij}^2 = \frac{k_{ij}^2}{1 - k_{ij}^2}\left(1 - \frac{K_s}{K_T}\right) \tag{5}$$

This approximate formula was used for active damage detection in composite beams using the modal effective EMCC [18, 19]. Formulas (4, 5) were also used in the following form that uses the eigen frequency of the beam with shorted patches and that (with a '*' subscript) of the beam with the *mass of the piezoelectric patch included but its stiffness neglected* [10]

$$K_{ij}^2 = \frac{k_{ij}^2}{1 - k_{ij}^2}\frac{\omega_{sc}^2 - \omega_*^2}{\omega_{sc}^2} \tag{6}$$

This expression has been evolved later to the following one that uses the eigen frequency of the beam structure (subscript s) and *neglects the added mass of the piezoelectric patches* [12]

$$K_{ij}^2 = \frac{k_{ij}^2}{1 - k_{ij}^2}\frac{\omega_{sc}^2 - \omega_s^2}{\omega_{sc}^2} \tag{7}$$

Another alternative evaluation procedure to formulas (4, 5) considers formula (2) but with the following definitions of the SC and OC eigen frequencies [9]

$$\omega_{sc}^2 = \frac{K_s + K_{jj}^{sc}}{M_s}, \quad \omega_{oc}^2 = \frac{K_s + K_{jj}^{oc}}{M_s} \tag{8}$$

Where, M_s is the modal stiffness of the beam only (*patches mass is neglected*) and

$$K_{jj}^{oc} = \frac{K_{jj}^{sc}}{1 - k_{ij}^2} \tag{9}$$

Formulas (8, 9) were used for transverse mode response-based PSD [17, 20].

3 Energy-Based Modal Effective EMCC Approximations

Seen as the ratio of convertible energy by the piezoelectric patch to the total energy stored in the piezoelectric structure, the EMCC can be defined by Ultiko's formula using OC and SC internal *energies* as [21]

$$k_e^2 = \frac{U_{conv}}{U^{oc}} = \frac{U^{oc} - U^{sc}}{U^{oc}} \tag{10}$$

This formula can be used to evaluate the modal effective EMCC by solving the following SC and OC *piezoelectric* eigenvalues problems, respectively [6]

$$\left([\mathbf{K}_s] + \left[\mathbf{K}_p^{sc}\right] - \omega_{sc}^2\,[\mathbf{M}]\right)\{\boldsymbol{\Phi}^{sc}\} = \{\mathbf{0}\} \tag{11a}$$

$$\left([\mathbf{K}_s] + \left[\mathbf{K}_p^{oc}\right] - \omega_{oc}^2\,[\mathbf{M}]\right)\{\boldsymbol{\Phi}^{oc}\} = \{\mathbf{0}\} \tag{11b}$$

Where, the structure ($[\mathbf{K}_s]$) and patch ($[\mathbf{K}_p^{sc}]$) stiffness contributions have been formally separated (but *not the total mass* $[\mathbf{M}]$). For mass normalized *ith mode*, these relations provide the expressions of the SC and OC MSE as

$$U_i^{sc} = \frac{1}{2}\{\boldsymbol{\Phi}_i^{sc}\}^t\left([\mathbf{K}_s] + \left[\mathbf{K}_p^{sc}\right]\right)\{\boldsymbol{\Phi}_i^{sc}\} = \frac{1}{2}\omega_i^{sc2} \tag{12a}$$

$$U_i^{oc} = \frac{1}{2}\{\boldsymbol{\Phi}_i^{oc}\}^t\left([\mathbf{K}_s] + \left[\mathbf{K}_p^{oc}\right]\right)\{\boldsymbol{\Phi}_i^{oc}\} = \frac{1}{2}\omega_i^{oc2} \tag{12b}$$

Hence, formulas (1, 2) can be recovered for the *ith mode* so that

$$K_{ie}^2 = \frac{U_i^{oc} - U_i^{sc}}{U_i^{oc}} = \frac{\omega_i^{oc2} - \omega_i^{sc2}}{\omega_i^{oc2}} \tag{13}$$

$$K_{in}^2 = \frac{U_i^{oc} - U_i^{sc}}{U_i^{sc}} = \frac{\omega_i^{oc2} - \omega_i^{sc2}}{\omega_i^{sc2}} \tag{14}$$

From these, relation (3) can also be recovered for the *ith mode* so that

$$\frac{K_{ie}^2}{K_{in}^2} = \frac{U_i^{sc}}{U_i^{oc}} = \frac{U_i^{sc}}{(K_{in}^2 + 1)U_i^{sc}} = \frac{1}{K_{in}^2 + 1}, \quad \text{or} \quad K_{ie}^2 = \frac{K_{in}^2}{K_{in}^2 + 1} \tag{15}$$

Now, using definition (10) for the piezoelectric patch material EMCC under *a given response mode* (k^2) provides

$$U_p^{oc} = \frac{U_p^{sc}}{1 - k^2} \tag{16}$$

Also, structure and patch contributions to OC/SC MSE can be separated into

$$U^{oc} = U_s + U_p^{oc}, \; U^{sc} = U_s + U_p^{sc} \tag{17}$$

Then, combining these relations with (16) provides

$$U^{oc} - U^{sc} = U_p^{oc} - U_p^{sc} = \frac{k^2}{1-k^2} U_p^{sc} = \frac{k^2}{1-k^2} (U^{sc} - U_s) \tag{18}$$

On the other hand, the base structure (without patches) eigenvalues problem is

$$\left([\mathbf{K}_s] - \omega_s^2 [\mathbf{M}_s] \right) \{\mathbf{\Phi}_s\} = \{\mathbf{0}\}, \tag{19}$$

This gives the base structure MSE as $U_{si} = \frac{1}{2} \{\mathbf{\Phi}_{si}\}^t [\mathbf{K}_s] \{\mathbf{\Phi}_{si}\} = \frac{1}{2}\omega_{si}^2$ for the *ith mode* which allows, together with (18), recovering formula (7) for the *ith mode* so that

$$K_{in}^2 = \frac{k^2}{1-k^2} \frac{U_i^{sc} - U_{si}}{U_i^{sc}} = \frac{k^2}{1-k^2} \frac{\omega_i^{sc2} - \omega_{si}^2}{\omega_i^{sc2}} \tag{20}$$

4 New Approximation of the Modal Effective EMCC

Consider the following *elastic* eigenvalues problems that use, respectively, only the SC (at constant electric field E, $[\mathbf{C}^{\mathrm{E}}]$) and OC (at constant electric displacement D, $[\mathbf{C}^{\mathrm{D}}] = [\mathbf{C}^{\mathrm{E}}] + [\mathbf{e}]^{\mathrm{t}}[\varepsilon^{\mathrm{s}}]^{-1}[\mathbf{e}]$) elastic properties of the piezoceramics

$$\left([\mathbf{K}_s] + \left[\mathbf{K}_p^E\right] - \omega^{E2} [\mathbf{M}] \right) \left\{\mathbf{\Phi}^E\right\} = \{\mathbf{0}\} \tag{21a}$$

$$\left([\mathbf{K}_s] + \left[\mathbf{K}_p^D\right] - \omega^{D2} [\mathbf{M}] \right) \left\{\mathbf{\Phi}^D\right\} = \{\mathbf{0}\} \tag{21b}$$

For mass normalized *ith mode*, these relations approximate SC and OC MSE as

$$U_i^{sc} \approx U_i^E = \frac{1}{2} \left\{\mathbf{\Phi}_i^E\right\}^t \left([\mathbf{K}_s] + \left[\mathbf{K}_p^E\right] \right) \left\{\mathbf{\Phi}_i^E\right\} = \frac{1}{2}\omega_i^{E2} \tag{22a}$$

$$U_i^{oc} \approx U_i^D = \frac{1}{2} \left\{\mathbf{\Phi}_i^D\right\}^t \left([\mathbf{K}_s] + \left[\mathbf{K}_p^D\right] \right) \left\{\mathbf{\Phi}_i^D\right\} = \frac{1}{2}\omega_i^{D2} \tag{22b}$$

This leads to the following *new approximation* of the modal effective EMCC

$$K_{iE}^2 = \frac{U_i^D - U_i^E}{U_i^E} = \frac{\omega_i^{D2} - \omega_i^{E2}}{\omega_i^{E2}} \tag{23}$$

This approximate formula has the major advantage to be obtainable with *general purpose structural FE codes*. However, it shares with the other approximate formulas (4–9, 20) the main disadvantage of *ignoring the physical equipotential constraint* on the patch electrode. Hence, care should be taken when using these approximate formulas for uncoupled modes since they will show them coupled [6]. Formula (23) is expected to be more accurate than above other approximate ones since it *does not assume uniaxial loading* and *does not assume similar mode shapes* of base and piezoelectric structures. Also it *considers the patches mass and stiffness*.

5 Modal Effective Viscoelastic Loss Factor-EMCC Analogy

From relations (12a, 12b), the following expressions hold

$$\omega_i^{sc2} = \{\boldsymbol{\Phi}_i^{sc}\}^t [\mathbf{K}_s] \{\boldsymbol{\Phi}_i^{sc}\} + \{\boldsymbol{\Phi}_i^{sc}\}^t \left[\mathbf{K}_p^{sc}\right] \{\boldsymbol{\Phi}_i^{sc}\} \tag{24a}$$

$$\omega_i^{oc2} = \{\boldsymbol{\Phi}_i^{oc}\}^t [\mathbf{K}_s] \{\boldsymbol{\Phi}_i^{oc}\} + \{\boldsymbol{\Phi}_i^{oc}\}^t \left[\mathbf{K}_p^{oc}\right] \{\boldsymbol{\Phi}_i^{oc}\} \tag{24b}$$

Using these expressions in (13) and assuming *similar OC and SC mode shapes* ($\{\boldsymbol{\Phi}_i^{oc}\} \approx \{\boldsymbol{\Phi}_i^{sc}\} = \{\boldsymbol{\Phi}_i\}$), transforms it into

$$K_{ie}^2 = \frac{\{\boldsymbol{\Phi}_i\}^t \left(\left[\mathbf{K}_p^{oc}\right] - \left[\mathbf{K}_p^{sc}\right]\right) \{\boldsymbol{\Phi}_i\}}{\{\boldsymbol{\Phi}_i\}^t \left([\mathbf{K}_s] + \left[\mathbf{K}_p^{oc}\right]\right) \{\boldsymbol{\Phi}_i\}} \tag{25}$$

For a piezoelectric given response mode, assumption (9) can be generalized $\left(\left[\mathbf{K}_p^{sc}\right] \approx (1 - k^2) \left[\mathbf{K}_p^{oc}\right]\right)$ so that (25) reduces to

$$K_{ie}^2 = k^2 \frac{\{\boldsymbol{\Phi}_i\}^t \left[\mathbf{K}_p^{oc}\right] \{\boldsymbol{\Phi}_i\}}{\{\boldsymbol{\Phi}_i\}^t \left([\mathbf{K}_s] + \left[\mathbf{K}_p^{oc}\right]\right) \{\boldsymbol{\Phi}_i\}} \tag{26}$$

Relations (25, 26) were first demonstrated in [7]. The form of (26) is similar to that of the MSE-based modal *effective* viscoelastic loss factor

$$\eta_i = \eta \frac{\{\boldsymbol{\Phi}_i\}^t \left[\mathbf{K}_v'\right] \{\boldsymbol{\Phi}_i\}}{\{\boldsymbol{\Phi}_i\}^t \left([\mathbf{K}_s] + \left[\mathbf{K}_v'\right]\right) \{\boldsymbol{\Phi}_i\}} \tag{27}$$

Where, η is the viscoelastic *material* loss factor and $\left[\mathbf{K}_v'\right]$ is the real part of the viscoelastic patch complex stiffness; these quantities present an analogy with, respectively, the squared material EMCC and piezoelectric patch OC stiffness of relation (26). This *analogy* provides an *interpretation* of the modal effective EMCC as the product of the squared material EMCC by the fraction of the MSE stored in the OC piezoelectric patch when the structure is vibrating in its ith mode. Hence, the popular MSE approach, often used for the evaluation of the effective viscoelastic loss factor, can be adapted [22] for the evaluation of the modal effective EMCC using *general purpose FE codes*.

6 Numerical Assessments and Validations

This section aims to assess the known (20) and new (23) approximate EMCC evaluations against its reference definition (14), used without (w/o or NE) and with the consideration of the equipotentiality (EP) constraint. For this purpose, the smart structure benchmark, shown in Fig. 1, is used; it consists of a cantilever steel (Young's modulus = 210 GPa, Poisson's ratio = 0.3 and mass density = 7,800 Kg/m^3) bonded with collocated PZT-4 patches (k_{31} = 0.33, mass density = 7,500 Kg/m^3, elastic (GPa): $C_{11}^E = C_{22}^E = 139$, $C_{13}^E = C_{23}^E = 74.285$, $C_{44}^E = C_{55}^E = 25.64$, $C_{12}^E = 77.84$, $C_{66}^E = 30.60$, piezoelectric (C/m^2): $e_{31} = e_{32} = -5.2$, $e_{33} = 15.08$, $e_{15} = e_{24} = 12.7$, free stress dielectric (nF/m): $\varepsilon_{33}^T = 11.5$ $\varepsilon_{11}^T = \varepsilon_{22}^T = 13.1$).

After a convergence study, the beam and patches are meshed, using ANSYS® FE commercial code, with 1 mm sized elastic SOLID45 and 0.25 mm sized piezoelectric SOLID5 (with activated *Extra Shape Functions* in order to avoid any numerical locking) eight nodes FE, respectively, leading to a total of 102,000 elements. The first ten modes were extracted using the Block Lanczos algorithm. Obtained frequencies and post-treated modal effective EMCC, using Eqs. (14), (20), (23), of the beam *transverse bending modes only*, are given in Table 1 and Table 2, respectively. They show that the new formula (23) behaves better than the well known one (20) since the latter can lead to negative squared EMCC.

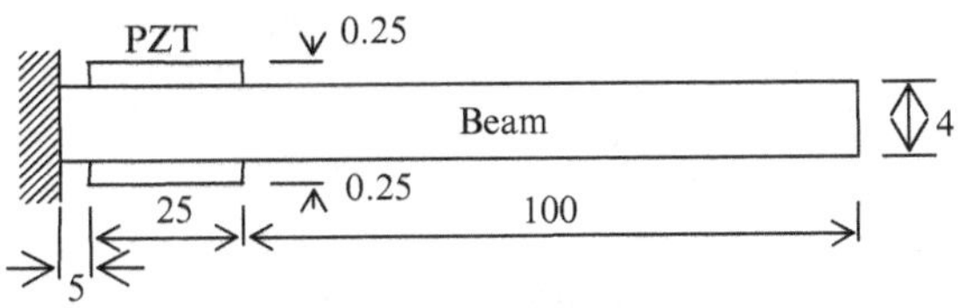

Fig. 1 EMCC evaluations benchmark (dimensions in mm and width = 25 mm)

Table 1 3D transverse (x-z) bending frequencies of the smart cantilever beam

Mode		Baseline	SC frequencies		OC frequencies		
3D	beam	f (Hz)	f_{SC} (Hz)	f_E (Hz)	f_D(Hz)	f_{NE}(Hz)	f_{EP}(Hz)
1	1	200.393	207.878	207.873	208.63	208.623	208.58
3	2	1,249.874	1,261.877	1,261.864	1,263.568	1,263.552	1,263.011
5	3	3,482.339	3,478.435	3,478.401	3,481.152	3,481.117	3,478.488
8	4	6,777.702	6,755.169	6,755.1	6,763.039	6,762.954	6,757.156

Table 2 Squared EMCC (%) of the smart cantilever beam

Formula Modes	(20)	(23)	(14) w/o EP (NE)	(14) with EP
1	0.86	0.72	0.71	0.67
2	0.23	0.27	0.26	0.18
3	−0.03	0.16	0.15	0.003
4	−0.08	0.23	0.23	0.06

7 Conclusions

This contribution first reviewed, and then derived in a unified manner using energy concepts available formulas in the smart structures literature for the approximate evaluation of the *modal effective electromechanical coupling coefficient* (EMCC). A *new approximate formula* was next proposed and assessed using *ANSYS*® 3D piezoelectric coupled finite elements (FE); it has the advantage to use only *short-circuit* and *open-circuit* elastic properties of the piezoelectric patches, thus allowing the use of *general-purpose* structural FE codes for the modal effective EMCC approximate evaluation. Besides, it avoids several usual assumptions (similar mode shapes, mass-less patches, etc.). An *interpretation* of the modal effective EMCC was also reached using an analogy with the modal strain energy-based effective viscoelastic loss factor, allowing also the use of classical FE codes. Its numerical and experimental validation can be found in [7].

References

1. IEEE (1988) IEEE Standards on piezoelectricity. ANS/IEEE Std 176-1987.
2. Benjeddou A, Ranger JA (2006) Use of shunted shear-mode piezoceramics for structural vibration passive damping. Comput Struct 84:1415–1425
3. Chevallier G, Ghorbel S, Benjeddou A (2009) Piezoceramic shunted damping concept: testing, modelling and correlation. Mécanique & Industrie 10:397–411
4. Al-Ajmi MA, Benjeddou A (2008) Damage indication in smart structures using modal effective electromechanical coupling coefficients. Smart Mater Struct 17:035023 (15pp)
5. Deü JF, Benjeddou A (2005) Free-vibration analysis of laminated plates with embedded shear-mode piezoceramic layers. Int J Solids Struct 42:2059–2088
6. Chevallier G, Ghorbel S, Benjeddou A (2008) A benchmark for free vibration and effective coupling of thick piezoelectric smart structures. Smart Mater Struct 17:065007 (11pp)
7. Trindade MA, Benjeddou A (2009) Effective electromechanical coupling coefficients of piezoelectric adaptive structures: critical evaluation and optimization. Mech Adv Mater Struct 16: 210–223
8. Naillon M, Coursant RH, Besnier F (1983) Analyse de structures piézoélectriques par une méthode d'éléments finis. Acta Electron 25:341–362
9. Hagood NW, von Flotow A (1991) Damping of structural vibrations with piezoelectric materials and passive electrical networks. J. Sound Vib 146:243–268
10. Hollkamp JJ (1994) Multimodal passive vibration suppression with piezoelectric materials and resonant shunts. J Intell Mater Syst Struct 5:49–57
11. Hollkamp JJ (1994) A self-tuning piezoelectric vibration absorber. J Intell Mater Syst Struct 5:559–566
12. Hollkamp JJ, Gordon RW (1996) An experimental comparison of piezoelectric and constrained layer damping. Smart Mater Struct 5:715–722
13. Agnes G (1995) Development of a modal model for simultaneous active and passive piezoelectric vibration suppression. J Intell Mater Syst Struct 6:482–487
14. Agnes G, Inman DJ (1996) Non linear piezoelectric vibration absorbers. Smart Mater Struct 5: 704–714
15. Viana FAC, Steffen V (2006) Multimodal vibration damping through piezoelectric patches and optimal resonant shunt circuits. J Braz Soc Mech Sci Eng 28: 293–310
16. Preumont A, de Marneffe B, Deraemaeker A, Bossens F (2008) The damping of a truss structures with a piezoelectric transducer. Comput Struct 86:227–239

17. Benjeddou A, Ranger JA (2005) Simple finite element modeling and performance evaluation of passive vibration damping using shunted piezoceramics. In Proc 1st Int Congr Des Model Mech Syst (Hammamet, Tunisia)
18. Benjeddou A, Vijayakumar S, Tawfiq I (2006) A new damage identification and quantification indicator for piezoelectric advanced composites. In Proc III Europ Conf Comput Mech.: Solids Struct Coupl Prob Eng (Lisbon, Portugal)
19. Benjeddou A (2006) First use of the shear piezoceramics and effective electromechanical coupling coefficient for damage detection and characterization. In Proc. III Europ Workshop Struct Health Monitoring (Granada, Spain)
20. Park CH, Kim YH, Park HC (2005) Dynamic formulations of plates with shunted piezoelectric materials. J Intell Mater Syst Struct 16:971–976
21. Ulitko (1977) Theory of electromechanical energy conversion in non uniform deformable piezoceramics. Soviet Appl Mech (cover-to-cover translation of Prikl Mekh) 13:1055–1062
22. Trindade A, Maio CEB (2008) Multimodal passive vibration control of sandwich beams with shunted shear piezoelectric materials. Smart Mater Struct 17:055015 (10pp)

Distributed Point Source Model for Wave Propagation Through Multi-phase Systems

Prashant Bhise, Abhijit Mukherjee, Sandeep Sharma, and Raghu Ram

Abstract Ultrasonic waves are used in identification of interfaces both in fluids and solids. It is imperative to numerically model these problems for their effective use. A newly developed technique, distributed point source method (DPSM), is gaining popularity in solution of these problems. The technique has been used to predict steady state conditions with a single frequency excitation. In this work, DPSM is extended to model transient waves propagation in multi-phase systems including multi-layered non-homogeneous fluid interfaces. Interface of two fluids is modeled using DPSM. Fast Fourier transformation (FFT) is used convert time domain signal into frequency domain and inverse FFT is used to transform the pressure and velocity signals back in time domain. Numerical results have been verified experimentally for transient wave propagation in homogenous and non-homogeneous fluids.

Keywords DPSM · FFT · Interface · Non-homogenous fluids · Transducer · Wave propagations

1 Introduction

Ultrasonics has been successfully used in a variety of biological and artificial systems consisting of multiple phases and layers of materials. Investigation of biological tissues, exploration of underground and undersea resources, characterization of materials consisting of gases, fluids and solids, are some examples. Reliable numerical models based on the principles of mechanics are imperative to successfully design the ultrasonic devices and interpret the images created by them. Conventional mathematical tools such as finite element method are not fit for such problems due to their deficiencies in modeling, and the magnitude of oscillations and frequencies

R. Ram, A. Mukherjee, and S. Sharma (✉)
Thapar University, Patiala 147004, India
e-mail: sksharma@thapar.edu

P. Bhise
Research student, Department of Civil Engineering, IIT Bombay, Mumbai 400076, India

B. Dattaguru et al. (eds.), *IUTAM Symposium on Multi-Functional Material Structures and Systems*, IUTAM Bookseries 19, DOI 10.1007/978-90-481-3771-8_32,

that are encountered in ultrasonic waves. This paper describes application of DPSM, a semi-analytical tool for modeling transient ultrasonic waves through multilayered systems.

DPSM has been used for computing pressure and velocity fields generated by the transducers [1, 2] immersed in fluid, with presence of an interface. In all these studies steady state response due to a single frequency has been reported. However, it is important to study propagation of transient waves through the field. Reference [3] describes transformation of non-periodic wave excitations of finite duration such as pulse into series of periodic signals using FFT. Present work describes DPSM modeling of transient ultrasonic waves through fluids with interfaces. Tone burst signal has been used to obtain the ultrasonic field in the fluid. The time response in velocity and pressure at different key points has been reported. The results have been validated with the available closed form solutions as well as with experimental results and have been found to be in good agreement.

2 Distributed Point Source Method (DPSM)

The basic principles of DPSM for the modeling of ultrasonic and magnetic transducers have been described in detail in [1,2]. Active transducer surface is discretized in to periodically pulsating source points. Ultrasonic field at any point of interest at a distance r from the surface of the transducer interface is computed by superposition of the contributions of all point sources (Fig. 1(a)). Reference [2] provides the methodology for the determination of the diameter of the spheres, strength of point sources etc.

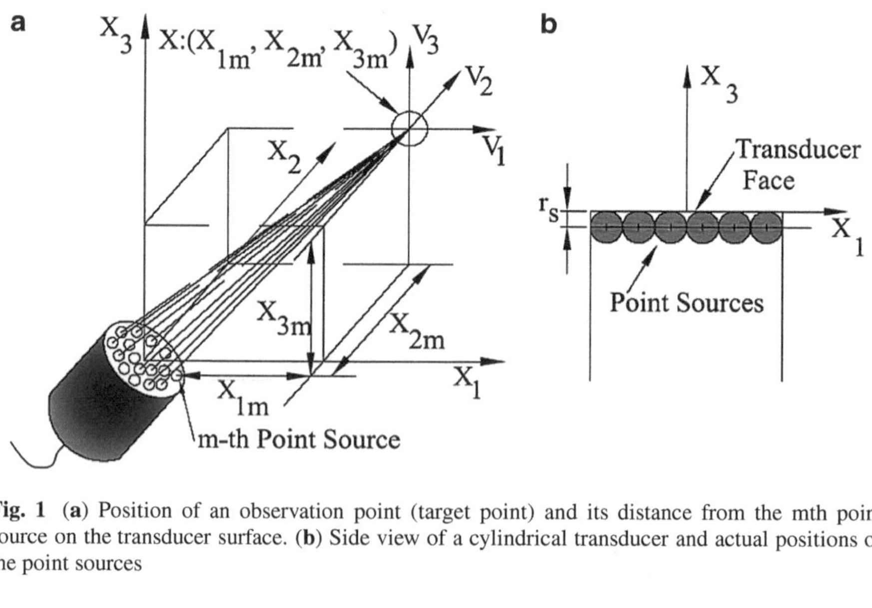

Fig. 1 (**a**) Position of an observation point (target point) and its distance from the mth point source on the transducer surface. (**b**) Side view of a cylindrical transducer and actual positions of the point sources

2.1 Ultrasonic Field in Homogeneous Fluid

Reference [1] provides the formulation for computing ultrasonic velocity ($\hat{V}_T$) and pressure ($\hat{P}R_T$) response at a given target point in homogeneous fluid due to N source points on transducer face as

$$\hat{V}_T = \hat{M}_{TS}.\hat{A}_s \quad \hat{P}R_T = \hat{Q}_{TS}.\hat{A}_s \tag{1}$$

where A_s is the source strength vector for point sources and it can be evaluated for given transducer face velocity v_0. $\hat{M}_{TS}$ and $\hat{Q}_{TS}$ are the matrices for coupling the geometrical relation between the target and source point as mentioned in [1].

2.2 Ultrasonic Field in Multi-Layered Fluids

Multiphase systems having interfaces formed by several fluids with different densities and acoustic properties have been modeled by using DPSM [4].

Each interface acts as a transmitter as well as a reflector of elastic wave energy, having two sets of point sources denoted by $\hat{A}$ and $\hat{A}^*$ (Fig. 2). Continuity conditions across the interfaces and boundary conditions at the transducer face are satisfied for computing ultrasonic field.

At transducer surface

$$\hat{\mathbf{M}}_{SS}\hat{\mathbf{A}}_S + \hat{\mathbf{M}}_{SI}\hat{\mathbf{A}}_I = \hat{\mathbf{V}}_{S0} \tag{2}$$

At interface, for continuity of the acoustic pressure and velocity

$$\begin{aligned} \hat{\mathbf{Q}}_{IS}\hat{\mathbf{A}}_S + \hat{\mathbf{Q}}_{II}\hat{\mathbf{A}}_I &= -\hat{Q}_{II^*}\hat{\mathbf{A}}_I{}^* \\ \hat{M}_{IS}\hat{\mathbf{A}}_S + \hat{M}_{II}\hat{\mathbf{A}}_I &= -\hat{M}_{II^*}\hat{\mathbf{A}}_I{}^* \end{aligned} \tag{3}$$

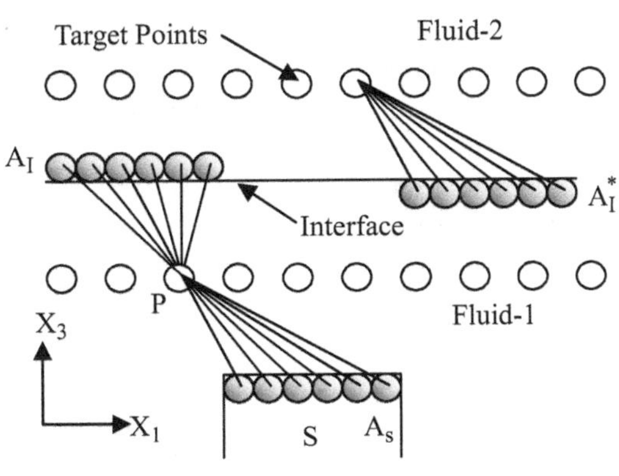

Fig. 2 Distribution of point sources for multi layered system

Equations (2) and (3) in matrix form can be written as

$$\begin{bmatrix} Mss & Msi & 0 \\ Mis & Mii & Mii^* \\ Qis & Qii & Qii^* \end{bmatrix} \begin{Bmatrix} \mathbf{A}_S \\ \mathbf{A}_I \\ \mathbf{A}_I{}^* \end{Bmatrix} = \begin{Bmatrix} \mathbf{V}_{SO} \\ \mathbf{0} \\ \mathbf{0} \end{Bmatrix} \tag{4}$$

$$\left[\mathbf{M\hat{A}T}\right].\left\{\mathbf{\hat{A}}\right\} = \left\{\mathbf{\hat{V}}\right\}$$

Point source strengths $\left\{\mathbf{\hat{A}}\right\}$ vector is evaluated for given transducer surface velocity vector. Thus, acoustic pressure at any point in fluid 1 and fluid 2 is obtained by

$$\text{For fluid-1} \quad P_T = Q_{TS} A_S + Q_{TI} A_I \tag{5}$$

$$\text{For fluid-2} \quad P_T = Q_{TI}{}^* A_I{}^* \tag{6}$$

2.3 Transient Wave Propagation Using Spectral Approach

Using DPSM formulation, steady state response of any periodic ultrasonic wave of infinite duration can be modeled. But real life non-periodic wave excitations of finite duration such as pulse can be converted into series of periodic signals using FFT. By superimposing response of all the periodic signals transient response can be obtained. Thus the problem of characterizing a signal is transformed into one of the determining set of combination coefficients known as Fourier coefficients.

$$u(x,t) = \sum_n \hat{F}_n \left\{ G_1(\hat{K}_{1n} x) + G_2(\hat{K}_{2n} x) + \cdots \right\} e^{i\omega_n t} = \sum_n \hat{F}_n . G_1 . (\hat{K}_{mn} x)\, e^{i\omega_n t} \tag{7}$$

where G is the analytically known transfer function of the problem.

In this case transfer function G obtained from DPSM formulation at given frequency. $\hat{F}_n$ is the amplitude spectrum, and known from input conditions. Thus, $\hat{F}_n G$ is recognized as the Fourier transform of the solutions.

The time input $F(t)$ is converted to its spectrum $\hat{F}_n$ through use of forward FFT. The transformed solution is then obtained by evaluating the product $\hat{u}_n = \hat{F}_n G\ (K_{mn})$ at each frequency. The last step in the analysis involves performing an inverse FFT (reconstructing the signal). It should be noted that while using FFT for inversion, the transformed solution in frequency domain is calculated up to Nyquist frequency and the rest is obtained by imposing the condition that it must be the complex conjugate of the initial part. This is done to ensure that the reconstructed history is real.

3 Numerical Results and Discussions

MATHCAD programs have been developed to model the ultrasonic field based on the DPSM formulation. Diameter of transducer (D) is taken as 0.1 inch (2.528 mm). For 1 MHz frequency, transducer face has been discredited in to 348 point sources. For convenience, x_1 and x_3 are taken as X and Z directions.

Using DPSM formulation, steady state ultrasonic field in a homogeneous fluid (water) generated by a flat circular transducer is computed. Numerically results of near field zone (N_F) and pressure (P) along X_3 direction are matching reasonably well with closed form results (Figure 3).

$$N_F = \frac{D^2}{4\lambda} = \frac{(2.528)^2}{4 \times 1.49} = 1.07\,mm \quad p(z) = \rho.c\ _f v_0 \left[e^{ik_f z} - e^{ik_f \sqrt{z^2+a^2}} \right] \tag{8}$$

Same technique is extended to predict transient wave propagation in homogeneous and non-homogeneous fluids using tone burst pulse of time duration of 250 μs and frequency content between 20 to 150 kHz. Refer Fig. 4, at Z = 0, pressure peak lies at 95 μs, same as the input velocity pulse peak time. Acoustic pressure pulse response at Z = 50 mm and Z = 150 mm arrives at 129 μs (having time lag r_1 = 34 μs) at 195 μs (having time lag r_2 = 100 μs).

Numerically computed time lags also match with experimental results (Fig. 5) for travel of ultrasonic wave in water column of varying heights. Also the pressure amplitudes are matching with the close form solution.

For modeling non-homogeneous fluids, two fluids (having P-wave speed 1.49 and 2 km/s, density 1 and 1.5 g/cc respectively) have been considered. Transducer is immersed in fluid 1 and the interface lies at Z = 250 mm. Acoustic pressure is

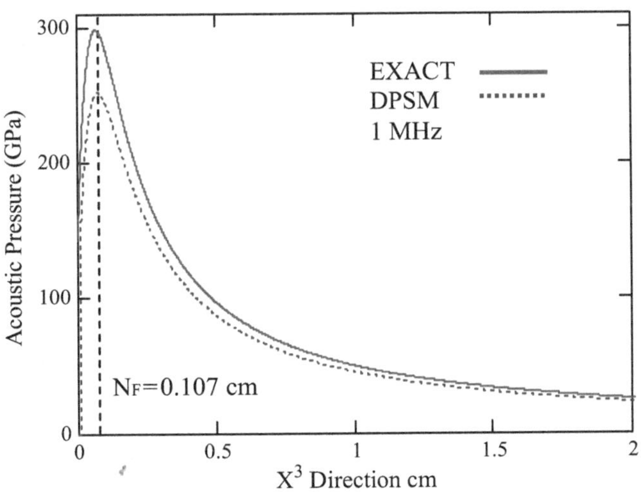

Fig. 3 Near zone (N_F) and acoustic pressure along Z axis

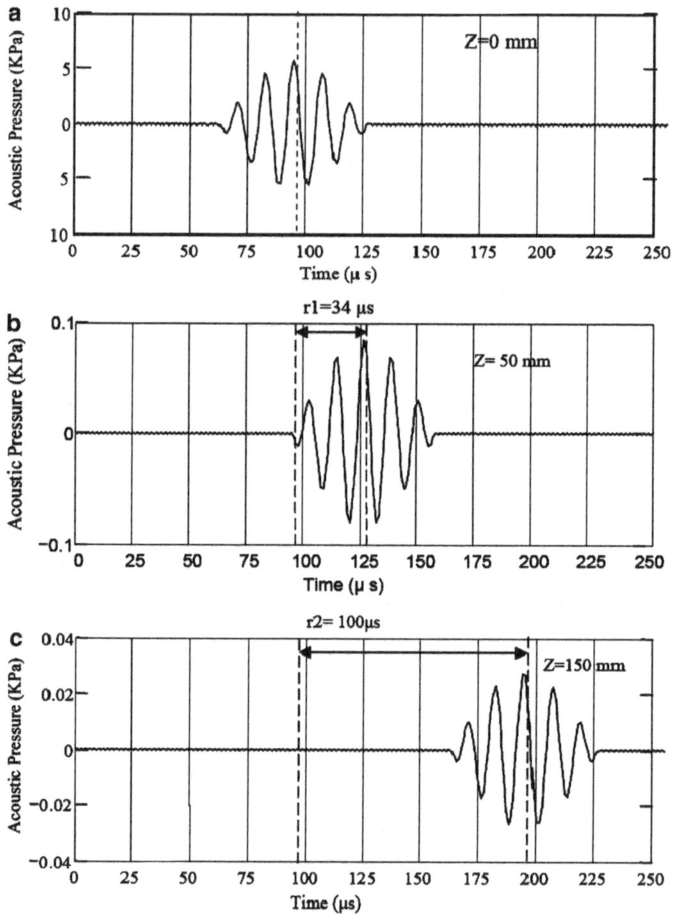

Fig. 4 Acoustic pressure at three different target points (**a**) At Z=0 mm (**b**) At Z=50 mm (**c**) At Z=150 mm)

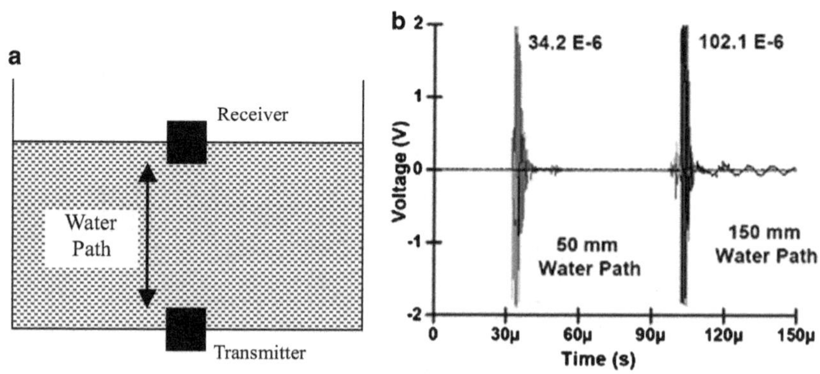

Fig. 5 (**a**) Experimental set up (**b**) V-t signatures for water path Z = 0 and 50 mm

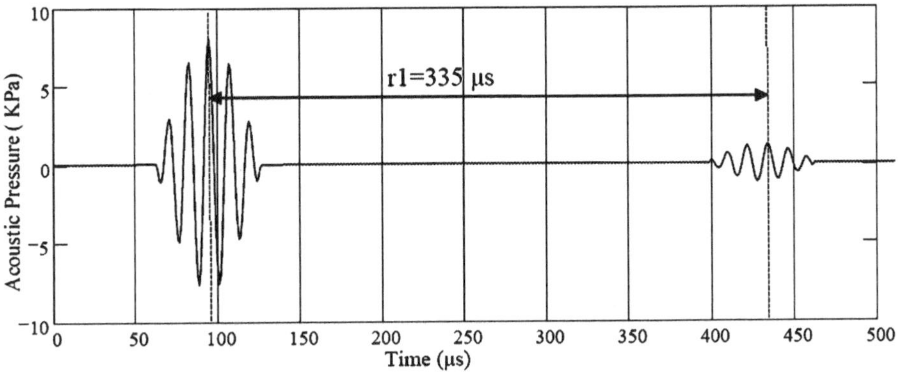

Fig. 6 Acoustic pressure response at Z = 0

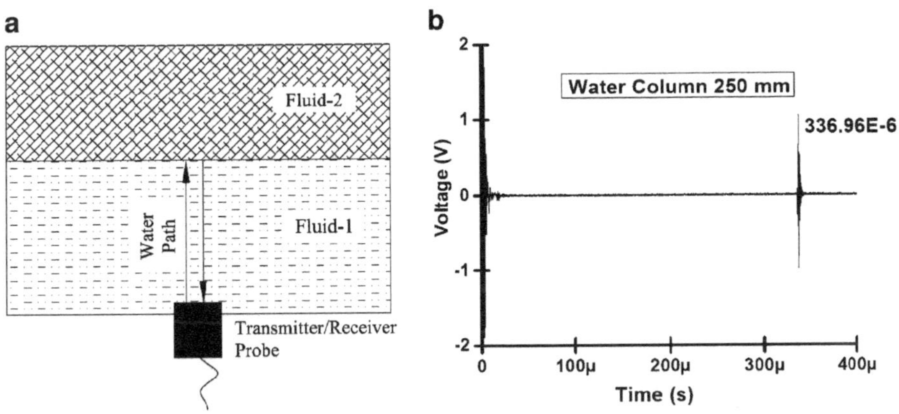

Fig. 7 (**a**) Experimental set up (**b**) V-t signature for water path Z = 250 × 2

recorded at Z = 0 mm, as shown Fig. 6. Pulse on left is the input pulse and pulse on right is the reflected pulse from interface. The time lag between the peaks 'r1' is 335 μs, which is matching with the time taken by wave to reach the interface and rebound back to the transducer face as shown in experimental results in Fig. 7. It may be noted that for clarity the reflected pulse is magnified by 10^5 times in Fig. 6.

4 Conclusions

The above developed model can be extended to inclined interface and solid-fluid interaction problem. From the time taken by reflected wave to reach original location one can determine the location of interface from transducer and the impendence ratio of far end fluid. Using multiple transducer and observation points it is also possible to determine the shape of simple interface.

References

1. Placko D. and Kundu T. (2004) Modeling of Ultrasonic Field by DPSM, Ultrasonic Nondestructive Evaluation: Engineering and Biological Characterization. CRC Press, Boca Raton, FL), Chap. 2:144–201.
2. Kundu T. et al. (2002) Theoretical Computation of Acoustic Pressure Generated by Ultrasonic Sensors in Presence of an Interface. SPIE 7th Annual International Symposium on NDE for Health Monitoring and Diagnostics. 4702:157–168.
3. Doyle, J. F., (1989) Wave propagation in structures, Springer, NY.
4. Banerjee S. et al. (2006) Ultrasonic Field Modeling in Multilayered Fluid Structures Using the DPSM. J. App. Mechanics. 73:1–12.

Intrinsic Localized Modes in Micro-scale Oscillator Arrays Subjected to Deterministic Excitation and White Noise

Subramanian Ramakrishnan and Balakumar Balachandran

Abstract The authors study the influence of noise on an array of micro-scale oscillators with a focus on the phenomenon of intrinsic localized modes (ILMs). It is shown numerically that a white noise excitation, by itself, is unable to produce ILMs. However, under a combined deterministic and random excitation, a noise strength threshold is observed beyond which the ILM at one location in attenuated whilst the localization at another location is strengthened. The results motivate an analytical approach wherein the Fokker-Planck equation and approximate moment evolution equations are derived for a typical cell of the array. Numerical solutions of the moment equations suggest that once a localization event occurs in the array, an excitation with noise strength above a threshold value contributes to the persistence of the event. The reported results can form the basis for developing a fundamental understanding of the influence of noise on ILMs in coupled arrays of nonlinear oscillators including micro-scale oscillator arrays.

Keywords Intrinsic localized modes · Micro-scale oscillators · Nonlinear vibration · White noise

1 Introduction

The combined influence of nonlinearity and stochastic effects on dynamical systems engender unique phenomena that are of interest from a fundamental standpoint as well as in practical applications. In this paper, the authors study the dynamics of an array of nonlinear micro-scale oscillators subjected to a combined harmonic and white noise excitation. The array under consideration is an example of a micro-electromechanical system (MEMS), the importance of which is well recognized

---

S. Ramakrishnan and B. Balachandran (✉)
Department of Mechanical Engineering, University of Maryland - College Park, MD 20742, USA
e-mail: subbu@umd.edu; balab@umd.edu

B. Dattaguru et al. (eds.), *IUTAM Symposium on Multi-Functional Material Structures and Systems*, IUTAM Bookseries 19, DOI 10.1007/978-90-481-3771-8_33,

for a broad spectrum of engineering applications including signal processing and communications (see, for instance [2, 7]). Noise effects such as those arising from thermal sources and fabrication irregularities, which are negligible in macro-scale systems, could become significant at the micro-scale [11]. Hence, it is important to account for stochastic effects in the analysis and design of micro-scale oscillator arrays. The aim of the paper is to provide an analytical framework for the study of white noise effects on the dynamics of coupled arrays of nonlinear oscillators and study the influence of noise on the formation and sustenance of intrinsic localized modes (ILMs) in micro-scale oscillator arrays.

Intrinsic localized modes (ILMs) are localizations of energy observed in the response of perfectly periodic lattices of nonlinear oscillators subjected to an external forcing [3, 10]. The necessary prerequisites for the existence of ILMs are strong nonlinearity and discreteness and, notably, ILMs exist independent of impurities or dislocations in a lattice. While localization has been studied extensively in condensed matter physics over the years (for instance, Anderson localization [1]), the study of ILMs in arrays of micro-scale oscillators subjected to harmonic excitations is a recent development [6, 8, 9]. Sato and co-workers [8] studied ILMs in a micro-cantilever array and reported a drive-induced locking effect in which the amplitudes of the ILMs became fixed and the modes became spatially pinned.

The existence of ILMs in MEMS arrays has important consequences. Arrays of micro-scale oscillators typically serve as channels for the transfer of energy and information and, evidently, localization phenomena could adversely impact their functionality. In particular, pinned ILMs characterized by large amplitudes, stability, and a high degree of localization, could cause significant damage both to the array itself as well as to the associated electronic circuitry. On the other hand, if the phenomenon is understood well enough for it to be controlled, it could enable entirely new technologies.

Dick, Balachandran, and Mote [4] recently considered arrays of piezoelectrically actuated micro-scale cantilevers. By analyzing the spatial form of an ILM using weakly nonlinear analysis, they showed that the amplitude profile of an ILM in a MEMS array can be predicted as a forced nonlinear vibration mode. They also considered the effects of noise by adding a “small” magnitude random signal with a uniform distribution to the excitation, and found that the ILMs could be still be initiated and sustained.

With respect to ILMs, the questions that naturally arise include the following: (i) Can ILMs be realized in a micro-scale oscillator array driven by white noise? (ii) How does noise, when superimposed on harmonic forcing, impact the emergence, stability and propagation of ILMs? In particular, a deeper understanding of the influence of noise on the formation and propagation of ILMs could lead to effective methods of controlling the phenomenon thereby opening avenues for improved designs of MEMS arrays.

Motivated by the above questions, the authors begin with the deterministic model of a micro-cantilever array considered by Sato et al. [8], where the array is represented by a system of Klein-Gordon equations. The first objective of the present work is to extend the results obtained by one of the authors and co-workers [4] to the stochastic regime. After examining the influence of white noise on localization, the authors develop a new analytical framework for the stochastic analysis of the response. The approach is based on deriving the Fokker-Planck equation, which governs the evolution of the probability density function associated with the stochastic dynamics of the array. The averaged dynamics is then represented by a system of ordinary differential equations that are satisfied by the moments of displacements and velocities of each element of the array. Typically, nonlinear systems give rise to an infinite hierarchy of moment equations, closure of which is achieved through a set of moment approximations. The closed system of moment equations thus obtained is then numerically solved to obtain the averaged stochastic dynamics and the influence of white noise on localization.

The rest of this article has been organized as follows. In section 2, the system of Klein-Gordon equations representing the array is introduced and numerical results obtained from direct simulation of the equations by passing the deterministic excitation signal through an additive white Gaussian noise channel for a range of values of the signal-to-noise ratio are presented. In the analytical framework introduced in section 3, the Klein-Gordon system is first converted to a system of stochastic differential equations in the presence of both white noise and deterministic excitation. The corresponding Fokker-Planck equation is then derived in the Ito sense. In the next step, the system of moment evolution equations associated with the Fokker-Planck equation is derived and moment closure is realized by imposing a set of approximations. Preliminary results obtained from numerical solutions of the system of moment equations are presented next. Concluding remarks are collected together and presented along with a discussion of the results in section 4.

2 Effects of Noise on ILMs: Computational Results

Consider an array of micro-cantilever beams coupled together by the overhang region between them [8]. Each cell consists of two beams of different lengths, different spring constants, and different masses. Ignoring the mass of the overhang and assuming a harmonic coupling between the cantilevers, the cell dynamics is governed by the nonlinear Klein-Gordon equations which, for a typical cell "i" are given by

$$\begin{aligned} m_a\ddot{x}_{ai} + (m_a/\tau)\,\dot{x}_{ai} + k_{2a}x_{ai} + k_{4a}x_{ai}^3 + k_I\,(2x_{ai} - x_{bi} - x_{bi-1}) &= m_a\alpha \\ m_b\ddot{x}_{bi} + (m_b/\tau)\,\dot{x}_{bi} + k_{2b}x_{bi} + k_{4b}x_{bi}^3 + k_I\,(2x_{bi} - x_{ai+1} - x_{ai}) &= m_b\alpha \end{aligned} \quad (1)$$

In Eq. (1), the subscripts a and b correspond to the different cantilever lengths, m_a and m_b are their respective masses, τ is a time constant, k_{2a} and k_{2b} are the associated linear spring constants, k_{4a} and k_{4b} are the associated cubic spring constants, and k_I is the linear coupling constant. The entire array is subjected to a uniform base acceleration α that is achieved by using a piezoelectric actuation. It is important to note that this actuation does not directly apply a force to the cantilevers but rather causes an acceleration of the cantilever's base frame.

Direct numerical simulations carried out by using Eq. (1) showed the existence of ILMs in the considered array. In particular, pinned ILMs where the localized oscillation frequency locks with the driver frequency have been observed [8]. In addition to observing ILMs in simulations, as previously mentioned, Dick et al. [4] were able to predict the amplitude profile of ILMs by using simplifications based on a symmetry argument and weakly nonlinear analysis.

Considering stochastic excitations, a natural way to characterize noise in a signal is the signal-to-noise ratio (SNR). SNR is a measure of noise intensity in a signal, and it is defined as the ratio of the power of the signal to the power of the noise. In the present case, a random excitation is achieved by passing the deterministic forcing signal through an additive white Gaussian noise channel (*awgn*) which superimposes the noise of desired SNR onto the forcing signal. Correspondingly, the equations of motion of the ith cell of the array now become

$$\begin{aligned} &m_a\ddot{x}_{a,i} + \frac{m_a}{\tau}\dot{x}_{a,i} + k_{2a}x_{a,i} + k_4x_{a,i}^3 + k_I\left(2x_{a,i} - x_{b,i} - x_{b,i-1}\right) \\ &\quad = m_a\, awgn(\alpha, SNR) \\ &m_b\ddot{x}_{b,i} + \frac{m_b}{\tau}\dot{x}_{b,i} + k_{2b}x_{b,i} + k_4x_{b,i}^3 + k_I\left(2x_{b,i} - x_{a,i+1} - x_{a,i}\right) \\ &\quad = m_b\, awgn(\alpha, SNR) \end{aligned} \tag{2}$$

Numerical results that elucidate the influence of noise intensity on the formation and persistence of ILMs in the array are presented next. In Fig. 1, the results obtained for pure white noise forcing are presented while results for the cases of SNR = 175, and SNR = 15 are presented in Figs. 2 and 3, respectively. The time evolution of the

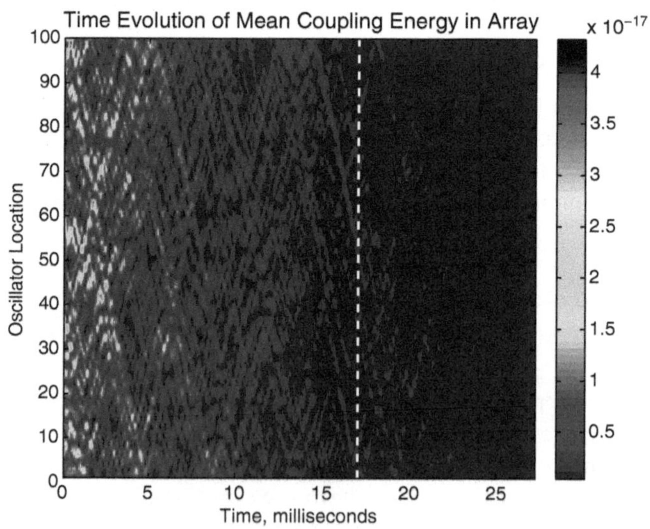

Fig. 1 Pure white noise forcing. No localization is observed

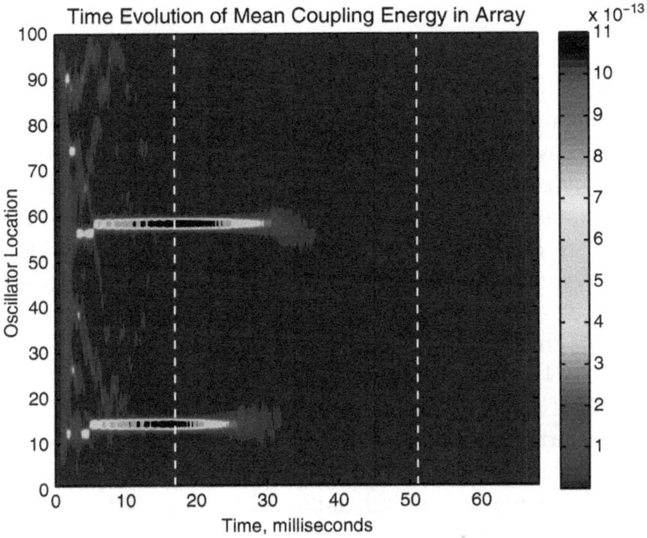

Fig. 2 Noise of intensity SNR = 175 superimposed on deterministic forcing which is harmonic in the window bracketed by the broken lines. Localization is observed

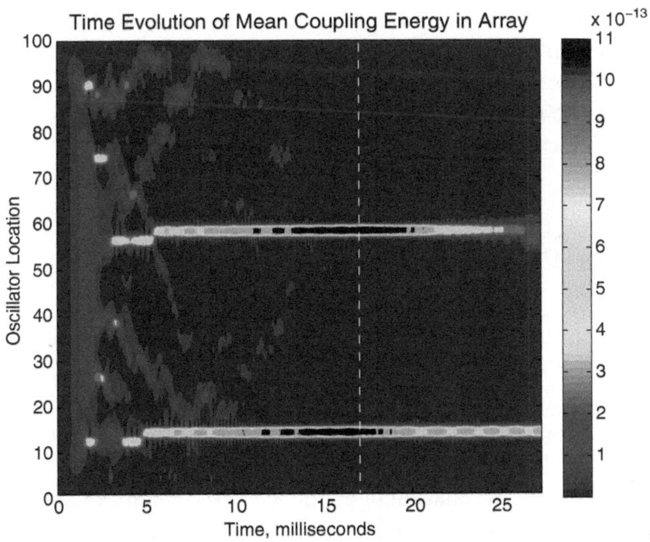

Fig. 3 Noise of intensity SNR = 15 superimposed on deterministic forcing

mean coupling energy is a measure of the degree of localization and this is shown for all oscillators in the array, in each of these figures. The color code represents mean coupling energy in units of Nm.

2.1 Discussion

The numerical results presented in Fig. 1 to Fig. 3 may be compared with the purely deterministic case considered by of Dick et al. [4] since the excitation profiles are identical in all the cases. Together, the results clearly indicate that passing the excitation signal through an additive white Gaussian noise channel has an appreciable effect on the formation and persistence of ILMs in the array. From Fig. 1, it is seen that white noise forcing by itself is unable to produce ILMs (a wide spectrum of noise intensities were used in the simulations). Next, the noise intensity was incrementally increased in the simulations and lower levels of noise intensity did not induce appreciable changes in the results. However, as seen in Fig. 2, a white noise of intensity SNR $= 175$ in the noise channel (when compared with the deterministic results of Dick et al. [4]) significantly reduces the time until which localization is sustained in the array at one location (oscillator 58) while there is no appreciable effect on the localization at the other location (oscillator 14). Interestingly, as the noise intensity is increased further, at the SNR value of 15, another threshold is reached. As seen from Fig. 3, at SNR $= 15$, the time for which the ILM is sustained at oscillator 58 decreases while the coupling energy at the other localization location begins to increase, even though these changes are of small magnitudes.

The above results indicate that, in the case of multiple localizations occurring in an array, noise could potentially be used to manipulate the spatial and temporal characteristics of an ILM. In addition, while the numerical values of threshold SNRs that induce significant changes are likely dependent on the particular characteristics of the considered array, the existence of the thresholds seems to indicate that white noise forcing may be advantageously employed to attenuate the persistence of ILMs in micro-cantilever arrays.

3 The Fokker-Planck Formalism

The authors refer to the vast literature on stochastic dynamical systems for the details of the approach briefly described here (see, for instance, Gardiner [5]). White noise (treated as the differential of a Weiner process) when introduced into the dynamical equations converts the latter into a stochastic differential system. Associated with the stochastic differential system is the Fokker-Planck equation satisfied by the probability density function that dictates the system dynamics. By solving for the probability density function, mean values of all dynamical variables may be evaluated for all times. However, Fokker-Planck equations for nonlinear systems seldom yield closed-form analytical solutions and approximate moment evolution equations (derived from the Fokker-Planck equation) can be solved to obtain the mean dynamics in such cases.

3.1 Stochastic Klein-Gordon Equations

Denoting the noise strength by σ, the white noise processes (the differential of the Weiner processes) by dW, the deterministic forcing function by f(t), the spring constants by $k_{2a}/m_a = K_{2a}$ and so on and defining a constant $c = 1/\tau$, the stochastically forced Klein-Gordon equations for cell i may be written in first order form as

$$
\begin{aligned}
dx_a &= y_a dt \\
dy_a &= \left(-cy_a - K_{2a}x_a - K_{4a}x_a^3 - K_{Ia}\left\{2x_a - x_b - x_{bi-1}\right\} - f(t)\right) dt + \sigma dW \\
dx_b &= y_b dt \\
dy_b &= \left(-cy_b - K_{2b}x_b - K_{4b}x_b^3 - K_{Ib}\left\{2x_b - x_{ai+1} - x_a\right\} - f(t)\right) dt + \sigma dW
\end{aligned} \tag{3}
$$

3.2 The Fokker-Planck Equation

The Fokker-Planck equation corresponding to Eq. (3) is derived as

$$
\begin{aligned}
\frac{\partial P}{\partial t} = &-\left[y_a \frac{\partial P}{\partial x_a} + y_b \frac{\partial P}{\partial x_b}\right] + \Big[cy_a + K_{2a}x_a + K_{4a}x_a^3 \\
&+ K_{Ia}\left(2x_a - x_b - x_{bi-1}\right) - f(t)\Big] \frac{\partial P}{\partial y_a} \\
&+ \left[cy_b + K_{2b}x_b + K_{4b}x_b^3 + K_{Ib}\left(2x_b - x_{ai+1} - x_a\right) - f(t)\right] \frac{\partial P}{\partial y_b} \\
&+ 2cP + \frac{\sigma^2}{2}\left[\left(\frac{\partial}{\partial y_a}\right)^2 + \left(\frac{\partial}{\partial y_b}\right)^2\right] P
\end{aligned} \tag{4}
$$

3.3 Moment Evolution Equations

Denoting the moments by using angled brackets, the moment evolution equations for the "a" cantilever at cell i may be derived from the Fokker-Planck Eq. (4) as

$$
\begin{aligned}
\frac{d}{dt}\left\langle x_a^n y_a^m \right\rangle = &\; n\left\langle x_a^{n-1} y_a^{m+1}\right\rangle - cm\left\langle x_a^n y_a^m\right\rangle - mK_{2a}\left\langle x_a^{n+1} y_a^{m-1}\right\rangle \\
&- mK_{4a}\left\langle x_a^{n+3} y_a^{m-1}\right\rangle - 2mK_{Ia}\left\langle x_a^{n+1} y_a^{m-1}\right\rangle + mK_{Ia}\left\langle x_a^n y_a^{m-1} x_b\right\rangle \\
&+ mK_{Ia}\left\langle x_a^n y_a^{m-1} x_{b_{i-1}}\right\rangle + mf(t)\left\langle x_a^n y_a^{m-1}\right\rangle + \frac{m(m-1)}{2}\sigma^2\left\langle x_a^n y_a^{m-2}\right\rangle
\end{aligned} \tag{5}
$$

By symmetry, a similar set of moment evolution equations can be obtained for the "b" cantilever as well. In order to obtain a closed system of moment evolution equations, the following assumptions are made: (i) moments of order three and higher can be neglected except for the third order moment of displacement, and (ii) moments of products involving variables in neighboring cells as well as different cantilever types are approximated as the product of moments of the respective variables. These approximations result in a system of eight moment evolution equations for each cantilever in the array that are numerically solved.

3.4 Numerical Solutions of Moment Evolution Equations

The following set of parameter values are assumed for the micro-cantilever array: $m_a = 5.46\text{E-}13\,\text{kg}$; $m_b = 4.96\text{E-}13\,\text{kg}$; $\tau = 8.75\text{E-}3\,\text{s}$; $k_{2a} = 0.303\,\text{N/m}$; $k_{2b} = 0.353\,\text{N/m}$; $k_{4a} = k_{4b} = 5\text{E}8\,\text{N/m}^3$; $k_I = 0.0241\,\text{N/m}$; $\alpha_0 = 10\text{E}4\,\text{m/s}^2$. The dynamics of a representative array of 50 cells (100 cantilevers) are simulated with random initial displacements of the order of magnitude 10E-7 m. The excitation profile, as in section 2, is identical to that used by Dick et al. [4]. The results are presented below with the color code representative of the mean coupling energy level in units of Nm.

3.5 Discussion

The numerical results obtained on the basis of the Fokker-Planck formalism provide interesting insights into the effects of noise on ILMs. It is noted here that localization in micro-cantilever arrays discussed in the literature so far arise from a direct simulations of the Klein-Gordon equations while the approach used here is from the viewpoint of averaged dynamics.

It is clear from Fig. 4 that noise of relatively low intensity ($\sigma = 10$) does not produce a localization. However, it is observed that the oscillator at the center of the array (oscillator 50 in this case) exists in states of higher coupling energy at later times (at 10 and 13 ms) relative to the other oscillators even when excited with a noise of low intensity.

The simulations were repeated for increasing levels of noise intensity. From the results presented in Fig. 5, it is observed that a noisy excitation of higher intensity ($\sigma = 250$) results in higher coupling energies for longer times for the oscillators around the center of the array. This effect is most pronounced for the oscillator 50 and its immediate neighbors at 17 ms. This suggests that once a localization occurs, an excitation with relatively high noise strength tends to sustain the event.

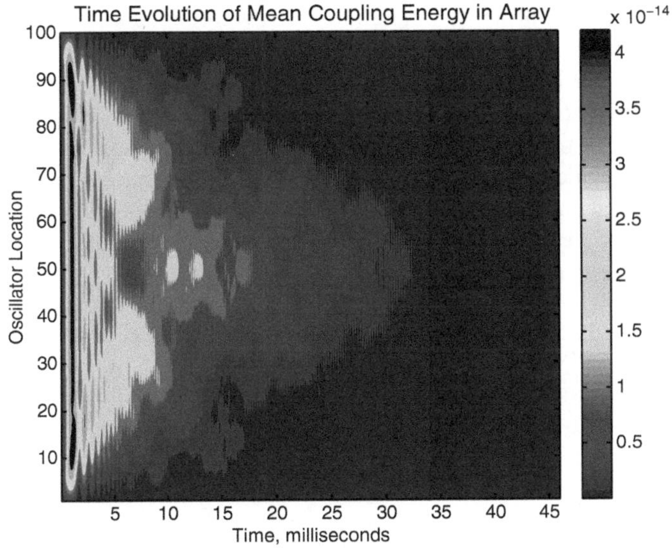

Fig. 4 Response for noise strength $\sigma = 10$ superimposed on harmonic excitation

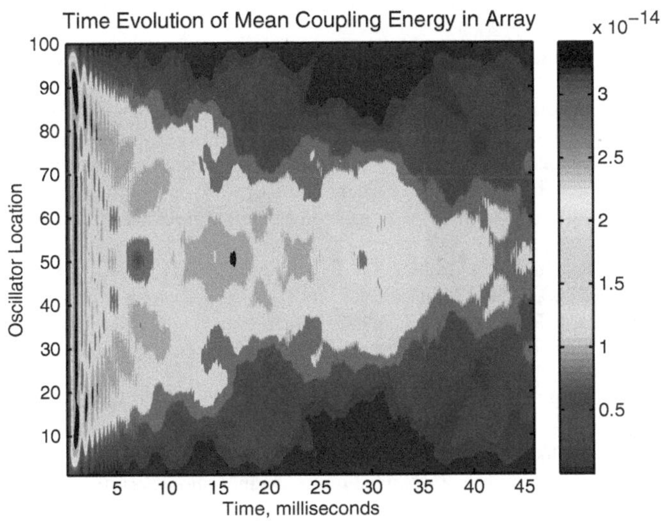

Fig. 5 Response for noise strength $\sigma = 250$ superimposed on harmonic excitation

4 Concluding Remarks

In this article, the authors addressed the effects of noise on nonlinear micro-scale oscillator arrays with an emphasis on the phenomenon of energy localization. An analytical framework based on the Fokker-Planck formalism was introduced and numerical simulations of approximate moment evolution equations indicated that, a white noise of higher intensity, in combination with a deterministic excitation, helps localize more energy into oscillators at the center of the array. In addition, localization in the array was also studied independently by direct numerical simulations. The results obtained by this method indicate that threshold levels of white noise strengths, while inhibiting the sustenance of ILMs in one location, tend to sustain localization at another location in the array. The implications of this for moving localizations spatially across the array as well as other aspects of the analytical approach are currently under investigation.

Acknowledgments Partial support received for this work through ARO Grant No. W911NF05100 76 and NSF Grant No. 0826173 are gratefully acknowledged.

References

1. Anderson, P.W.: Absence of diffusion in certain random lattices. Phys. Rev.**109,**1492-1505 (1958).
2. Balachandran, B., Li, H.: Nonlinear phenomena in micro-electromechanical resonators. In: Proc. Of the IUTAM Symposium on Chaotic Dynamics and Control of Systems and Processes in Mechanics, pp. 97-106 (2003).
3. Campbell, D.K., Flach, S., Kivshar, V.S.: Localizing energy through nonlinearity and discreteness. Phys. Today **57**, 43-49 (2004).
4. Dick, A.J., Balachandran, B., Mote, C.D.: Intrinsic localized modes in microresonator arrays and their relationship to nonlinear vibration modes. Nonlinear Dynamics (2008).
5. Gardiner, C.W.: Handbook of Stochastic Methods. Springer, Heidelberg (1983).
6. Lifshitz, R., Cross, M.C.: Response of parametrically driven nonlinear coupled oscillators with applications to micromechanical and nanomechanical resonator arrays. Phys. Rev. B **67**, 134302 (2003).
7. Nguyen, C.T.C., Katehi, L.P.B., Rebeiz, G.M.: Micromachined devices for wireless communications. Proc. IEEE **83**, 1756-1768 (1998).
8. Sato, M., Hubbard, B.E., Sievers, A.J., Ilic, B., Czaplewski, D.A., Craighead, H.G.: Observation of locked intrinsic localized vibrational modes in a micromechanical oscillator array. Phys. Rev. Lett. **90**, 044102 (2003).
9. Sato, M., Hubbard, B.E., English, L.Q., Sievers, A.J., Ilic, B., Czaplewski, D.A., Craighead, H.G.: Study of intrinsic localized vibrational modes in micromechanical oscillator arrays. Chaos **13**, 702-715 (2003).
10. Sievers, A.J., Takeno, S.: Intrinsic localized modes in anharmonic crystals. Phys. Rev. Lett. **61**, 970-973 (1988).
11. Vig, J.R, Kim, Y.: Noise in micromechanical system resonators. IEEE Trans. Ultrasonics, Ferroelectrics and Freq. Control. **46**, 1558-1565 (1999).

A Theoretical and Computational Framework for Modeling Diffusion-Driven Boundary Motion Without Remeshing

Kaushik Mysore and Ganesh Subbarayan

Abstract Diffusion-driven phase boundary motion influences both shape and topologies in many microstructural systems. In this paper, we discuss two aspects. First we present a theoretical framework, in which we permit in the models (i) bulk fields such as displacements and bulk mass density (ii) surface fields such as surface energy and surface mass density. The resulting equation for the free energy rate contains appropriate coupling between physical fields. We postulate a minimization principle and consequently obtain updated boundary locations. Second, we use the hierarchical partition of unity meshless compositional procedure as the computational framework for modeling evolving phases. Non-Uniform Rational B-Splines or NURBS are used for discretizing the geometrical model, distribution of material properties, and the accompanying behavioral field. This is a geometry-based alternative to the finite element method for modeling moving boundary problems. Within this framework, computational geometry inspired ideas such as parametric embedding and boolean compositions enable modeling of moving boundaries and changing topologies without remeshing. We demonstrate evolution towards optimal slit and circular shapes of single void systems. We also demonstrate dissolution, growth, coalescence, break-up and disappearance in multi-void systems. In all cases energetic trade-offs between surface and strain energies and their interactions with surface and bulk diffusions dominate the final shapes.

Keywords Diffusion · Finite element method · Interconnects · Meshless methods · Moving boundaries · Multi-physics phenomena · NURBS · Voids

1 Introduction

Copper line and via interconnect structures commonly used in microelectronic systems serve multiple roles such as electrical connection, mechanical integrity and thermal energy transport. These structures experience high current densities coupled

---

G. Subbarayan (✉) and K. Mysore
Purdue University, West Lafayette, IN 47907, USA
e-mail: skaushik@purdue.edu; ganeshs@email.address

B. Dattaguru et al. (eds.), *IUTAM Symposium on Multi-Functional Material Structures and Systems*, IUTAM Bookseries 19, DOI 10.1007/978-90-481-3771-8_34,

with thermomechanical stress during use environments and as a result fail due to formation and growth of voids. In this paper, motion of voids in stressed solids with direct applications to copper interconnect structures is modeled. In prior published work, motion of voids in solids has been studied through experiments [1, 2], and using analytical-and-computational models [3–8]. Mechanisms for such motions include diffusion on boundary surfaces, in bulk volumes of solids and also vapor transport inside pores [3]. In this study, we limit out discussion to diffusion mechanisms in solids. The stress- and electro-migration failures in microelectronic chips are examples of diffusion-driven failures induced by void formation and growth. The transport processes that create these voids are not completely understood. However, different aspects of the complex phenomena have been addressed in the past [4–7]. The above survey of literature is not intended to be comprehensive, but a listing of research relevant to the present paper. From a geometry modeling viewpoint, a challenge to modeling void evolution are the geometrical changes as an outcome of the applied stresses. Sharp interface descriptions of void boundaries are natural for modeling evolving voids. Constructing the evolving geometry is a Computer-Aided Design (CAD) step that may be represented procedurally as shown in Fig. 1. The CAD and FE procedures are arguably poorly integrated as they remain separated for most of the modeling phase. Meshless schemes on the other hand are alternatives that can alleviate problems of remeshing during iterative geometry changes. Such procedures enable better integration of the modeling process. Even closer integration is possible if the analysis procedures are aware of geometry construction so the effects of geometry changes can be localized to regions undergoing change. The above issues are discussed in detail in [9] where a unified, meshless representational framework based on hierarchical compositions of geometry, material, and behavioral fields is developed. Here, we adapt such a framework for solving the force, mass and energy balance equations for modeling void evolution. Details of both the theoretical and computational framework presented in this work can also be found in [8, 10]. Other examples of multi-functional structures where the procedures developed here can be employed are discussed in the summary section.

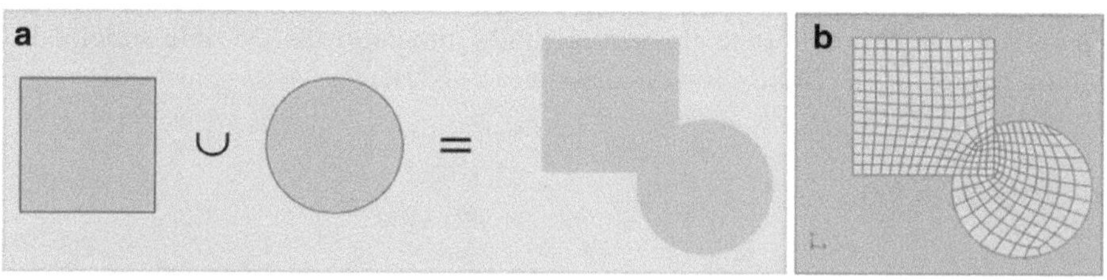

Fig. 1 Schematic shows the process of geometrical model and mesh construction: (**a**) shows a schematic for a Boolean union operation; (**b**) shows mesh on the final domain used for analysis. Three challenges commonly encountered are (i) representing complex geometrical shapes (ii) integrating design and analysis procedures (iii) reducing cost associated with remeshing for moving boundary problems

2 Theoretical Framework

A goal of this section is to present the final results of a mechanics framework for modeling evolution of voids in metal lines. Stresses, lattice and surface diffusion, and surface energy effects are all considered. Details of the derivation is not presented here in the interest of space, but is expected to be published later [10]. A schematic of the system under consideration is shown in Fig. 2 - such a system naturally describes modern interconnect structures. Divergence and Reynolds transport theorems are commonly used for stating balances of bulk fields. Similarly, surface divergence and transport theorems can be employed for expressing balance of surface fields [11]. We consider both densities in the bulk and on surfaces for mass balances. We also consider a scalar surface energy, and stress equilibrium assuming a quasi-static process. We make assumptions of linear elastic material behavior, ignore explicit time and spatial variations of surface energy and surface densities. We also consider coupling of different physical fields through energy balance, see Eq. (1).

$$\frac{d}{dt}\int_{\Omega} \rho u d\Omega = \int_{\Omega} \sigma : D d\Omega + \textit{other contributions} \tag{1}$$

where, ρ is the bulk mass density, and σ, D are the stress and rate of deformation tensors respectively. Employing suitable thermodynamic potentials $u = \psi - s\theta$ (where u, ψ are internal, helmholtz energies, and s, θ are entropy and temperature) and considering simplifying assumptions (as mentioned in the paragraph above) the final form of the energy balance (see Eq. (2)) is

$$\begin{aligned}\frac{d}{dt}\int_{\Omega} \rho\psi d\Omega &= \int_{\Gamma} \frac{1}{2}(\varepsilon : E : \varepsilon)(v \bullet n)\, d\Gamma - \int_{\Gamma} 2K\gamma_s (v \bullet n)\, d\Gamma \\ &+ \int_{\Gamma} J_s F_s d\Gamma + \int_{\Gamma} J_n F_n d\Gamma \end{aligned} \tag{2}$$

where, E is Young's modulus, ϵ is the strain tensor, K is the mean curvature, n is the unit normal, v is material velocity, γ_s is the scalar surface energy coefficient, J_s and

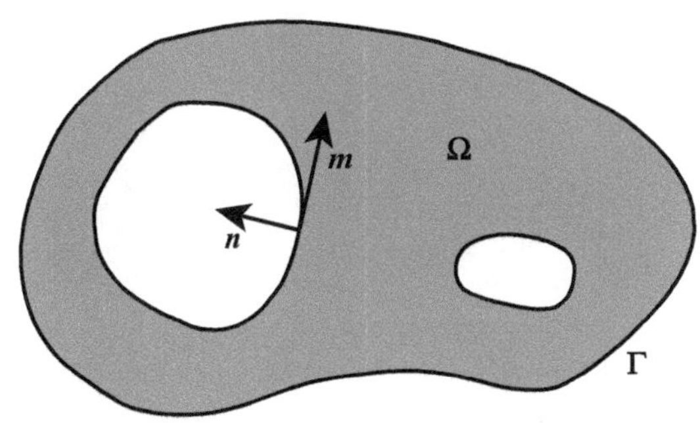

Fig. 2 A two dimensional schematic of a homogeneous body Ω with mutiple voids, and unit normals n, m along with the boundary Γ

J_n are surface and normal fluxes, F_s and F_n are their respective power conjugates. We further postulate that the system evolves to minimize the Helmholtz free energy. This postulate provides the necessary equations to solve for the unknown boundary locations.

3 Computational Framework

The compositional procedure is discussed briefly in this section. The procedure derives inspiration from Constructive Solid Geometry procedures in CAD, and from the partition of unity finite element method [9]. In partition of unity field compositions, each primitive is associated with a design state (representing both the approximation to the geometry and the distribution of the material property within the geometry) and a behavior state (for representing different physical fields such as displacements, temperatures, and electric potentials). The notion of constructing geometries through Boolean compositions is carried over to the analysis of behavioral fields as well by defining a scalar weight field on each primitive that is a measure of the contribution of the primitive to the overall behavior. Mathematically such a composition becomes valid if the weights are constructed in such a manner that they satisfy the three properties that are listed to the right on Fig. 3. NURBS basis functions are used for representing geometry (Fig. 4), material and behavior. NURBS

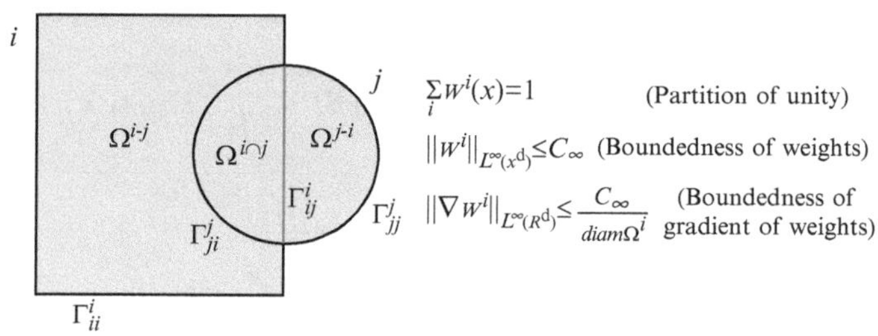

Fig. 3 Illustration of the composition of primitives (*left*) with domains $\Omega's$ and Boundaries (*both external and internal*) $\Gamma's$. A scalar weight w is defined on each primitive i, j. The equations to the right describe three properties of the scalar weights that enable partition of unity approximations

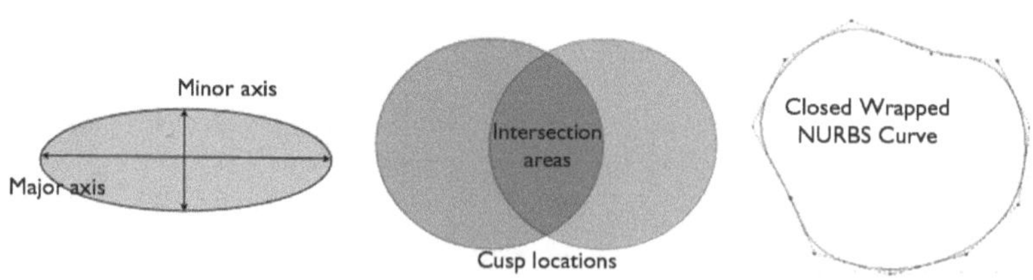

Fig. 4 Elliptical, circular and arbitrary free-form NURBS primitives

approximations can be constructed using the NURBS-basis functions in a manner analogous to finite element *interpolations* as shown below:

$$\tilde{f}(\xi,\eta,\zeta) = \frac{\sum_{i=0}^{l}\sum_{j=0}^{m}\sum_{k=0}^{n} N_{i,p}(\xi)\, N_{j,q}(\eta)\, N_{k,r}(\zeta)\, w_{ijk} f_{ijk}}{\sum_{i=0}^{l}\sum_{j=0}^{m}\sum_{k=0}^{n} N_{i,p}(\xi)\, N_{j,q}(\eta)\, N_{k,r}(\zeta)\, w_{ijk}} \quad 0 \le \xi,\eta,\zeta \le 1 \qquad (3)$$

where, f_{ijk} are the values of the control points, w_{ijk} are the weights, and the $N's$ are the basis functions. In a compositional procedure, the same basis functions are used for geometry, material and behavior providing desirable local properties. For the analysis of surface fields, in this work, the discussion is limited to elliptical families of geometries. For the case of multi-void systems the cusps and intersection areas require more care during quadrature (see [10] for details).

4 Results and Discussions

Energetic considerations determine the favorable void shapes for both single and multi-void systems. The specific shape, however, depends on the competition between strain energy densities and surface energies. It also depends on whether lattice or bulk diffusion mechanisms dominate. When the only active mass transport is on the surface, we find that a void prefers a circular shape if surface energy dominates while it collapses to a slit if strain energy dominates [5, 10]. Here, we discuss a limited set of results. First, we consider a case of combined surface and lattice diffusion (Fig. 5). To facilitate comparison between the two mechanisms, we define a parameter $f = D_l/D_s$ which is a ratio of diffusivities of lattice and surface diffusion. We further assume lattice diffusion to be shape conserving. In order to differentiate the two mechanisms, we consider void volumes. Surface diffusion is volume conserving, unlike lattice diffusion. Therefore a plot of void area (volume in three dimensions) will suffice to deduce the dominant diffusion mechanism. Figure 5a is a plot where the area remains almost a constant, and therefore surface diffusion has taken over as a dominant mechanism guiding an elliptical void towards a stable circular configuration. The plot shown in Fig. 5b describes the effect of a dominant lattice diffusion mechanism. In the beginning, both surface and lattice diffusion influence shape changes. However, as the voids evolve, the contributions of surface diffusion wane away. Lattice diffusion dominates and the void eventually disappears.

Topological changes have been observed in interconnect structures. The voids nucleate and sometimes also disappear. Such changes are dramatic at a microstructural scale, and require significant transport of atoms between regions within the domain. Surface diffusion alone cannot lead to connectivity changes in the domain

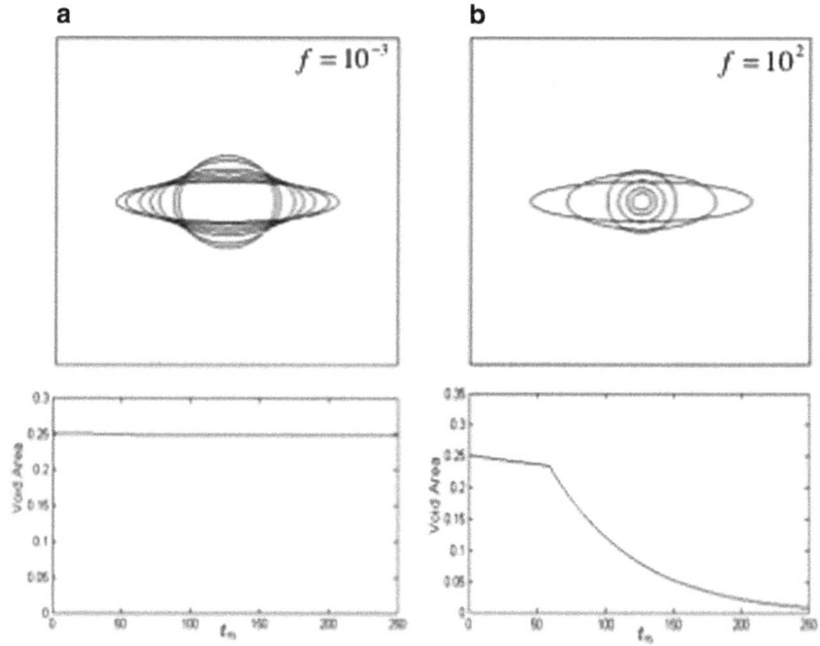

Fig. 5 Void areas and evolving primitive configurations for dominant surface diffusion (*left top and bottom*) and dominant lattice diffusion (*right top and bottom*). In both cases, the starting configuration was an ellipse

as they are volume conserving mechanisms. Consequently, diffusion in the bulk (also referred to as lattice diffusion) causes to bring about these changes. Therefore, a second set of results focuses on two void systems. The primary intent here is to demonstrate the ease of describing a variety of topological changes that are possible (due to the multiple forces and mechanisms) when using a partition of unity compositional procedure. The details are described elsewhere [10]. A parameter $\lambda = L_c/L$ which is a ratio of center-to-center distance L_c of voids to the cell size L. λ together with $\Lambda = \sigma^2 A/E\gamma_s$ describes a variety of topological changes (see Fig. 6).

5 Summary

The developed modeling procedure is an alternative to mesh-based finite element scheme and becomes attractive when geometry changes are frequent. Such situations occur in applications involving moving boundaries, and also in shape and topology optimization. Apart from interconnect structures, other examples include modeling shape memory alloys, piezoelectric, MEMS actuators, and woven composites to name a few.

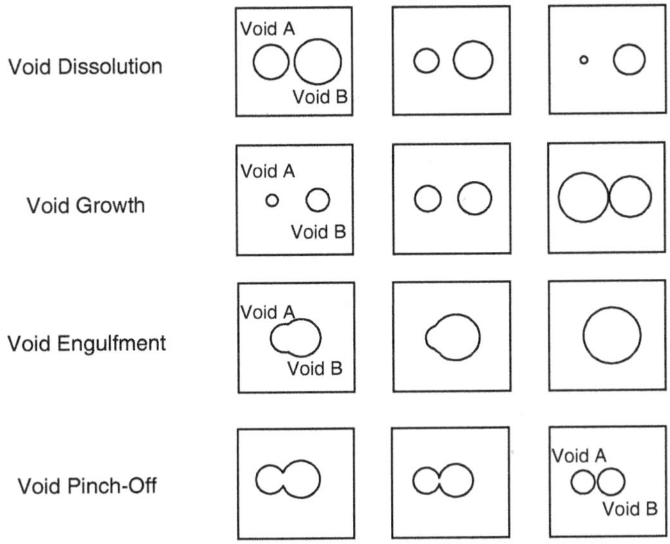

Fig. 6 Topological changes due to lattice and surface diffusion. Multi-void systems exhibit dissolution, growth, coalescence, and pinch-off owing to mass transport through the bulk and on surfaces. λ and Λ values for the above cases are 0.4, 0.0021 (void dissolution), 0.4, 3.199 (void growth), 0.18, 0.841 (void engulfment), 0.3, 0.000087 (void pinch-off)

Specifically, we discussed (a) the final results of mechanics framework for modeling boundary motion (b) bulk and surface effects that were separated and systematically accounted for (c) thermodynamic potentials and a energy minimization postulate to govern boundary evolution (d) compositional procedure for modeling moving boundary problems (e) NURBS basis functions as isoparametric (or isogeometric) approximations making mesh generation procedures unnecessary (f) combined effects of lattice and surface diffusion (g) a variety of topological changes; the compositional procedure describes topological changes consistent with the geometry construction procedures, this together with NURBS approximations makes the entire modeling procedure simpler compared to the mesh-based schemes.

References

1. W. Diez and J. Biersack. Experiments on the motion of bubbles in solids by sublimation in a temperature gradient. *Phys. Status Solidi*, 32(1), 1969.
2. JH Rose. Fatal electromigration voids in narrow aluminum-copper interconnect. *Applied Physics Letters*, 61:2170, 1992.
3. F. Nichols. Kinetics of diffusional motion of pores in solids (a review). *Journal of Nuclear Materials*, 30(1–2):143–165, 1969.
4. A. Needleman and JR Rice. Plastic creep flow effects in the diffusive cavitation of grain boundaries. *Acta Metall.*, 28(10):1315–1332, 1980.

5. Z. Suo. Motions of microscopic surfaces in materials. *Advances in Applied Mechanics*, 33: 193–294, 1997.
6. D.N. Bhate, A. Kumar, and A.F. Bower. Diffuse interface model for electromigration and stress voiding. *Journal of Applied Physics*, 87:1712, 2000.
7. A.F. Bower and S. Shankar. A finite element model of electromigration induced void nucleation, growth and evolution in interconnects. *Modeling and Simulation in Material Science and Engineering*, 15(8):923, 2007.
8. K. Mysore and G. Subbarayan. Arbitrary sharp-diffuse interface computational modeling of phase evolution in multi-component systems. *TMS Annual Symposium Invited Talk*, 2007.
9. M. Rayasam, V. Srinivasan, and G. Subbarayan. CAD inspired hierarchical partition of unity constructions for NURBS-based, meshless design, analysis and optimization. *Int. J. Numer. Meth. Engng*, 72:1452–1489, 2007.
10. K. Mysore. *Arbitrary Implicit-Explicit Interface Modeling for Evolving Boundary Problems with Applications to Stress and Electromigration*. PhD thesis, Purdue University, 2009. (expected).
11. J.C. Slattery. *Interfacial Transport Phenomena*. Springer, 2006.

Multiscale Simulation of Metal/Ceramic Interface Fracture

Amir Siddiq and Siegfred Schmauder

Abstract Interface failure in metal/ceramic composites plays an important role in modern materials technology, as evident by their use in a variety of applications. High-strength materials, such as metal-matrix composites consist of internal interfaces between ceramic (e.g. SiC or Al_2O_3) particles or filaments within a metallic host. In microelectronics packaging, interfaces between metallic (Cu and/or Al) interconnects and SiO_2, carbide/nitride (TiCN) or oxide (Al_2O_3) ceramics are commonplace, and impact the performance and longevity of solid state devices. Despite their widespread use, a basic understanding of these interfaces has been elusive. For example, given a particular metal/ceramic interface, it is not yet possible to accurately predict such fundamental properties as its fracture energy. In most of the cases, improvements in interface properties proceed via a costly and time consuming trial-and-error process in which numerous materials are evaluated until suitable performance is obtained. Computational methods provide a wide range of possibilities to study the fracture behaviour of such metal/ceramic interfaces.

In the first part of the presented work, the deformation behaviour of niobium single crystals has been simulated using crystal plasticity theory. An automatic identification procedure has been proposed to identify the crystal plasticity parameters for each family of slip systems and simulation results of the mechanical behaviour of single crystal niobium are compared with the experiment. Good agreement between the experimental and simulation results was found.

The second part presents effects of the different niobium single crystalline material orientations on crack initiation energies of the bicrystal niobium/sapphire four-point-bending-test specimens for a stationary crack tip. The trends of crack initiation energies are found to be similar to those observed during experiments.

In the third part, crack propagation analyses of niobium/alumina bicrystal interface fracture have been performed using a cohesive modelling approach for three different orientations of single crystalline niobium. Parametric studies have been performed to study the effect of different cohesive law parameters, such as work

---

A. Siddiq and S. Schmauder (✉)
Institut für Materialprüfung, Werkstoffkunde und Festigkeitslehre (IMWF), Pfaffenwaldring 32
70569 Stuttgart, Germany
e-mail: amir.siddiq@gmail.com; siegfried.schmauder@mpa.uni-stuttgart.de

B. Dattaguru et al. (eds.), *IUTAM Symposium on Multi-Functional Material Structures and Systems*, IUTAM Bookseries 19, DOI 10.1007/978-90-481-3771-8_35,

of adhesion and cohesive strength, where work of adhesion is the area under the cohesive law curve while cohesive strength is the peak stress value of the cohesive law. The results show that cohesive strength has a stronger effect on the macroscopic fracture energy as compared to work of adhesion. Cohesive model parameters are identified for different combinations of cohesive strength and work of adhesion by applying a scale bridging procedure.

In the last part, a correlation among the macroscopic fracture energy, cohesive strength, work of adhesion and yield stress of niobium single crystalline material will be derived.

Keywords Crystal plasticity · Cohensive model · Fracture mechanics · Metal/ceramic interface · Multiscale modelling · Work of adhesion

1 Introduction

The combination of metal and ceramic is well known and has many applications. For instance, alumina has been used in electronic industry for many years as a ceramic insulator, whereas niobium, aluminium and copper are three of the best electrodes, having good conductivity. Powders and fibres of alumina are also used to reinforce aluminium alloys, one of the most common lightweight materials for components in automobile engines and other transportation systems.

The safety of such components inevitably requires a thorough understanding of their behaviour under load. Therefore, an effort has been put to devise a procedure which can simulate realistic fracture behaviour of niobium/alumina interface. Effects of crystal orientations of niobium single crystalline material on the fracture energy have been studied using crystal plasticity theory. The ultimate goal of the investigation is to theoretically clearing-up the connection between local adhesion capacity and macroscopic fracture energy.

The mechanical properties of single crystalline sapphire/niobium/sapphire joints were investigated in [1, 2] for different orientation relationships between metal and ceramic. The orientation of the single crystalline materials was found to have significant effect on the fracture energies of the Nb/Al_2O_3 interface. The influence of the chemical composition of regions at or close to the interface on the strength of a metal/ceramic interface was studied in [3] It was found that the strength of a metal/ceramic interface depends upon the density of the metal-oxide bonds.

Gibbesch et al. [3] found that the fracture energy of niobium/sapphire and copper/sapphire interfaces change dramatically with the relative crystallographic orientation of the two constituents. Korn et al. [4] observed that the amount of plasticity initiated in the metal during fracture of such interfaces is strongly influenced by both, the interface chemistry and the orientations of slip systems relative to the crack surface. Kysar et al. [5, 6] studied the copper/sapphire interface and the directional dependence of the crack for various orientations.

Continuum mechanical studies have been done by many researchers to study and understand the fracture behaviour of metal/ceramic interfaces.

Kohnle [7, 8] analyzed polycrystalline Niobium/Sapphire interfaces. The influence of the plastic properties of the metal part on the interface strength and on the energy release rate is examined. It was shown that low yield stress results in high plastic deformation in the metal part and consequently higher energy is stored before the critical stress value for fracture is reached.

In reality, engineers are interested in validating the analysis results with experiments which has not been considered in the majority of these works as how these theories can be used to quantitatively validate the experimental results, such as, experimentally computed fracture energies, etc.

Therefore, an effort is required to devise a procedure based on multiscale simulation which can simulate the fracture behaviour of bcc metal/ceramic interfaces and quantitatively validate it with experiments.

The aim of the present work is to study the interface fracture behaviour of niobium/alumina bicrystal specimens using crystal plasticity theory [9, 10] in conjunction with a cohesive modelling approach [11]. Parameter studies have been done to study the effect of different parameters of the cohesive model [11], such as interface strength and cohesive energy.

An insight into the role of cohesive strength and work of adhesion in macroscopic fracture is also presented which can be used by experimentalists to design better bimaterials by varying cohesive strength and work of adhesion.

2 Material Behaviour at Sub Micron Level

Crystal plasticity theory [9, 10] has been used to simulate the deformation behaviour of niobium single crystalline material. Crystal plasticity theory is based on the assumption that plastic deformation is the sum of the crystalline slip in all activated slip systems. The theory is based on the fact [9, 10] that plastic slip occurs when resolved shear stress onto a crystallographic plane in the direction of slip reaches a critical value. The precise theory was formulated by Hill and Rice [9, 10].

Through out this work Bassani and Wu [12] model has been used to model three stage hardening of single crystalline materials during deformation.

Bassani and Wu [12] is based upon the analytical characterization of the hardening moduli at any stage during deformation. Their expression for self and latent hardening depends on the shear strains γ^α of all slip systems

$$h_{\alpha\alpha} = \left\{(h_o - h_s)\sec h^2\left[\frac{(h_o-h_s)\gamma^\alpha}{\tau_s-\tau_o}\right] + h_s\right\} G(\gamma^\beta; \beta \neq \alpha)$$

$$h_{\alpha\beta} = q h_\alpha \alpha \quad (\alpha \neq \beta)$$

$$G(\gamma^\beta; \beta \neq \alpha) = 1 + \sum_{\beta\neq\alpha} f_{\alpha\beta} \tanh(\gamma^\beta/\gamma_o)$$

where, h_0 is the initial hardening modulus, τ_0 the initial yield stress, τ_s the saturation stress, g^α the total shear strain in system αh_s the hardening modulus during stage

I deformation, $f_{\alpha\beta}$ the interaction strength between slip system α and β and γ^{β} the total shear strain in slip system $\tilde{\beta}$.

The hardening moduli are described with an initial hardening (h_0) which saturates after reaching resolved shear stress (τ_s). After the diminishing of the hyperbolic secant term, the saturation hardening (h_s) term specifies that each slip system has a finite hardening rate. The function G deals implicitly with cross-hardening that occurs between slip systems during stage II and III.

The above mentioned hardening law by Bassani and Wu [12], respectively has been implemented numerically within the UMAT written by Huang [13]. This UMAT subroutine has been used with ABAQUS for the present work.

The crystal plasticity model has been implemented for finite deformation and rotation. Jaumann Stress rate is used in the formulation, which is the corotational stress rate on axes that rotates with the crystal lattice.

The orientation of the slip systems are updated during every call to the UMAT and stresses are updated using the updated orientation.

3 The Cohesive Model

Cohesive models have found increasing interest to simulate fracture in metallic, polymeric, and ceramic materials and their composites. Using cohesive models, the fracture can be simulated for any structure with or without an initial crack. A versatile cohesive law is proposed by Scheider [11], in which the initial stiffness of the cohesive element can be varied. A region can be defined, where the traction in the cohesive element is kept constant. This has been achieved by using two additional parameters δ_1 and δ_2 (Fig. 1), leading to the following formulation for the function T(δ):

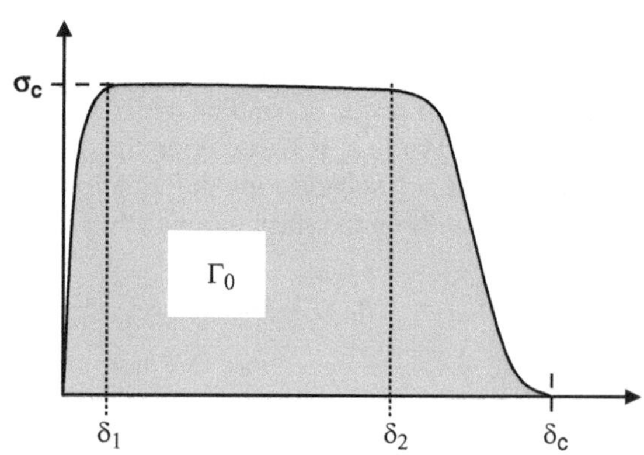

Fig. 1 Cohesive laws proposed by Scheider [87]

$$\sigma = \sigma_c \begin{cases} 2\left(\dfrac{\delta}{\delta_1}\right) - \left(\dfrac{\delta}{\delta_1}\right)^3 & \delta \leq \delta_1 \\ 1 & \delta_1 \leq \delta \leq \delta_2 \\ 2\left(\dfrac{\delta - \delta_2}{\delta_c - \delta_2}\right)^3 - 3\left(\dfrac{\delta - \delta_2}{\delta_c - \delta_2}\right)^2 + 1 & \delta_2 \leq \delta \leq \delta_c \end{cases}$$

In general all the cohesive models consist of a traction-separation law. The traction-separation relation for interfaces is such that with increasing interfacial separation, the traction across the interface reaches a maximum, then decreases and eventually vanishes, permitting a complete decohesion. There is common belief that cohesive models can be described by two independent parameters. These parameters may be two of the three parameters, namely the cohesive energy, and either of the cohesive strength or the separation length.

4 Finite Element Analysis

The finite element model is based on the experiments performed in [4], in which the influence of orientation and impurities on the fracture behaviour of Nb-sapphire interfaces were studied using notched bending tests. Single crystals were diffusion bonded in ultra high vacuum furnace for different interface orientations. For undoped bicrystals bonded at $1,400°\text{C}$, the computed interfacial fracture energy ranged from 77 to $2,100\,\text{J/m}^2$ depending on the interface planes of Nb and sapphire. For fracture evaluation, interfacially notched bending test specimens of dimensions $2 \times 4 \times 32\,\text{mm}^3$ were prepared. The specimen is shown in Fig. 2. The notch length is 0.4 mm. The specimen is loaded to the fracture load F_c in a four-point bending tests device at a crosshead speed of $96.8\,\mu\text{m/min}$. The load F and the crosshead deflection at the load points v are simultaneously recorded.

Finite element simulations of a stationary crack tip were done using a three-dimensional model; the finite element mesh consisted of 20-noded quadratic reduced integration elements. The total number of elements in the model is 17,208. Also, crack propagation analyses were performed using two dimensional plane strain elements.

For all simulations, both outer Alumina shanks (ceramic) and Alumina single crystal at the middle of the specimen were treated as purely elastic with a Young's

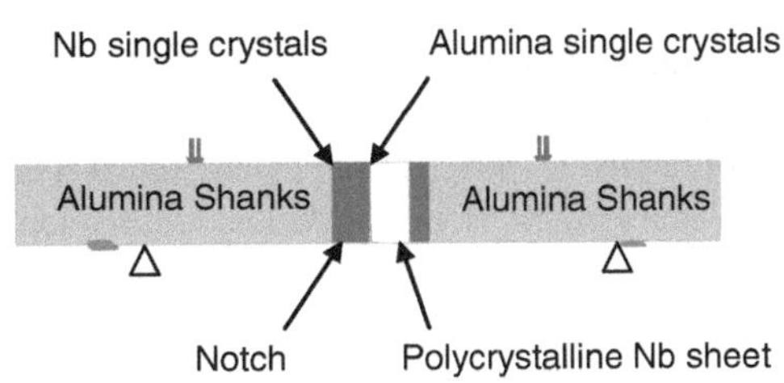

Fig. 2 Four point bending test specimen

Table 1 Hardening parameters for Bassani and Wu hardening law

	τ_o (MPa)	τ_s (MPa)	h_o (MPa)	h_s (MPa)	γ_o	γ_{oI}	$f_{\alpha\beta}$	$f_{\alpha\beta I}$	q	q_I
(110)[111]	13.7	16.4	292.262	1.4	0.25	0.25	10.0	9.9	0.0	0.0
(112)[111]	13.07	16.344	49.03325	39.2266	0.1	0.1	0.34	0.3	0.0	0.0

modulus of 390 GPa and a Poisson's ratio of 0.27. The polycrystalline niobium sheet is always modelled with an elastic-plastic constitutive law. The Young's modulus and the Poison's ratio were the same for all simulations ($E = 104.9$ GPa, $\nu = 0.397$). These elastic data are adjusted to alumina and niobium, respectively [7, 8]. The plastic behaviour of the stress-strain curve of the polycrystalline niobium sheet is approximated by a Ramberg-Osgood relation [14], which is described in the one-dimensional case by the following equation:

$$E_c = \sigma + \alpha \left(\frac{\sigma}{\sigma_0} \right)^{n-1} \sigma$$

Here, n denotes the hardening exponent, α the yield offset and σ_0 the yield stress. This material law is nonlinear from the beginning, but for commonly used hardening exponents ($n \geq 5$) the divergence from linearity is only slight for stresses below σ_0. The chosen plasticity theory is the deformation plasticity theory (for details see [20] and references therein), which describes not a plastic material behaviour, but a nonlinear elastic material. This means, that no unloading criterion exists. The parameters of the above equation are adjusted to the niobium stress-strain curves in [15]. The parameters used are $n = 6$, $\sigma_0 = 180$ MPa and $\alpha = 0.3$.

For the case of crystal plasticity the single crystal niobium is modelled using the hardening law of Bassani and Wu [12]. The hardening parameters for each slip system are derived in [15] and are given below (Table 1):

A four point bending test has been simulated with the boundary conditions as shown in Fig. 2. The displacement of 30 μm is applied stepwise (increasing linearly with time) at the loading points.

5 Results and Discussion

5.1 Stationary crack

The energy release rates for three different orientations denoted as Orientation I (Nb(100)[001]||Sp(11 − 20)[0001]), II (Nb(110)[001]||Sp(11 − 20)[0001]) and III (Nb(111)[−1 − 1 − 2]||Sp(11 − 20)[0001]) have been plotted against the strain at the stationary crack tip in Fig. 3. The results depict that the change in crystal orientation influences the energy release rates only after the plastic deformation has started. The start of the plastic deformation is the point where the resolved shear stress of any slip system exceeds the initial yield stress.

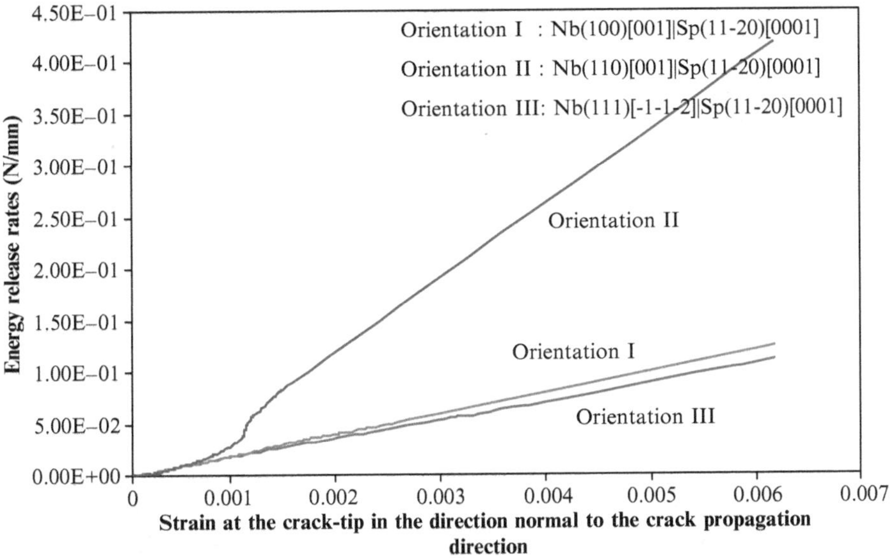

Fig. 3 Comparison of energy release rates for three orientations I, II and III

It was found during the four point bending experiments [4] that the energy release rate of orientation II (370 J/m^2) was always higher than that of the other two orientations I (115 J/m^2) and III (112 J/m^2). Figure 3 shows the similar trends with highest energy release rate for orientation II. While orientation I and III having energy release rate values very close with orientation I having slightly higher values. Based on crystal plasticity the stress-strain curves of the uniaxial tension test simulations for the three orientations I, II and III are plotted in Fig. 4 which shows that the orientation of niobium single crystals changes the stress-strain behaviour and it can be observed that for a specific strain value, higher stresses are induced for the case of orientation II than orientation I and III. These higher stresses cause the higher total energy of the niobium single crystals which is given by the area under the stress-strain curve. This ultimately ends up with a higher J-integral value as shown in Fig. 3. It was also observed during the simulation [Cames] that orientation I and III always higher plastic strain than orientation II for the same loading conditions.

5.2 *Crack Propagation Using Cohesive Modelling Approach*

In what follows next, we have performed crack growth analyses using different sets of values of cohesive law parameters, such as, cohesive strength and work of adhesion. We firstly study the effect of cohesive strength on the J-resistance curves of niobium (110)[001]/alumina (11-20)[0001] bicrystal specimen. Later we also compare the results for the case when the cohesive strength is kept constant while the

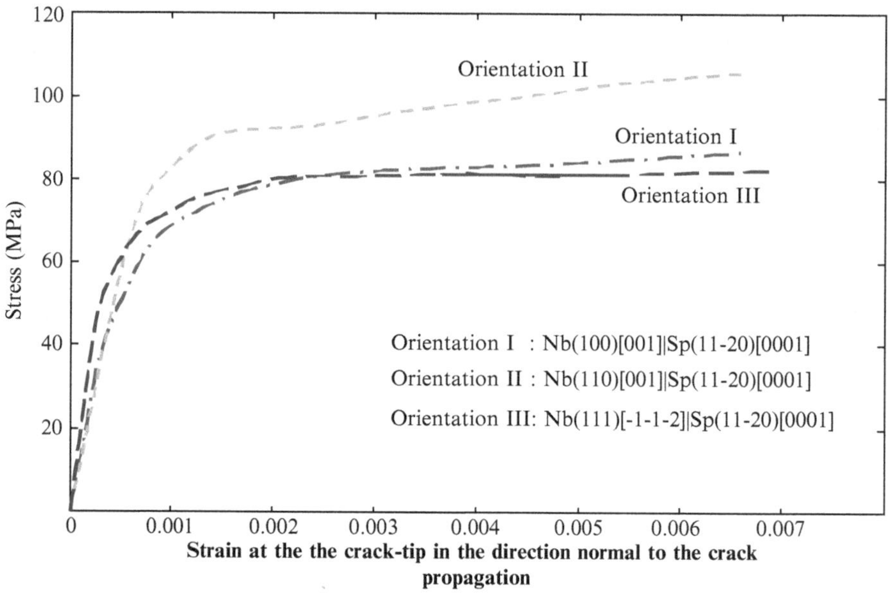

Fig. 4 Stress-strain curves for orientations I, II and III of uniaxial simulation

cohesive energy is varied. The ultimate goal of this parametric study is to find a correlation between the local adhesion capacity and the macroscopic fracture energy.

Three different values of cohesive energy (work of adhesion) reported in [16–20] have been used, i.e., 1, 4, 9.8 J/m^2.

The effect of cohesive strength has been studied by keeping the work of adhesion constant. The cohesive law curves for various values of cohesive strength at constant work of adhesion are plotted in Fig. 5. The J-resistance curves for the work of adhesion of 4 J/m^2 are shown in Fig. 6. The results clearly show that by increasing the cohesive strength value while keeping the cohesive energy constant, the fracture resistance increases, this is due to the higher plastic dissipation around the crack tip. It is also found that the amount of crack length to reach the steady state crack growth increases with increasing cohesive strength. This behaviour can also be explained in terms of the stress level or traction ahead of the crack tip, i.e., for higher cohesive strength values, the traction required to advance the crack is high, and hence a larger applied load is required to attain higher traction. On the other hand, a lower cohesive strength corresponds to a smaller applied load required to reach lower traction ahead of the crack tip.

The plastic dissipation rate as a function of the normalized crack length for various cohesive strength values is plotted in Fig. 7. The plastic dissipation rate increases as the cohesive strength increases. It is also shown in Fig. 7 that the plastic dissipation rate is initially high and decreases as the steady state crack growth is reached. This decrease in plastic dissipation increases as the cohesive strength increases. This is due to the fact that during the initial stage of crack growth when the crack tip is

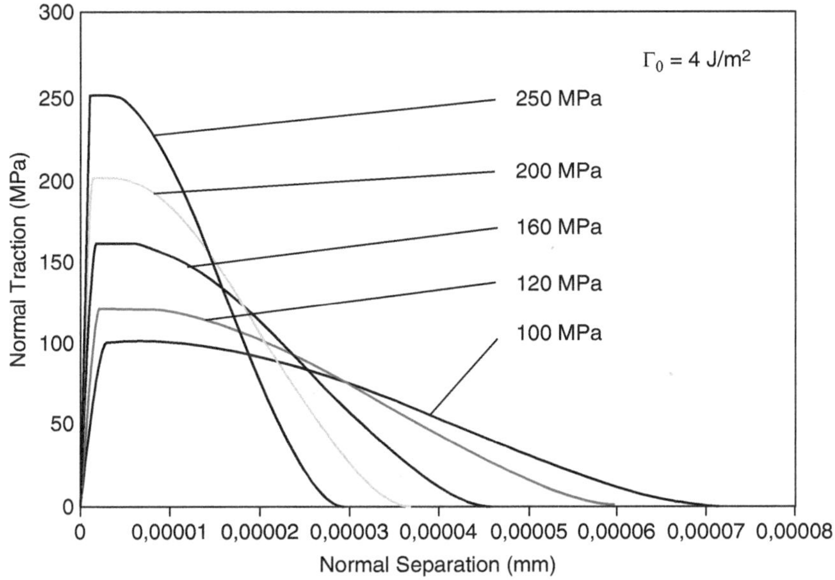

Fig. 5 Cohesive law curve of various values of cohesive strength for constant work of adhesion ($\Gamma_0 = 4\,\mathrm{J/m^2}$)

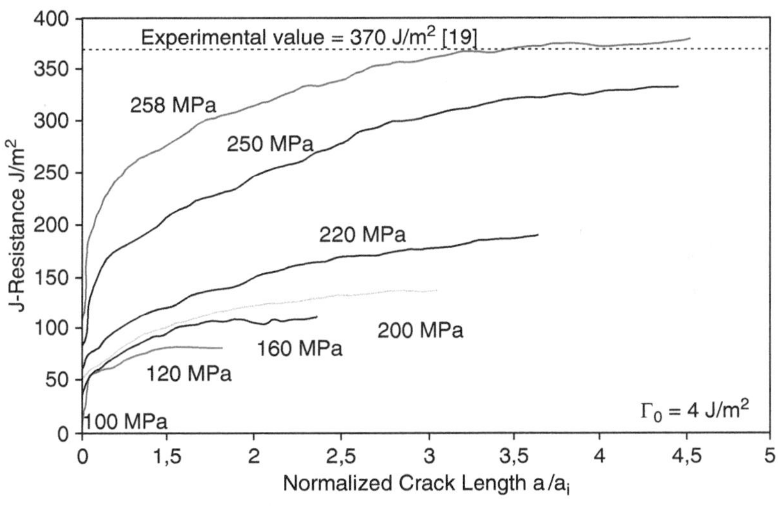

Fig. 6 J-resistance curves for different values of cohesive strength ($\Gamma_0 = 4\,\mathrm{J/m^2}$)

sharp, a higher amount of plastic dissipation is present which decreases eventually as the crack tip starts to blunt. Also, in terms of stress level, the higher the cohesive strength the higher is the amount of stress singularity at the crack tip during the initial stages causing higher plastic dissipation rates during the initial stages of the crack growth.

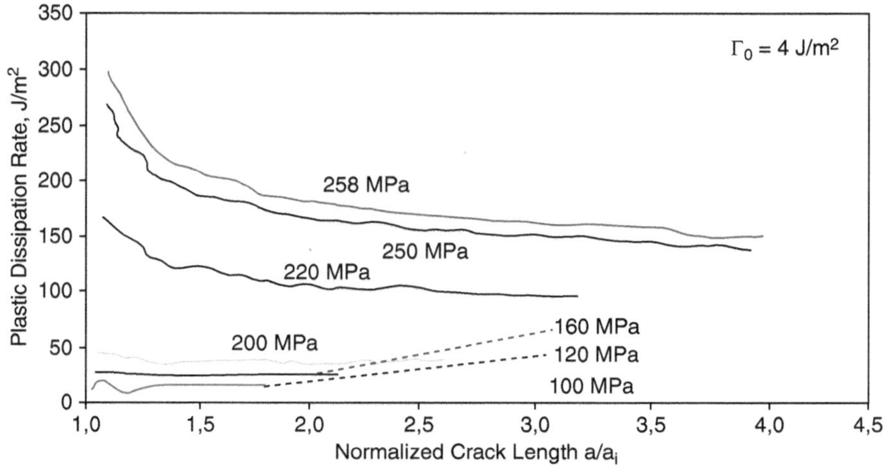

Fig. 7 Plastic dissipation rate vs normalized crack length ($\Gamma_0 = 4\,\mathrm{J/m^2}$)

Table 2 Cohesive model parameters for Nb(110)[001]||Al_2O_3 (11-20)[0001]

Work of adhesion ($\mathrm{J/m^2}$)	Cohesive strength (MPa)	Steady state fracture energy ($\mathrm{J/m^2}$)
1	283	370.6
4	258	371
9.8	225	372

As shown in Fig. 6, the experimental fracture energy value of $370\,\mathrm{J/m^2}$ is reached when the cohesive strength was assumed to be 258 MPa for the work of adhesion value of $4\,\mathrm{J/m^2}$.

Similar studies are done for the work of adhesion value of 1 and 9.8 $\mathrm{J/m^2}$. The cohesive law parameters identified for the three different values of work of adhesion are given in Table 2.

Fracture energies as a function of cohesive strength have been plotted in Fig. 8 for three different work of adhesion values at steady state crack growth. These curves show that as the cohesive strength increases the fracture energy at the steady state crack growth also increases. This increase is gradual till the cohesive strength value is approx. 220 MPa. As soon as this value is reached the slope of the fracture energy curve increases this is due to the high plastic dissipation. In a similar fashion the fracture energy is plotted as a function of the work of adhesion for different values of cohesive strength at steady state crack growth, as shown in Fig. 9. It shows that a linear relation exists between fracture energy and work of adhesion.

Similar studies have been done for the other two orientations, i.e., when niobium (100)[001] is bonded with alumina (11-20)[0001] and niobium (111)[-1-12] is bonded with alumina (1-20)[0001].

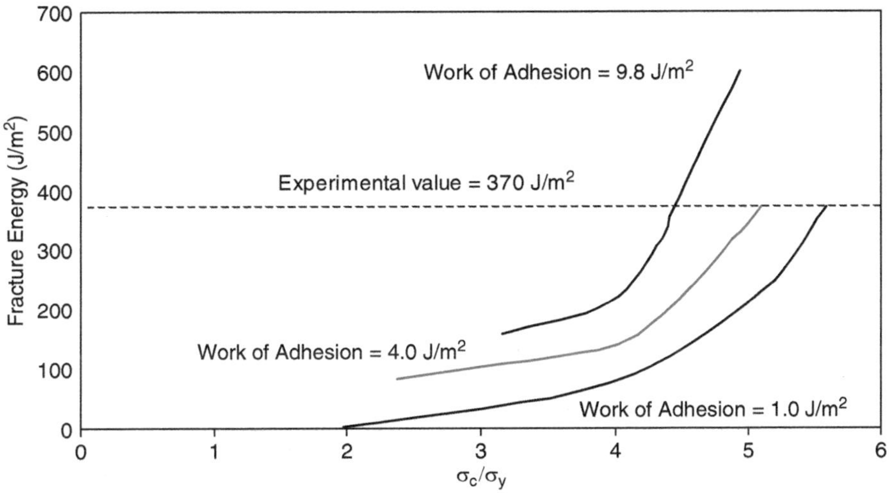

Fig. 8 Fracture energy vs cohesive strength (σ_c) to yield stress (σ_y) ratio at steady state crack growth for different work of adhesion, Γ_0 (Nb(110)[001]||Al_2O_3(11-20)[0001])

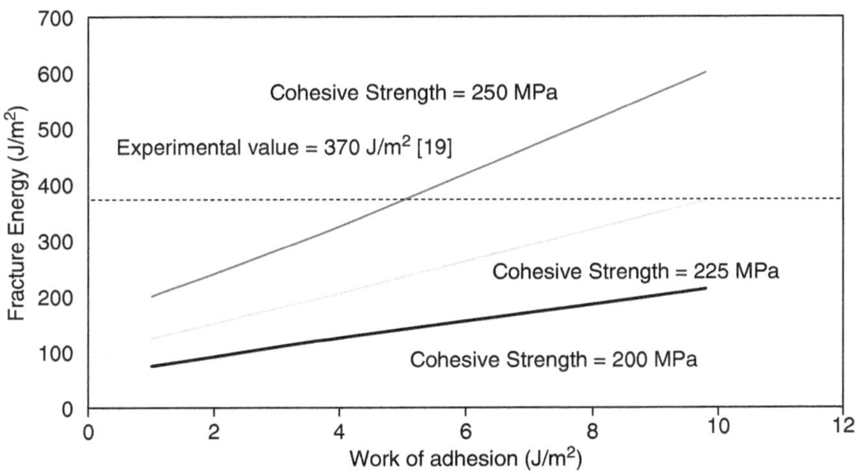

Fig. 9 Fracture energy vs work of adhesion at steady state crack growth (Nb(110)[001]||Al_2O_3(11-20)[0001])

5.3 Correlation between Local Adhesion Capacity and Macroscopic Fracture Energy

In previous section, the influence of different cohesive model parameters, such as, cohesive strength and work of adhesion on the fracture energy was presented. This section is devoted to find a correlation between the local adhesion capacity and macroscopic fracture energy for two different orientations of niobium single crystals in niobium/alumina bicrystal specimen. Results plotted in the previous section show that a cubic relation exists between cohesive strength (σ_c) to yield stress (σ_y) and fracture energy. It is also found that a linear relation exists between work of adhesion (W_{adh}) and fracture energy (J_c).

Based on the above conclusions following relation has been used to describe the correlation between the local adhesion capacity and macroscopic fracture energy.

$$J_c = \left(2.98\left(\alpha_i \cdot \frac{\sigma_c}{\sigma_y}\right)^3 - 34.8\left(\alpha_i \cdot \frac{\sigma_c}{\sigma_y}\right)^2 + 135.24\left(\alpha_i \cdot \frac{\sigma_c}{\sigma_y}\right) - 164.89\right) W_{adh}$$

Where α_I is the orientation parameter, its value is found to be 1.0 for orientation I, 1.35 for orientation II and 0.853 for orientation III.

The relations found above can be used to predict the fracture energy of niobium/alumina systems, if the rest of the material parameters are available. The relation not only takes into account the dependence of the fracture energy on cohesive model parameters, such as, cohesive strength and work of adhesion, but, also material properties, such as, the yield stress which is strongly dependent on the crystal orientation [21].

6 Conclusion

Interface fracture analyses of niobium/alumina bicrystal specimen are presented in this work using a multiscale approach. It is found that for the case of stationary crack tip, fracture energy of the bicrystal specimen changes for different orientations of single crystalline niobium. The results obtained are in conjunction with the experimental results.

Interface crack growth analyses of bicrystal niobium/alumina specimen are also presented in the presented work using cohesive modelling approach. It was found that cohesive strength has a strong influence on the fracture energy of the bicystal niobium/alumina interface for different niobium orientations. It was also found that as the cohesive strength increases the fracture energy also increases this is due to the higher plastic dissipation. The relation between the cohesive strength and the fracture energy is found to be non-linear cubic.

It was also shown that work of adhesion also has a influence on the fracture energy, i.e., as the work of adhesion increases the fracture energy also increases. The relation for various orientations of niobium single crystalline material in the niobium/alumina bicrystal specimen is found to be almost linear.

In the last part of this section a correlation was derived among the fracture energy, cohesive strength, work of adhesion and yield stress. And parameters for these correlations were identified for different orientations of niobium single crystalline material in the bicrystal niobium/alumina specimen.

Acknowledgements The presented work is funded by the Deutsche Forschungsgemeinschaft within the Graduiertenkolleg "Internal interfaces in Crystalline Materials" and within the project SFB 716-B2. The financial support is gratefully acknowledged.

References

1. G. Soyez, "Plastische Verformung und Rißbildung in druckbeanspruchten Niob/a-Al_2O_3-Verbunden", PhD Dissertation, University of Stuttgart, Stuttgart, 1996.
2. G. Soyez, G. Elssner, M. Ruhle and R. Raj, *Constrained yielding in niobium single crystals bonded to sapphire*, Acta Mater. **46** (1998), 3571–3581.
3. H. F. Fischmeister, W. Mader, B. Gibbesch and G. Elssner, *Joining of ceramic, glass and metal*, Mat. Res. Soc. Symp. Proc. **122** (1988), 529–540.
4. D. Korn, G. Elssner, R. M. Cannon and M. Rühle, *Fracture properties of interfacially doped Nb-Al_2O_3bicrystals: I, fracture characteristics*, Acta Mater. **50** (2002), 3881–3901.
5. J. W. Kysar, *Continuum simulations of directional dependence of crack growth along a copper/sapphire bicrystal interface: Part i, experiments and crystal plasticity background*, J. Mech. Phys. Solids **49** (2001), 1099–1128.
6. J. W. Kysar, *Continuum simulations of directional dependence of crack growth along a copper/sapphire bicrystal interface: Part ii, crack tip stress and deformation analysis*, J. Mech. Phys. Solids **49** (2001), 1129–1153.
7. C. Kohnle, O. Mintchev, D. Brunner and S. Schmauder, *Fracture of metal/ceramic interfaces*, Comp. Mater. Sci. **19** (2000), 261–266.
8. C. Kohnle, O. Mintchev and S. Schmauder, *Elastic and plastic fracture energies of metal/ceramic joints*, Comp. Mater. Sci. **25** (2002), 272–276.
9. R. J. Asaro, *Crystal plasticity*, J. Appl. Mech. **50** (1983), 921–934.
10. R. Hill and J. R. Rice, *Constitutive analysis of elastic-plastic crystals at arbitrary strain*, J. Mech. Phys. Solids **20** (1972), 401–403.
11. I. Scheider and W. Brocks, *Simulation of cup-cone fracture using the cohesive model*, Engng. Fract. Mech. **70** (2003), 1943–1961.
12. T. Y. Wu and J. L. Bassani, *Latent hardening in single crystals I. Theory and experiments*, Mathematical and Physical Sciences **435** (1993), 21–41.
13. Y. Huang, *A user material subroutine incorporating single crystal plasticity in the Abaqus finite element program*, 1991.
14. W. Ramberg and W. R. Osgood, "Description of stress-strain curves by three parameters, " *NASA Technical Note No 902*, 1945.
15. A. Siddiq and S. Schmauder, *Simulation of hardening in high purity niobium single crystals during deformation*, Steel Grips, Journal of steel and related materials **3** (2005), 281–286.
16. I. G. Baitrev, A. Alavi and M. W. Finnis, *First-principle calculations of the ideal cleavage energy of bulk niobium(111)/alpha-alumina(0001) interfaces*, Phys. Rev. Lett. **82** (1999), 1510–1513.
17. R. M. Cannon, D. Korn, G. Elssner and M. Rühle, *Fracture properties of interfacially doped Nb-Al_2O_3bicrystals: II, relation of interfacial bonding, chemistry and local plasticity*, Acta Mater. **50** (2002), 3903–3925.
18. D. F. Duffy, J. H. Hardening and A. M. Stoneham, *A calculation of the structure and energy of the niobium/alumina interface*, Acta Mater. **44** (1996), 3293–3301.
19. M. W. Finnis, C. Kruse and U. Schönberger, *Ab initio calculations of metal/ceramic interfaces: What have we learned, what can we learn?* Nanostructured Mat. **6** (1995), 145–155.
20. W. Zhang and J. R. Smith, *Stoichiometry and adhesion of Nb/Al_2O_3*, Phys. Rev. B **61** (2000), 16883–16889.
21. A. Siddiq and S. Schmauder, *Modelling of crystal plasticity effects on the crack initiation energies of a bicrystal interface*, Computer Assisted Mechanics & Engng. Science **14** (2007), 67–78.

Editor's Bio-Sketches

Prof. B. Dattaguru
IUTAM-MMSS 08, Chair

B. Dattaguru received Ph.D. in Aerospace Engineering from the Indian Institute of Science (IISc) in 1972. He excelled in aero-structures teaching, research and industrial consultancy for more than 3 decades at IISc. He has also worked outside India at the then Fatigue & Fracture Branch, NASA Langley Research Center in USA. His fields of interest are Aerospace Structures covering Composite Structures, Fracture Mechanics and Damage Tolerance, Finite Element Methods and Smart Structures. Rustom Choksi award for excellence in Engineering Research from IISc, Academic Excellence award, DRDO, 2002 and Padmashree from the Govt. of India in 2005 are among several awards received by him. He retired as the Chairman of the Department of Aerospace Engineering from IISc and he is currently a Consultant to Aerospace group at M/s. Mahindra Satyam, at Benguluru.

Dr. V. K. Aatre

V. K. Aatre received PhD. (Electrical Engineering) from University of Waterloo, Canada in 1967. He worked at the Technical University of Nova Scotia, Halifax, Canada as Professor of Electrical Engineering from 1968 to 1980. Back in India, he served Defense Research and Development Organization in various capacities and retired as Scientific Adviser to Defense Minister in 1996. He is currently Visiting Professor at the Indian Institute of Science. As an academic, he specialized in Signal Processing, extensively published and authored a popular

UG text book. Among several of his outstanding achievements at DRDO, that of direct relevance to this symposium has been the leadership provided by him in the area of Smart Materials and MEMS. He is the Convener of National Program on Smart Materials, and founder President of the Institute of Smart Structures and Systems (ISSS) and been involved in productizing MEMS in various industries. Vasavik award for Electronic Sciences and Technology (1992), Technology Leadership, DRDO (1998) and Padma Bhushan (2000) are among several awards conferred on him.

Prof. S. Gopalakrishnan
IUTAM-MMSS 08, Co-Chair

S. Gopalakrishnan received his Ph.D. (Aeronautics and Astronautics) from Purdue University, West Lafayette USA and is currently a Professor in the Department of Aerospace Engineering at the Indian Institute of Science, Benguluru, India. He is an acknowledged expert in the area of Wave Propagation, Structural Health Monitoring and Smart Structures. He has visited Ecole de Technologie Superieure (ETS), Canada, Penn State University, USA and Georgia Institute of Technology, USA as Visiting Professor. He currently heads the Aerospace application and Structural Health Monitoring group of the National Program on Micro and Smart Systems, Government of India. He has published two books, 102 journal papers and 75 international conference papers. He is in the editorial board of seven journals and was invited to be the part of the editorial team for the new five volume book on Encyclopedia on Structural Health Monitoring. He is also a Fellow of the Indian National Academy of Engineering.

FSC
www.fsc.org
MIX
Papier | Fördert gute Waldnutzung
FSC® C083411

Kolibri 360 GmbH
Ferdinand-Jühlke-Straße 7
99095 Erfurt, Deutschland
produktsicherheit@kolibri360.de